Contents

Acknowledgments

We should like to extend our sincere thanks to the following organisations which have assisted us with technical information, drawings and advice. Without such assistance, the task of preparation would have been impossible.

Ambac Industries Incorporated, USA
Australian Institute of Petroleum Ltd, Australia
British Leyland (B L) Plc England
Bryce Berger Limited, England
Caterpillar Inc, Australia
Clayton Dewandre Company Limited, England
Consolidated Pneumatic Tool Co Australia Pty Ltd, Australia
Cummins Diesel Australia
Detroit Diesel Allison Division of General Motors Holdens, Ltd, Australia
Deutz Australia Pty Ltd, Australia
Donaldson Australasia Pty Limited, Australia
Dynamometers Australia
English Electric Diesels Limited, Ruston Engine Division, England
English Electric Diesels, Ruston & Hornsby (Aust) Pty Ltd, Australia
General Motors Corporation, Detroit Diesel Allison Division, USA
The Glacier Metal Co Ltd, England
Hastings Deering Queensland Pty Ltd, Australia
Lab Volt, Australia
Lucas CAV Limited, England
Lucas Industries Australia Ltd, Australia
Mack Trucks Australia Pty Ltd, Australia
Mack Trucks Inc, USA
Mazda Motors Pty Limited, Australia
Perkins Engines Pty Limited, Australia
Perkins Engines Pty Limited, England
Robert Bosch GmbH, German Federal Republic
Robert Bosch (Australia) Pty Ltd, Australia
Society of Automotive Engineers — Australasia
Stanadyne Inc, USA
Volvo Australia Pty Ltd, Australia

We would like to extend our sincere thanks for their assistance, in so many ways, to the TAFE teachers of Queensland; in particular to those from the Bundaberg and Mt Gravatt Colleges of TAFE. Our thanks are also extended to those interstate teachers who have given us advice and guidance.

Diesel Engines and Fuel Systems

A.F. ASMUS
Technical Teacher
Bundaberg College of Technical and
Further Education

B.F. WELLINGTON
Technical Teacher
Mt Gravatt College of Technical and
Further Education

Pitman

Melbourne   London   Toronto   Wellington

PITMAN PUBLISHING PTY LTD
A Longman Company
Kings Gardens, 95 Coventry Street
Melbourne 3205 Australia
Offices in Sydney, Brisbane,
Adelaide and Perth

First published 1979
Reprinted 1981, 1982, 1984
Second edition 1988

© A.F. Asmus, B.F. Wellington 1988

Designed by John van Loon
Cover design by John van Loon
Illustrations by Jane Pennells and Marion Buckton
Set in 10/11 pt Cheltenham Book (Linotron 101)
Printed in Malaysia
by Percetakan Mun Sun Sdn. Bhd., Shah Alam, Selangor Darul Ehsan

National Library of Australia
Cataloguing in Publication data

Asmus, A.F. (Alan F.).
    Diesel engines and fuel systems.

    Bibliography.
    Includes index.
    ISBN 0 7299 0013 4.

    1. Diesel motor—Fuel systems. I. Wellington, B.
    (Barry). II. Title.

621.43'6

Preface

It is with a great deal of pleasure that we present this new edition. There have been quite a few major changes, with some totally new sections, some expanded to include new developments in the diesel industry, and some sections extensively pruned of obsolete detail. Some material that has been retained does not relate to equipment in current production, but meets the needs of various training syllabi.

Turbochargers may now be considered a major component of the work, while Roosa Master and Bosch rotary fuel injection pumps have been added to the section on distributor pumps. Cummins Fuel Systems has been extensively rewritten and the chapters dealing with jerk-type injection pumps and governors have been upgraded.

Two new sections have been added—Engine Emission Controls because of the increasing impact this is having on engine and fuel system design as well as on the serviceman, and Alternative Fuels because of the wide spread interest and extensive research being undertaken.

Once again, response to requests for assistance from manufacturers, their agents, servicemen—in fact, the industry—has been both positive and prompt, and has made the compilation of this volume possible. The interpretation of the data has been our own, however, and is presented in the belief that it is accurate and correct.

As in the previous edition, the use of the Imperial system of measurement has been restricted to only those instances where it is the industry standard—for example, the Fahrenheit scale of temperature has been used in the American Petroleum Institute (API) specifications for lubricating oils.

A.F. ASMUS
B.F. WELLINGTON   June, 1987

1
Some important terms

Inertia is the resistance offered by a body to a change in its state of rest or uniform motion.

Force is that which changes, or tends to change a state of rest or uniform motion of a body.

The **unit of force** is the **newton**, which is the force required to give a mass of one kilogram an acceleration of one metre per second squared (m/s^2), hence the force due to gravity acting on one kilogram is 9.806 65 (usually accepted as 9.8) newtons.

Work is done when a force overcomes resistance, and the point of application of the force moves. The unit of **work** is the **joule**, which is the work done when a force of one newton is applied and moves a body one metre in the direction of the application of the force, eg if a jack lifts a mass of one kilogram through a distance of one metre, then 9.8 joules of work have been done, ie 9.8 newtons (the force necessary to overcome gravity) multiplied by one metre (the distance) is equal to 9.8 joules.

Power is the rate of doing work. The **unit of power** is the **watt**. When one joule of work is done in one second, one watt of power is consumed, eg 9.8 watts are required to lift a mass of one kilogram through a distance of one metre in one second—again the 9.8 factor because the force due to gravity acting on one kilogram is 9.8 newtons.

Indicated mean effective pressure (IMEP) is the average pressure in the engine cylinder during a complete engine cycle. Since the intake stroke pressure and the exhaust stroke pressure are close to atmospheric, they may be neglected, so that IMEP may be considered equal to the average pressure on the piston on the power stroke (which acts on the piston) minus the average pressure on the compression stroke (which represents work done by the piston).

Note In practice, IMEP is found by means of an indicator diagram (see Chapter 5).

Brake mean effective pressure (BMEP) is that portion of IMEP that is converted to usable power. It is obvious that some of the IMEP must be consumed in overcoming internal engine friction. Therefore, only portion of the IMEP is available to be converted to usable power at the flywheel.

If friction could be eliminated, then IMEP would equal BMEP. However friction exists inside an engine. Therefore BMEP = IMEP – the pressure consumed in overcoming friction.

Indicated power is the theoretical power of an engine, calculated from the indicated mean effective pressure.

If the indicated mean effective pressure is multiplied by the cross-sectional area of the piston, the total force acting on the piston may be found.

That is,
force on piston = P_M (pascals) × A (metres^2)
    where P_M is indicated mean effective
                pressure
    and A is the cross-sectional area of the
                piston
But 1 pascal = 1 newton/metre^2

Therefore:
force on piston
$$= P_M \left(\frac{\text{newtons}}{\text{metres}^2}\right) \times A(\text{metres}^2)$$
$$= P_M \times A \left(\frac{\text{newtons}}{\text{metres}^2} \times \text{metres}^2\right)$$
$$= P_M \times A \text{ newtons}$$

Now, work done (joules) is equivalent to the applied force (newtons) multiplied by the distance through which it acts.

Therefore:
work done per working stroke
    = force × length of stroke
    = $P_M \times A$ (newtons) × L (metres)
    where L is the length of stroke
    = $P_M \times A \times L$ (newton metres)
But 1 newton metre = 1 joule

Therefore:

work done per working stroke

$$= P_M \times A \times L \text{ joules}$$

Now, power (work done per second)

= work done per working stroke ×
    number of **working strokes per
    second**

$$= P_M \times A \times L \left(\frac{\text{joules}}{\text{strokes}}\right) \times N \left(\frac{\text{strokes}}{\text{second}}\right)$$

where N is the number of **working strokes
per second**

$$= P_M \times A \times L \times N \left(\frac{\text{joules}}{\text{strokes}} \times \frac{\text{strokes}}{\text{second}}\right)$$

$$= P_M \times A \times L \times N \text{ (joules/second)}$$

But 1 joule per second = 1 watt

Therefore:

indicated power $= P_M \times A \times L \times N$ watts,
which is usually written as:

Indicated power $= P_M LAN$

Brake power is a measure of the engine
power available, at the engine flywheel, to do
work. It is less than **indicated power** by an
amount equal to the power necessary to
overcome the engine's internal friction. The
brake power of an engine is usually found by
means of a dynamometer, which applies a
breaking force to the turning effort of the
engine in question. The reaction causes a
turning effort to be applied to the dynamometer,
and this turning effort is measured. Its value,
together with the engine's speed, is then
substituted in a formula quoted by the
manufacturer of the dynamometer for that
particular dynamometer. The result is the brake
power of the engine.

Torque is turning effort. The crankshaft of an
engine turns because of the force applied to the
piston on the power stroke, which is transferred
to the crankshaft via the connecting rod. At that
engine speed at which the cylinder pressure is
highest, the greatest turning effort is applied to
the crankshaft. Usually, because of various
inefficiencies at high speed, the maximum
cylinder pressure does not occur at peak rpm,
but at much lower speed. Therefore the
maximum torque will occur at much less than
peak rpm.

Torque must not be confused with brake
power, which involves a rate (or speed) factor.
Thus, while torque will be less than maximum
at peak rpm, the speed factor will result in
brake power being maximum at or near peak
revs.

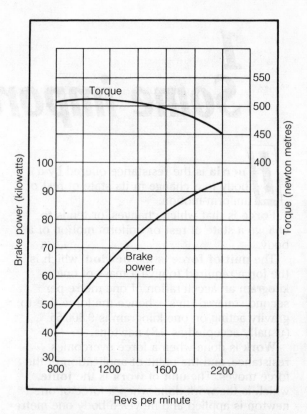

Fig 1.1 Typical brake power and torque curves for a
large automotive diesel engine

The **mechanical efficiency** of an engine is
the ratio of its **brake power** to its **indicated
power**. An engine that is mechanically efficient
will not waste much power overcoming friction,
and the brake power will not be much less than
indicated power.

Mechanical efficiency
$$= \frac{\text{brake power}}{\text{indicated power}} \times 100\%$$

Indicated thermal efficiency is the
efficiency of an engine with regard to how
efficiently it converts the chemical energy of the
fuel to indicated power.

Indicated thermal efficiency
$$= \frac{\text{indicated power}}{\text{power available from fuel used}} \times 100\%$$

Every fuel contains a specific amount of heat
energy, which is released when the fuel is burnt—
this is known as the **calorific value of the
fuel**, and is expressed in joules per kilogram of
fuel. By measuring the amount of fuel used per
minute or per hour, it is a simple matter to

calculate the amount of heat energy supplied
per second.

Heat energy supplied (joules) per second
= fuel used (kilograms) per second ×
calorific value of fuel (joules per kilogram)
But joules per second = watts

Therefore:

power available from fuel
= fuel used per second
× calorific value of fuel

Brake thermal efficiency is found by
comparing the power output at the flywheel
with the theoretical power available from the
combustion of the fuel.

Brake thermal efficiency
$$= \frac{\text{brake power}}{\text{power available from fuel used}} \times \mathbf{100\%}$$

The **volumetric efficiency** of a naturally
aspirated engine is the ratio of the volume of air
induced into the engine cylinder to the piston
displacement, that is, how completely the
cylinder fills with air (or air–fuel mixture). No
unsupercharged engine has a volumetric
efficiency of 100 per cent, and volumetric
efficiency falls off at high rpm.

The following factors influence volumetric
efficiency:

- valve size and lift
- port size
- intake and exhaust manifold design
- valve timing, and
- exhaust system restrictions

If an engine is supercharged, then the
volumetric efficiency exceeds 100 per cent,
since the air (or air–fuel mixture) is forced into
the engine under pressure. If the quantity of air
forced into the cylinder was allowed to regain
normal pressure, then its volume would be
greater than the piston displacement.

2
Comparing petrol and diesel engines

Both petrol and diesel engines are internal combustion (IC) engines working on either the two- or four-stroke cycle. The basic design of both engines is similar, the main difference between the two being the method of introducing the fuel charge into the combustion chamber, and the means employed to ignite it. However, the engine features and performance characteristics may differ greatly. In the table of comparisons set out below, we have endeavoured to illustrate some of the differences between petrol and high-speed diesel engines of comparable size.

Admission of the fuel to the combustion chamber

In the diesel engine, air only is compressed on the compression stroke. The fuel charge required to produce the power stroke in each cylinder is accurately metered and pressurised by the fuel injection pump. It passes to the high-pressure injectors via pipes, and is sprayed into the combustion chamber, where it mixes with hot compressed air, and is ignited.

In the petrol engine, fuel and air have traditionally been mixed in the throat of the carburettor, which is situated outside the engine cylinder. The fuel-air mixture then passes to the engine cylinder through the inlet manifold and the cylinder head inlet ports. Modern design has seen the introduction of fuel injection into the inlet ports of the cylinder head, air only passing through the inlet manifold. However, in either case, the fuel-air mixture passes through the inlet ports into the combustion chamber of the engine, where it is ignited after being compressed.

Compression ratio

Before dealing with compression ratio, it is necessary to become familiar with the terms below:

- **Swept volume**—This is the volume of air displaced when the piston moves from bottom dead centre (BDC) to top dead centre (TDC) on its compression stroke.

- **Clearance volume**—This is the volume of air between the top of the piston and the cylinder head when the piston is at TDC at the end of the compression stroke.

Compression ratio is a comparison of the volume of air in the cylinder when the piston is at BDC, with the volume of air present in the cylinder when the piston is at TDC. The volume of air in the cylinder when the piston is at BDC is equal to the swept volume plus the clearance volume, and the volume present when the piston is at TDC is equal to the clearance volume.

Therefore:

$$\text{Compression ratio} = \frac{\text{swept volume} + \text{clearance volume}}{\text{clearance volume}}$$

It should be mentioned that the higher the compression ratio used, the hotter the gas being compressed will become.

Since the fuel and air are compressed in the cylinders of petrol engines, the maximum compression ratios used are limited, since high compression ratios, with their subsequent high air temperatures, would cause detonation of the mixture. Detonation is a dangerous condition, which inflicts severe mechanical stress on engine components, as well as reducing engine performance.

Table 2.1 Comparison between diesel and petrol engines

Feature	High-speed diesel engine	Petrol engine
Admission of fuel	Directly from fuel injector	From carburettor via the manifold, or injected into the inlet port
Compression ratio	From 14:1 to 24:1	From 7:1 to 10:1
Ignition	Heat due to compression	Electric spark
Torque	Varies little throughout the speed range	Varies greatly throughout the speed range
Brake thermal efficiency	35–40%	25–30%
Exhaust gases	Non–poisonous, but may cause suffocation	Poisonous
Engine construction	Robust	Relatively lighter than the diesel engine
Maximum crankshaft rpm	From 2500 to 5000 rpm	From 4000 to 6000 rpm
Compression pressure	Actual, 3100–3800 kPa Theoretical at 16:1 CR, 4254 kPa	Actual, 750–1400 kPa Theoretical at 8:1 CR, 1728 kPa
Compression temperature	Actual, 425–550°C Theoretical at 16:1 CR, 525°C	Actual, up to 230°C Theoretical at 8:1 CR, 375°C
Fuel used	Automotive distillate	Petrol

However, diesel engines rely on the heat of the compressed air to ignite the fuel as quickly as possible when it is sprayed into the engine cylinders, and so high compression ratios are very necessary. Thus, diesel engines invariably have much higher compression ratios than petrol engines.

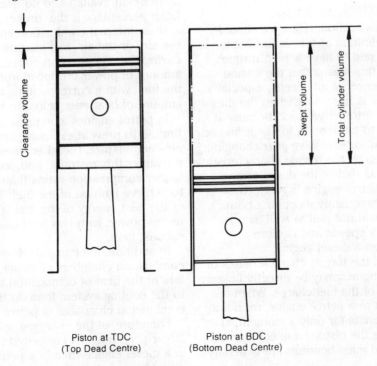

Piston at TDC
(Top Dead Centre)

Piston at BDC
(Bottom Dead Centre)

Fig 2.1

Ignition of the charge

It is well-known that when air is compressed, heat is generated. Because of the high compression ratios used in diesel engines (14:1 to 24:1), the air taken into the engine cylinder is highly compressed (up to 3800 kPa). When the fuel charge is injected into and mixed with this highly compressed air at a predetermined point in the cycle, the heat in the air is sufficient to cause the fuel to ignite. This is called **compression ignition.**

With regard to the petrol engine, however, it must be remembered that the fuel and air are mixed external to the engine cylinder, and the mixture is taken into the engine cylinder and compressed. This would make compression ignition, as used in the diesel engine, impractical, because the petrol vapour would ignite explosively during the compression stroke, causing very severe physical damage to engine components. It is therefore necessary to make use of lower compression ratios, and to ignite the air–fuel mixture in the combustion chamber by means of an electric spark at the desired point in the cycle.

Torque

Before studying this section, reference should be made to the definition of torque in Chapter 1. A diesel engine is said to have a 'high torque'. Broadly speaking, this means that the engine has good pulling power at all speeds, especially at low speeds. This is important when the diesel engine is used for automotive work, because it allows more work to be done in top gear and so eliminates the need for excessive gear changing.

With a petrol engine, the torque characteristic is not as constant as that of the diesel engine. As the speed of a petrol engine is reduced, so is its torque (and consequently its pulling power).

A petrol engine will not pull as well as a diesel engine at low speeds and is more inclined to stall than a diesel engine.

The difference in the torque characteristics of petrol and diesel engines may be directly linked to the combustion of the fuel charge. When combustion occurs in a petrol engine, maximum cylinder pressure exists for only a moment, at or near TDC, where the piston, connecting rod, big-end bearing and main bearings are generally 'in line'.

In a diesel engine, maximum cylinder pressure is sustained for a considerable time after the piston passes through the TDC position. As the crankshaft turns and the big-end journal moves past the 'in line' position, the force exerted on the piston by the gas pressure is transferred more effectively to the crankshaft via the connecting rod.

This effect is greatest when the connecting rod forms a right angle with a line through the centres of the main bearings and the big-end journal, that is, the line of action of the force is the greatest distance from the centre of rotation of the crankshaft. Thus, the longer the maximum cylinder pressure is sustained, the greater the torque developed.

Thermal efficiency

It is usual to use the general term thermal efficiency to mean brake thermal efficiency, and it is used in that sense here. Very broadly, of the chemical energy of the fuel converted to heat energy in the engine, approximately 30 per cent is lost to the cooling system, 30 per cent to the exhaust, 10 per cent is used in overcoming the internal friction of the engine, leaving about 30 per cent available to do useful work. This latter percentage is the (brake) thermal efficiency.

In an internal combustion engine, the more the air (or the air–fuel mixture) is compressed during the compression stroke, the greater the amount of power produced from combustion of the fuel, with a corresponding reduction in the amount of heat energy lost to the exhaust gases.

In petrol engines, compression ratios must be limited to prevent explosive ignition of the air–fuel mixture. Diesel engines, however, do not suffer this restraint, and, indeed, need much higher compression ratios than petrol engines to achieve ignition of the fuel. Therefore, more of the heat energy of the fuel is converted to useful power, with less heat energy wasted as exhaust heat.

In addition, the compact, low-volume combustion chambers of diesel engines permit less of the heat of combustion to be conducted to the cooling system than do the larger combustion chambers of petrol engines.

Therefore, of the energy supplied as fuel, a larger percentage is converted to usable power in a diesel engine than in a petrol engine—the diesel has a higher thermal efficiency.

Exhaust gases

Because the compression ratio of a diesel engine is high and there is more than adequate air for combustion, the burning of the fuel charge is very nearly complete. This results in harmless carbon dioxide instead of the dangerous carbon monoxide associated with petrol engines. The amount of carbon monoxide present in the exhaust gases of direct-injection diesel engines is around 17 per cent of that from a petrol engine, and this figure drops to approximately 5 per cent with some other combustion chamber designs.

The exhaust gases from a diesel engine also contain unburnt hydrocarbons (unburnt fuel), nitrogen oxides and oxides of sulphur. Although the exact cause of the characteristic diesel odour is not yet known, it has been found that engines with low hydrocarbon emissions do not usually have a strong exhaust odour. The oxides of nitrogen and sulphur combine with the water vapour resulting from combustion to form acids, which attack the cylinder liners, cause the lubricating oil to become acidic, and may lead to lung irritation.

Because the compression ratio of a petrol engine is considerably less than that of a diesel engine, the fuel charge does not burn with the same efficiency. This causes a greater percentage of lethal carbon monoxide to be present in the petrol engine exhaust gases—approximately 7 per cent operating normally, but higher when idling, in some instances. It is for this reason that diesel engines should be used in preference to petrol engines for mines, tunnels or any other undertakings where ventilation is inadequate.

Engine construction

Due to the high compression ratios and the combustion process employed in diesel engines, a diesel engine has much higher cylinder pressures than a petrol engine of similar power. This means that the internal stresses set up in the diesel engine components will be greater than those in a petrol engine. Consequently, the crankshaft, bearings, pistons, connecting rods and so on must be made stronger to withstand these stresses, and so enable the engine to function efficiently. With rapid advances being made in metallurgy, however, the weight of the modern diesel compares favourably with that of a petrol engine of similar power.

Crankshaft rpm

To understand why the crankshaft revolutions of a diesel engine are lower than those of a petrol engine of comparable size, it is necessary to understand the term 'inertia' (see Chapter 1). A force has to act on a body to overcome its inertia and so cause it to move, or, if the body is in motion, to cause it to accelerate or decelerate. Now, if the body decelerates or comes to a complete stop, it is obvious that force exerted by the inertia of the body must be overcome.

The piston of an IC engine must be brought to rest at both TDC and BDC positions, before its direction of motion is changed. This means that the inertia force exerted by the piston and connecting rod must be absorbed by these and other engine components. As the pistons and connecting rods used in a diesel are of heavier construction than those used in a petrol engine, the inertia force acting in a diesel engine is greater than that in a petrol engine, doing the same number of rpm. The major components of a diesel engine such as connecting rods, connecting rod 'big-end' bolts and crankshaft must therefore be subjected to greater stresses. Mechanical failures such as broken connecting rod 'big-end' bolts and broken connecting rods may result from high inertia forces. In order to minimise these stresses, the crankshaft revolutions of a diesel engine are kept below that of a petrol engine.

Compression pressure

Compression pressure may be defined as the maximum air pressure created in the cylinder of an IC engine on the compression stroke. When the compression pressure in an engine cylinder is being checked, action should be taken to ensure that combustion does not occur. The actual compression pressure of the air in the cylinder of an engine is dependent on a number of factors such as engine rpm, cylinder wear, ring wear and valve condition, but the theoretical compression depends on three factors:

- the pressure of the air in the cylinder before it is compressed
- the compression ratio used, and
- the gas constant for the pressure–volume relationship of the air under the conditions existing in the engine. (As pressure increases, so the volume decreases; the mathematical relationship of the two for a particular gas is known as the gas constant.)

The theoretical compression pressure of an engine is found by using the formula:

$$P_2 = P_1 r^n$$

where P_1 = pressure of the air in the cylinder before compression

P_2 = theoretical compression pressure

r = compression ratio,

n = the characteristic gas constant.

As has been stated previously, a petrol engine has a lower compression ratio than a diesel engine. Also, when the air–fuel mixture enters the cylinder of a petrol engine, it is cooled to a certain extent as the petrol vaporises. This improves the pressure–volume relationship, and so the value for n is slightly higher for a petrol engine than for a diesel. Because of the lower compression ratio used and the higher value for n, it should be obvious that the theoretical compression pressure in the cylinder of a petrol engine is much lower than that in the cylinder of a diesel engine.

Compression temperature

When a mass of air is compressed, the speed of the molecules in the air is increased, heat is generated, and so the temperature of the mass of air rises. Compression temperature may be defined as the temperature attained by the air, which is compressed in the cylinder of an IC engine, on the compression stroke. It should be noted here that compression temperature is the temperature attained by the air due to compression in the cylinder, and is not the temperature of the gases in the cylinder when combustion is in progress.

The actual compression temperature of the air in the cylinder of an IC engine is mainly dependent on the compression pressure, which is, in turn, dependent on the compression ratio

and the mechanical fitness of the engine. An increase in engine rpm, however, will increase the compression pressure and subsequent compression temperature.

The theoretical compression temperature of the air in the cylinder of an IC engine depends on three factors:

- the temperature of the air entering the cylinder before it is compressed
- the compression ratio used, and
- the value of the characteristic gas constant, air being the gas used.

The theoretical compression temperature of an engine may be found by using the following formula:

$$T_2 = T_1 r^{n-1}$$

where T_1 = temperature of the air in the cylinder before compression begins. (In the above formula, absolute temperature must be used; this is found by adding 273° to the Celsius temperature.)

T_2 = theoretical compression temperature

r = compression ratio

n = the characteristic gas constant.

As has been discussed previously, petrol engines have a much lower compression ratio than diesel engines, while the value of n is slightly higher, than for a diesel. Calculations based on typical values show that the theoretical compression temperature of the air in the cylinder of a diesel engine with a compression ratio of 16:1 is approximately 40 per cent higher than that of a petrol engine with an 8:1 compression ratio.

Note   It must be stressed here that the theoretical compression pressure and the theoretical compression temperature of the air that is compressed in the cylinder of an IC engine are influenced by such things as engine rpm, altitude, pressure–volume relationship, and climatic conditions. Also, the theoretical figures are considerably higher than the actual compression pressure and compression temperature found in the engine working under normal conditions. This is due to such things as valve overlap, faulty engine valves, piston and ring wear, liner wear, and loss of heat to the cooling system via the cylinder head and liners.

Fuel used

The fuels used in both high-speed diesel engines and petrol engines are refined from petroleum. Automotive distillate is most commonly used in modern, high-speed diesel engines and petrol is used in petrol engines. Although the cost of producing distillate and petrol is approximately the same, some governments choose to subsidise distillate production, so making it a cheaper fuel for consumers.

Distillate has a higher flashpoint than petrol, so there is less fire risk involved with a diesel engine. It cannot be too highly stressed that petrol must never be used in a diesel engine. Petrol is more volatile than distillate, and if it were injected into the combustion chamber of a diesel engine and burnt, a high pressure rise would result. Extremely severe knocking would follow, and bearing failure, crushed pistons and broken connecting rods would probably result.

Some advantages of the diesel engine

- Because the thermal efficiency of a diesel engine is high, it uses less fuel than a petrol engine of comparable size, and is therefore more economical to operate.
- The diesel is more suitable for marine work because it does not use spark ignition, and consequently does not require electrical equipment, which may be damaged by water.
- Diesel fuel may be handled and stored with a greater degree of safety than petrol, as it has a higher flashpoint and there is, therefore, less risk of explosion or fire.
- The torque characteristic of a diesel engine does not vary greatly at either high or low speeds. This means that diesel engines, when used for automotive applications, will pull better in top gear.
- Diesel engines can be used underground in mines and tunnels with comparative safety, whereas a petrol engine cannot. The reason for this is that the amount of carbon monoxide present in the exhaust gas from a petrol engine is much greater than that present in the exhaust gas from a diesel, and could prove fatal for persons working in the vicinity.

Some disadvantages of the diesel engine

- A diesel engine used to be harder to start in cold weather than a petrol engine. However, direct injection and the use of thermostart devices have improved the cold starting ability of the diesel to the point where it is equal to, or better than, the petrol engine in this regard.
- As the working stresses in a diesel engine are greater than those in a petrol engine, the construction of a diesel is generally heavier than that of a petrol engine of comparable output.
- The initial cost of a diesel engine is greater than that of a petrol engine of similar size. However, substantial savings in operating costs more than offset this.
- Because of the high operating pressures and the efficiency and precision required of the fuel injection system, the system is extremely sensitive to dirt and water and the fuel must be maintained scrupulously clean.

3
The mechanical cycles

Although they range from minis to monsters in size, almost all reciprocating piston IC engines work on one of two mechanical cycles—the four-stroke cycle or the two-stroke (Clerk) cycle. These cycles designate, in correct sequence, the mechanical actions by which (a) the fuel and air gain access to the engine cylinder, (b) the gas pressure (due to combustion) is converted to power, and (c) the burnt gas is expelled from the engine cylinder.

There has been considerable contention regarding the Otto cycle, which is generally considered synonymous with the four-stroke cycle. More correctly, the Otto cycle is a four-stroke cycle **with combustion occurring at constant volume** (see Chapter 5.).

The basic four-stroke-cycle diesel engine

From the name, it is fairly obvious that there are four strokes in one complete engine cycle. A stroke is the movement of the piston through the full length of the cylinder, and, since one such movement causes the crankshaft to rotate half a turn, it follows that there are two crankshaft revolutions in one complete engine cycle. The four strokes, in correct order, are as follows:

1 **The inlet stroke**—With the inlet valve open and the exhaust valve closed, the piston moves from TDC to BDC, creating a low-pressure area in the cylinder. Clean, filtered air rushes through the open inlet valve to relieve this low-pressure area, and the cylinder fills with air.
2 **The compression stroke**—With both valves closed, the piston moves from BDC to TDC, compressing the air. During this

stroke the air becomes heated to a temperature sufficiently high to ignite the fuel.
3 **The power stroke**—At approximately TDC, the fuel is injected, or sprayed, into the hot, compressed air, where it ignites, burns and expands. Both valves remain closed, and the pressure acts on the piston crown, forcing it down the cylinder from TDC to BDC.

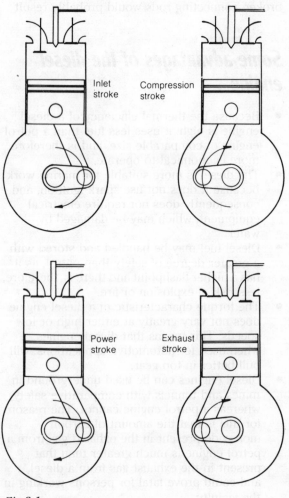

Inlet stroke   Compression stroke

Power stroke   Exhaust stroke

Fig 3.1

4 **The exhaust stroke**—At approximately BDC the exhaust valve opens and the piston starts to move from BDC to TDC, driving the burnt gas from the cylinder through the open exhaust valve.

At the completion of the exhaust stroke, the exhaust valve closes, the inlet valve opens and the piston moves down the cylinder on the next inlet stroke. Since there are three non-working strokes to one working stroke, some means of keeping the engine turning over must be provided, particularly in single-cylinder engines. It is for this reason, and to ensure smooth running, that a heavy flywheel is fitted to the crankshaft.

Scavenging the four-stroke cycle diesel engine

It is necessary, since efficient combustion is desired, to completely clear the burnt gas from the cylinder on the exhaust stroke, so providing a full cylinder of fresh air by the completion of the inlet stroke. Air that has burned with fuel has had almost all its oxygen consumed, and cannot be used again. The clearing of the exhaust gas from the cylinder of an internal combustion engine is known as 'scavenging'.

Both the upward movement of the piston on the exhaust stroke and the valve timing contribute to the scavenging of the burnt gas in the four-stroke engine.

It is usual for the exhaust valve to be opened before BDC, thus allowing a puff of high-pressure gas to escape at high velocity through the exhaust system. The upward movement of the piston ensures that the burnt gas continues this high-speed movement.

At the end of the exhaust stroke, the gas will continue to move through the exhaust system because of its momentum alone. This continued movement will cause a low-pressure area to develop behind the fast-moving exhaust gas, and this can be used to draw fresh air into the cylinder if the inlet valve opens at the instant the low-pressure area develops.

In practice, it is found that good scavenging can be achieved by opening the inlet valve just before TDC, and closing the exhaust valve just after TDC, thus allowing the exhaust gas to 'draw' fresh air into the cylinder before the inlet stroke actually begins.

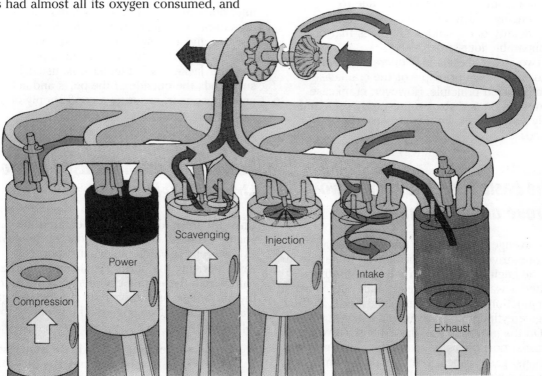

Fig 3.2 Stages of operation of a six-cylinder, naturally aspirated diesel engine

The period of crankshaft rotation when both valves are open together is known as **valve overlap**, and occurs only at TDC.

Suitable positioning of the valves, usually one at each side of the combustion chamber, helps to guide the gases on their correct paths, and assists in achieving complete scavenging.

It is worthwhile to note at this stage that supercharged engines usually have more valve overlap than naturally aspirated engines.

The two-stroke (Clerk) cycle diesel engine

*I*n engines of this type, there are obviously only two strokes, or one crankshaft revolution, to one complete engine cycle. This means that there are twice as many working strokes per minute in a two-stroke diesel as there are in a comparable four-stroke, working at the same engine speed. Theoretically then, the two-stroke engine should develop twice the power of a four-stroke of similar bore and stroke and the same number of cylinders, but, due to scavenging difficulties, the power output is in the vicinity of one-and-a-half times that of a comparable four-stroke.

Two-stroke diesels may operate on either the scavenge blown principle or the crankcase compression principle. However, crankcase compression two-strokes are rarely seen and only scavenge blown two-strokes will be discussed here.

The basic scavenge blown two-stroke diesel engine

A scavenge blown engine makes use of an engine-driven air pump, or blower, to supply air to the engine cylinder. An inlet port in the cylinder wall is used instead of an inlet valve, in conjunction with either an exhaust port in the engine cylinder or an overhead exhaust valve.

On the power stroke, the exhaust valve or exhaust port opens first, and the high-pressure exhaust gas escapes. The inlet port then opens, allowing air (at between 15 and 60 kPa pressure) to sweep through the cylinder,

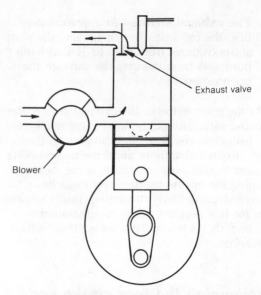

Exhaust valve

Blower

Fig 3.3 The scavenge blown two-stroke diesel engine (schematic diagram)

clearing the exhaust gas and recharging the cylinder with fresh air. It is usual to close the exhaust valve or port first, allowing a slight pressure to build up in the cylinder before compression begins.

It is general practice to have a number of inlet ports machined in the cylinder liner. These are cut at an angle or tangent to the cylinder liner, to give the incoming air a spiralling action, and are known as tangential ports. A large air jacket, called an air box, usually surrounds the outside of the ports and acts as an air reservoir, ensuring a good supply of air to all ports at all times.

Scavenging methods used in two-stroke diesels

Once the fresh air charge has entered the engine cylinder, it may act in one of three ways as it sweeps towards the exhaust port (or valve), driving the burnt gas ahead of it. The path within the engine cylinder through which the air sweeps as it drives out the exhaust gas classifies the scavenging system as one of the following three:

- **Cross scavenging**—This is the simplest system, and is used on many small petrol engines. However, due to its relative inefficiency with regard to scavenging, it is becoming increasingly rare in modern diesel engines.

Scavenging is achieved by situating the inlet and exhaust ports on opposite sides of the engine cylinder. The piston, reciprocating in the cylinder, opens or closes these ports. In some cases, the piston crown is specially shaped to deflect the incoming air upwards throughout the cylinder, in an attempt to obtain complete scavenging.

- **Uniflow scavenging**—Uniflow scavenging is considered the most efficient system of all, since the incoming air enters at one end of the cylinder and spirals throughout the entire length of the cylinder, to pass out through the exhaust port or the exhaust valve (depending on the design of the particular engine).

piston automotive engine and small Detroit Diesels Allison engines, to large marine engines up to and exceeding 15 000 kW.

- **Loop scavenging**—There are a number of loop-scavenging systems, but the scavenge air paths are basically the same in all cases. Loop scavenging is more efficient than cross scavenging, since the incoming air moves from the inlet port to the top of the cylinder, and down to the exhaust port, driving out the burnt gas.

Piston-controlled ports are used, the exhaust port (or ports) being the upper. The inlet port (or ports) may be directly beneath the exhaust, or may be some distance around the cylinder liner. Regardless of their actual position in relation to the exhaust ports, the inlet ports are so shaped that they direct the incoming air upwards, thus ensuring complete scavenging in the form of a loop.

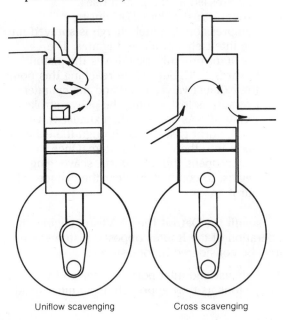

Uniflow scavenging   Cross scavenging

Fig 3.4

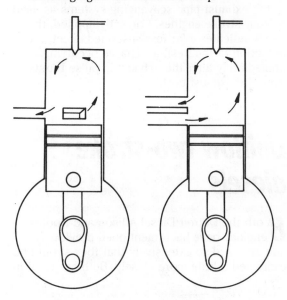

Fig 3.5 Loop scavenging

Mechanically operated exhaust valves in the cylinder head are usually used with this system, although opposed piston engines use one piston to control the exhaust ports and one to control the inlet ports. Again, the inlet ports are specially shaped to direct the incoming air in the required direction, and are known as tangential ports.

Because of its efficiency, the uniflow scavenging system is used on an extremely wide range of engines, from small high-speed engines such as the Rootes opposed-

It is not uncommon for all loop-scavenging systems to be referred to as the 'Schnuerle loop'. However, the Schnuerle loop system is a specific system in which the air is admitted to the cylinder through two inlet ports, one on each side of the single exhaust port.

The Kadenacy system

This is a system of exhaust pulse scavenging that applies to two-stroke engines of all types, and consists of opening the exhaust valves or ports on the power stroke while there is still a pressure of a few atmospheres in the cylinder.

As the exhaust gas escapes from the cylinder at high velocity, a partial vacuum is created in the cylinder. The inlet ports then open and fresh air rushes in to relieve the low-pressure area thus created.

Engines using this system may either be blown (the partial vacuum and blower pressure combining to give efficient scavenging and high volumetric efficiency), or be dependent on the partial vacuum alone to induce the fresh air charge into the cylinder. In the latter case, the system can only be applied to engines working at fairly constant speed. When used in conjunction with a blower, the system is claimed to give efficient scavenging, with increased power and better fuel economy as a result.

The Kadenacy system has often been likened to the exhaust-pipe scavenging systems as used on four-stroke engines. Correctly applied, the name indicates a far less sustained effect, the induction of the fresh-air charge occurring immediately after the exhaust impulse bursts from the exhaust port.

Uniflow two-stroke diesels

Both the Detroit Diesel Allison and Rootes engines have been mentioned previously, and are used so extensively that they should be examined a little more closely. Both are used in

automotive applications, while the Detroit Diesel Allison engine is manufactured in a large number of variations and sizes, which power a diversity of equipment ranging from boats to locomotives, bulldozers to tanks.

Although it is a two-stroke engine, the operation of the Detroit diesel is considered in four stages:

1   **Scavenging**—With the exhaust valves in the head open, and the piston crown below the ports, air from the blower passes into the cylinder and swirls towards the exhaust ports, driving out the burnt gases and leaving the cylinder filled with clean air.
2   **Compression**—As the piston begins to move towards TDC, the exhaust valves close and the piston then covers the inlet ports. The air trapped in the cylinder is increasingly compressed as the piston rises.
3   **Power**—Just before TDC during compression, the fuel charge is sprayed into the intensely hot air and ignites. The resultant pressure rise drives the piston towards BDC, but before reaching this point, the exhaust valves open and the cylinder pressure is lost, ending the power stroke.
4   **Exhaust**—As soon as the exhaust valves open, burnt gases pass through the exhaust ports into the exhaust system. Once the piston opens the inlet ports, scavenging begins, to completely clear the cylinder of burnt gases.

As with the Detroit Diesel Allison engine, the operation of the Rootes opposed-piston engine may be considered in four stages:

1   Exhaust and inlet ports are sealed by pistons as they approach mean inner dead

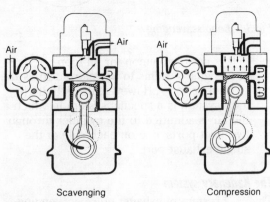
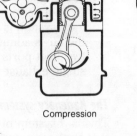
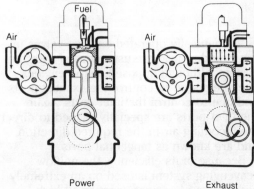

Scavenging   Compression   Power   Exhaust

Fig 3.6 Detroit Diesel operation

centre. Fuel is then injected into the cylinder and combustion takes place.

2   The high pressures generated by combustion force the pistons outwards on their working stroke, so actuating the linkage to turn the crankshaft.

3   Towards the end of the working stroke the phasing of the pistons and the location of ports enable the exhaust ports to open before the inlet ports. Exhaust ports are so designed that burnt gases are rapidly expelled from the cylinder into the exhaust system.

4   Shortly afterwards the inlet ports are uncovered and air, supplied to the air chest from the low-pressure blower, rushes into the cylinder and sweeps remaining exhaust gases out through the exhaust ports.

4
Turbochargers and blowers

Turbocharging

The turbocharger was invented by a Swiss named Buchi in 1906, and has been seen from time-to-time in various versions ever since. However, it is only in the past two decades that it has been developed to such a degree of reliability and performance that it is now being fitted to a continually increasing percentage of new IC engines.

The turbocharger is essentially an exhaust-driven supercharger, its primary purpose being to pressurise the intake air, so increasing the quantity entering the engine cylinders on the inlet stroke, and allowing more fuel to be burned efficiently. In this way, the torque and power output of an engine can be increased by up to 35 per cent by the addition of a turbocharger.

Turbocharger construction

The turbocharger is made up of three sections, the centre bearing housing assembly, the turbine housing and the compressor housing, as shown in Fig 4.2. The bearing housing contains

Turbo-Compressor

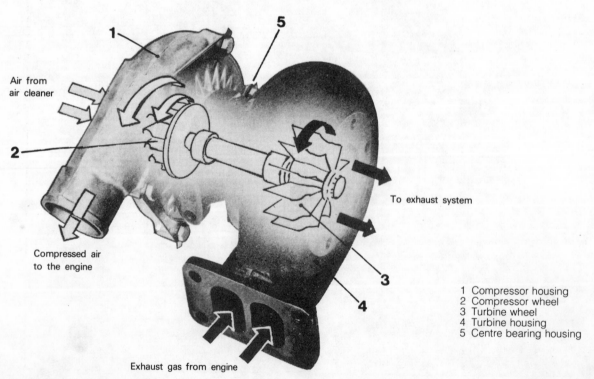

Air from air cleaner

Compressed air to the engine

To exhaust system

Exhaust gas from engine

1 Compressor housing
2 Compressor wheel
3 Turbine wheel
4 Turbine housing
5 Centre bearing housing

Fig 4.1 Exhaust-driven turbocharger

Fig 4.2 Turbocharger sub-assemblies

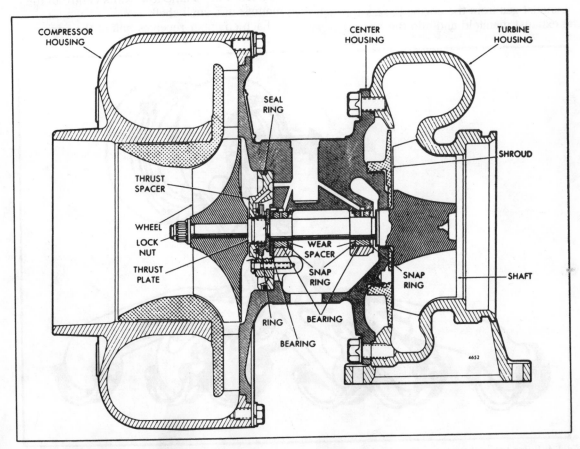

COMPRESSOR
HOUSING

CENTER
HOUSING

TURBINE
HOUSING

SEAL
RING

SHROUD

THRUST
SPACER

WHEEL

LOCK
NUT

WEAR
SPACER

THRUST
PLATE

SNAP
RING

SNAP
RING

SHAFT

RING

BEARING

BEARING

4652

Fig 4.3 Sectional view of a typical turbocharger

two plain bearings, piston-ring-type seals, retainers and a thrust bearing. There are also passages for the supply and dumping of oil to and from the housing.

The turbine wheel turns in the turbine housing, and is usually integral with the turbine shaft, which is carried in the plain bearings in the bearing housing. The compressor wheel, which is fitted to the opposite end of the turbine shaft forming a combined rotating assembly, turns in the compressor housing. Turbocharger rotational speeds can reach 120 000 rpm on some small high-performance units.

Turbocharger operation

In general terms, there are two types of turbocharger—the pulse type and the constant pressure type—each with its own operating characteristics. However, both operate in the same basic way.

Exhaust gas from the engine passes through the exhaust manifold and into the turbocharger turbine housing, where it impinges on the turbine blades causing the turbine, shaft and compressor wheel assembly to rotate.

As the compressor rotates, air is pressurised by centrifugal force and passes from the compressor housing to the engine inlet manifold, the quantity and/or pressure of the air being proportional to the speed of rotation.

The pulse-type turbocharger requires a specially designed exhaust manifold to deliver high-energy exhaust pulses to the turbocharger turbine. This design, with its individual branches as shown in Fig 4.4, prevents interference between the exhaust gas discharges from the separate cylinders, thus promoting a high-speed pulsing flow not achieved with other designs.

In some applications, a split-pulse turbine housing can be used to further aid in the excitation of the rotating assembly. This design has two volute chambers instead of one. The term 'volute chamber' is used in reference to the spiral-shaped turbine housing, which decreases in volume toward its centre in the manner of a snail shell.

Each chamber receives half of the engine

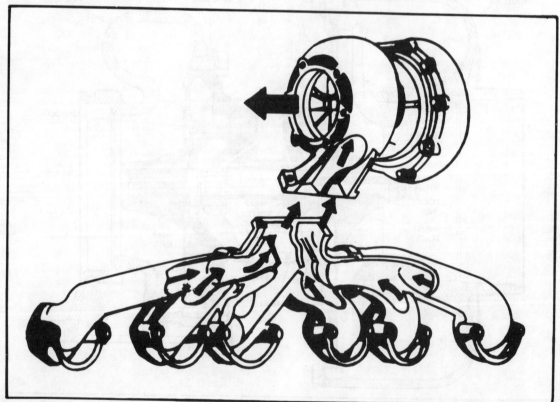

Fig 4.4 Pulse-type exhaust manifold

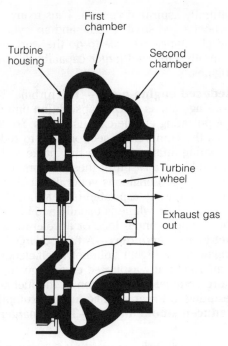

Fig 4.5 Split-pulse turbine housing

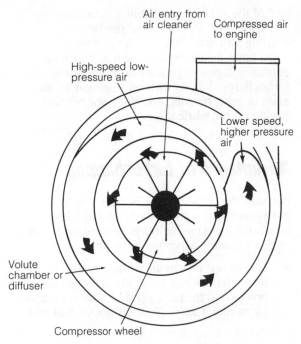

Fig 4.6 Turbocharger compressor housing

exhaust flow; for example, in a four-cylinder engine the front two cylinders are fed into the first chamber, the back two are fed into the second chamber as shown in Fig 4.5.

With the constant-pressure type of turbocharger, the exhaust gas from all cylinders flows into a common manifold, where the pulses are smoothed out, resulting in exhaust gas entering the turbine housing at an even pressure.

With both types of turbocharger, the exhaust gas then enters a volute-shaped annular ring in the turbine housing, which accelerates it radially inwards at reduced pressure and increased velocity onto the turbine blades. The blades are so designed that the force of the high-velocity gas drives the turbine and its shaft assembly.

The compressor assembly is of similar design and construction in both pulse and constant-pressure turbochargers (see Fig 4.6). The compressor consists of a wheel and a housing incorporating a single volute or diffuser. Air in the compressor chamber mainly lies between the blades of the compressor wheel, and is thrown out radially by centrifugal force into the volute during rotation of the wheel. Here the air velocity decreases and a corresponding increase in air pressure results. As the air progresses around the volute its velocity decreases further

and the pressure increases as the cross-sectional diameter of the chamber increases.

In summary, the pulse-type turbocharger offers a quick excitation of the rotating assembly due to the rapid succession of the exhaust gas pulses on the turbine assembly. It is predominantly used in automotive applications, where acceleration response is important.

Constant-pressure turbochargers are used mainly on large diesel engines in earthmoving equipment and in marine applications. In these applications, acceleration response is not as critical.

Turbocharger lubrication

In most applications, turbochargers are lubricated by the lubrication system of the engine to which they are fitted. Oil under pressure from the engine oil pump enters the top of the bearing housing and flows around the shaft and to the thrust bearings and oil seals. The oil flows both inside and around the outside of the shaft bearings, which fully float in oil during operation. The oil also flows to the piston-ring-type oil seals at either end of the rotating shaft to aid in sealing and lubrication. The thrust bearing located at the compressor end of the rotating assembly is lubricated by the same oil before it leaves the bearing housing and flows back to the engine sump.

On large diesel engines such as those used in marine and power-generation applications, the turbocharger has its own oil reservoir in the main bearing housing and does not rely on engine oil for lubrication.

Advantages of turbocharging

Because the turbocharger is driven by exhaust-gas energy that would otherwise have been lost, a turbocharged engine offers several advantages over a naturally aspirated version:

- **Increased power-to-weight ratio**—A turbocharger can generally increase the power and torque output of a diesel engine by as much as 35 per cent above that of a naturally aspirated version. Thus many turbocharged smaller four- and six-cylinder diesel engines are able to do the work of naturally aspirated larger capacity V8 engines.

- **Reduced engine noise**—The turbine housing acts as a noise absorbtion unit for the pulsating engine exhaust gases. So, too, does the compressor section help to reduce pulsating intake noises in the intake manifold. As a result of these factors, a turbocharged engine is generally quieter than a naturally aspirated unit, although a characteristic whine is usually audible when the engine is under load or accelerating.

- **Better fuel economy**—A turbocharged engine has a higher volumetric efficiency than a naturally aspirated engine, giving more complete combustion of the fuel and resulting in lower specific fuel consumption.

- **Reduced smoke output**—Turbochargers

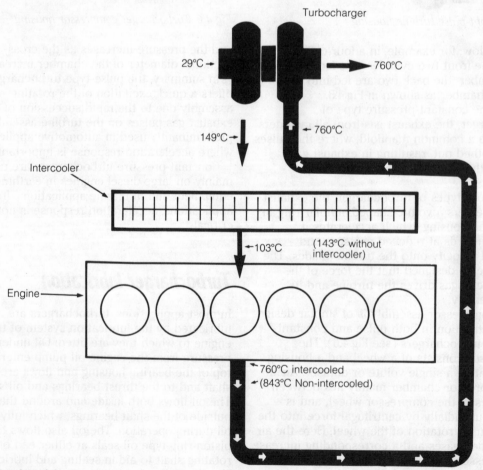

Fig 4.7 Intake/exhaust temperature comparisons—intercooled and non-intercooled engine

supply a surplus amount of air during medium- to high-speed operation, resulting in a much cleaner and efficient combustion phase, which reduces smoke output considerably.

Intercooling

Intercoolers are used to reduce the high temperature of the air leaving the turbocharger, so increasing the density of the intake air. The denser the air, the greater the quantity of fuel that can be burned efficiently, and the greater the subsequent power and torque output of the engine.

When air is compressed, it becomes hot. When the air charge leaves the compressor section of the turbocharger, it is at a much higher temperature than ambient air temperature. When it is heated, the air expands and becomes less dense, so that less oxygen is available in the engine cylinder for combustion. Further, the addition of heated air to the engine can increase engine operating temperature.

These effects become noticeable when the charge-air pressure exceeds 140 kPa, and the high temperature of the air starts to have an adverse effect on the engine performance and operating temperature. Charge-air temperature on non-intercooled turbocharged engines can reach temperatures of 100°C and above.

Intercooling or charge-air cooling is the process of cooling the heated compressed air before it enters the engine cylinders. In so doing, the air charge becomes more dense, allowing additional fuel to be efficiently burned, resulting in increased engine power and torque above that possible with a non-intercooled turbocharged engine. Fig 4.7 shows typical air and exhaust gas temperatures for intercooled and non-intercooled engines.

There are two types of intercoolers in current use, namely the air-to-water and the air-to-air intercooler. Both are heat exchangers, devices that bring a hot medium (in this case, the charge air) into close contact with a cooler medium (either water or air), allowing heat to be conducted from the hot to the cold.

Air-to-water intercooler

This type of intercooler operates by passing the charge air through a water-cooled heat exchanger mounted in the intake manifold

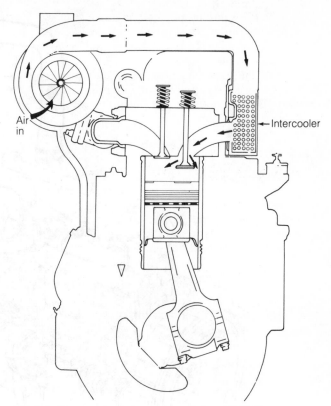

Fig 4.8 Air-to-water intercooler—inlet manifold installation

beside the cylinder head, as shown in Fig 4.8. Because the charge air is hotter than the engine cooling water, which runs through the intercooler, some heat transfer will take place. This transfer of heat reduces the charge-air temperature to a (possible) 85°C (engine operating temperature), if the cooling system is operating efficiently.

Air-to-air intercooler

With air-to-air intercooling, the charge air is passed through a finned heat exchanger (like water in an engine radiator), and the vehicle's forward movement causes air to flow across the fins of the heat exchanger thus cooling the charge air. A typical system is shown in Fig 4.9.

This type of intercooler can reduce charge-air temperature to as low as 15°C above ambient air temperature. With charge-air temperatures as low as this and under pressure between 175–189 kPa, it is possible to provide three times as much air for combustion as is possible in a naturally aspirated engine. Air-to-air intercoolers are used on mobile machines and are mounted in front of the engine radiator.

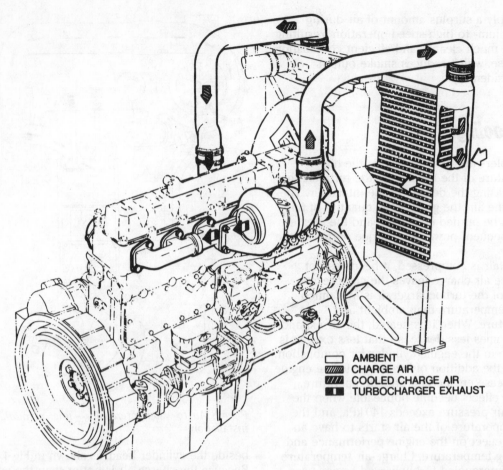

Fig 4.9 Layout of an air-to-air intercooler system

AMBIENT
CHARGE AIR
COOLED CHARGE AIR
TURBOCHARGER EXHAUST

Engine design changes when turbocharging

Because turbocharged engines operate under higher stress and temperature than naturally aspirated engines, original equipment

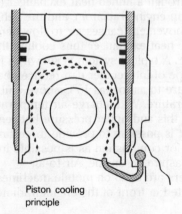

Piston cooling
principle

manufacturers design such engines to tolerate these conditions. Engine oil coolers are practically a standard fitment on turbocharged diesels; so too are oil sprays (as shown in Fig 4.10), which spray engine oil on the underside of the piston crown for efficient cooling. To cater for the extra oil-flow requirements of these additional features, larger capacity oil pumps are fitted. Further, stronger pistons are fitted to handle the increased loads.

Effects of altitude on turbocharged diesel engines

When an internal combustion engine is operated at high altitude where the air is less dense than at sea level, the quantity of air (and

◄ *Fig 4.10 Oil-spray piston cooling*

oxygen) entering the engine cylinder on induction stroke is insufficient for combustion of the normal fuel charge. As a result, the performance of the engine falls in proportion to the altitude at which it is being operated.

Turbocharged engines are not affected to the same degree. As the air becomes less dense with altitude, the turbocharger spins faster due to the reduced pumping load, producing a compensating effect. However, there is still a decrease in engine performance, although this is much less than for naturally aspirated engines.

On turbocharged engines, power output is reduced by approximately 1 per cent per 300 m rise in altitude above sea level. When the operating altitude is in the vicinity of 2000 m, the fuel delivery to the engine must be decreased according to engine specifications to prevent damage to the turbocharger due to overspeeding.

Turbocharger controls

In certain applications where fast acceleration is needed, engines are fitted with large-capacity turbochargers that require a speed-control device. This device, commonly referred to as a wastegate (as shown in Fig 4.11), prevents the turbocharger from over-speeding and subsequently over-boosting and damaging the engine.

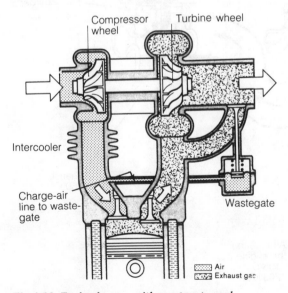

Fig 4.11 Turbocharger with wastegate and intercooler

On the other hand, the smaller turbochargers fitted to stationary engines or slow moving diesel-powered equipment are self regulating in their maximum speed and charge pressure by the design of the turbine and compressor housings. Because these engines operate in a narrow speed band with constant high charge pressures, there is no need for controls that regulate charge pressure relative to changing engine speeds.

Therefore it is essential that the turbocharger installed be matched to the engine and performance requirements. Fig 4.12 shows a typical performance comparison between a standard turbocharger and a turbocharger fitted with a wastegate.

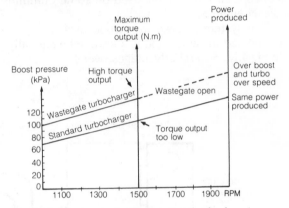

Fig 4.12 Wastegate turbocharger/standard turbocharger performance comparison

The standard turbocharger with its smaller output capacity supplies sufficient charge air for combustion of a full fuel charge only when the engine is operating in the high torque and speed ranges. But down in the medium speed range the charge-air pressure is considerably lower, generally resulting in incomplete combustion, black exhaust smoke, poor acceleration and lack of power.

The wastegate turbocharger, however, is of a higher output capacity and capable of delivering sufficient charge air for complete combustion of the fuel during acceleration as well as in high-torque situations. As the engine speed and exhaust-gas energy increase, so the turbocharger speed increases and the charge-air pressure rises. Without the wastegate, charge pressure would continue to rise, with considerable risk to both the engine and the turbocharger.

However, the increasing air pressure acts on the diaphragm in the wastegate until, at a predetermined pressure, the resulting force is sufficient to compress the spring and open the exhaust bypass passage. This allows sufficient exhaust gas to bypass the turbine to prevent any further rise in turbocharger speed and subsequent charge-air pressure.

Wastegate turbochargers are generally fitted to faster moving earthmoving equipment, for example dump trucks and road scrapers, as well as high-performance automotive vehicles.

Series turbocharging

Series boost turbocharging is a new design concept currently being used on some Cummins diesel engines.

The system utilises two turbochargers connected in series, and is shown schematically in Fig 4.13. Several advantages over the

conventional design using one larger capacity turbocharger are claimed for this concept:

- better specific fuel consumption
- reduced acceleration smoke
- improved torque peak
- higher overall pumping efficiency.

Precautions when operating turbocharged engines

Engine starting procedure

A turbocharged engine should always be allowed to idle when it is started until the engine oil pressure has built up to normal operating pressure. Starting an engine with the throttle wide open will result in the turbocharger operating at high speed with very

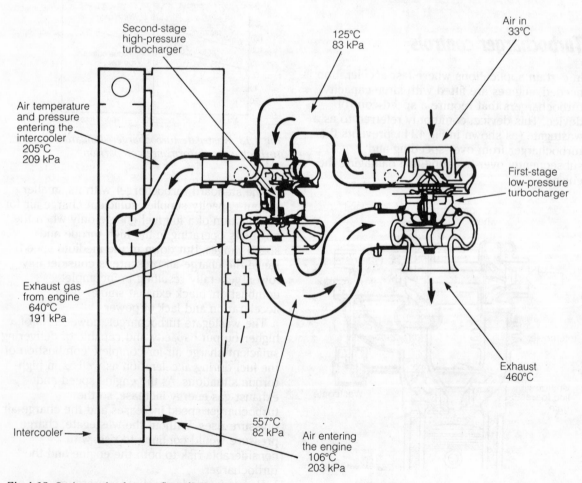

Fig 4.13 Series turbocharger flow diagram

little oil being circulated through its bearings, with resultant accelerated wear on the rotating assembly and bearings of the turbocharger.

Engine shutdown

Before a turbocharged diesel engine is shut down, the engine should be run at idle speed for 3–4 minutes. This will allow the high-speed rotating assembly to slow down, allow the engine operating temperature to normalise, and allow excessive heat to be dissipated from the turbocharger.

If a turbocharged engine is shut down while operating at high speeds or under load, the turbocharger rotating assembly will continue to rotate for some time without oil for essential lubrication and cooling. Because the exhaust turbine shaft operates at high temperature during engine operation, once the oil flow to the bearing housing stops, the heat in the shaft and housing is sufficient to decompose the oil to form gums and varnish, leaving no lubricating residue and causing premature wear to the rotating shaft, its support bearings and the bearing housing.

There are now ways of protecting the turbocharger against sudden engine shut down. An automatic timer unit can be fitted to the engine shutdown system, which over-rides the stop control and allows the engine to idle for a number of minutes before stopping.

Another method utilises an oil accumulator mounted on the engine, which is charged by the engine lubrication system during operation. When the engine is shut down, oil is forced from the accumulator via a check valve, to the turbocharger bearing housing, and lubricates the bearings for approximately 30 seconds.

Turbocharger service

Being unlike any other component of the engine, turbochargers have the need for specific service procedures. Further, because of the high operating temperatures and high operating speeds, turbochargers are susceptible to heat cracking and unbalance to a degree seldom seen in engine ancillary equipment.

Engine lubrication

Although not specifically a turbocharger service item, it is of the utmost importance to carry out regular oil and filter changes on turbocharged engines. It is just as important to monitor engine oil pressure and the quality of engine oil used. Turbocharged diesel engines should only operate on the lubricating oil recommended by the engine manufacturer, which is usually of a different classification from that required for naturally aspirated engines.

Inspection and cleaning

Many engine manufacturers and/or turbocharger manufacturers recommend periodic disassembly, inspection and cleaning of the compressor housing and turbine. A small deposit on the turbine wheel can seriously affect turbocharger performance and should be removed. The usually recommended cleaning procedure is to use a solvent and soft (not wire) brush, taking care to ensure that solvent does not enter the turbocharger bearing housing.

In addition to inspecting for deposits, the components should also be inspected for physical damage, paying particular attention to the turbine and compressor wheels and housings.

Foreign-object damage

The high-speed rotating assembly of a turbocharger is balanced to exacting standards. For this reason, a turbocharger should not be put back in service if any part of the rotating assembly or housings is damaged in any way. If damage by foreign objects does occur to the rotating members, the unit should be disassembled and the damaged component replaced, in consultation with the appropriate workshop manual.

Turbocharger installation

When a turbocharger is installed on an engine, and before the engine is started, engine oil should be poured into the oil intake hole in the bearing housing. This prelubricates all the bearings and sealing rings. The oil intake pipe can then be refitted and the engine started.

Turbochargers that have not operated for long periods will lose their residual lubrication. Therefore the engine should be cranked over in the 'no-fuel' position (stop lever actuated) until oil pressure registers on the gauge—this will prelubricate the turbocharger prior to start-up.

Operating checks

With experience, it is possible to gain a good indication of turbocharger operation from the sounds it produces in operation. The engine should be operated through all speed and load ranges while paying particular attention to unusual noises coming from the turbocharger. Generally, the only noise that should be heard is a high-pitched whine that occurs when the engine is placed under load or accelerated.

Turbocharger overhaul

Overhauling a turbocharger should not be attempted without reference to the appropriate workshop manual. However, the following provides a general overall description of the disassembly, inspection and reassembly of a typical turbocharger.

Disassembly

Clean the exterior of the turbocharger with a non-caustic cleaning solvent. Mount the unit on to a special fixture or in a vice as shown in Fig 4.14.

Before removing the housings use a scribing tool to mark the relative location of compressor and turbine housings to the bearing housing.

Remove the clamp or bolts securing the compressor housing and lift off the housing. Next remove the turbine housing clamp or bolts and lift the bearing housing clear of the turbine housing.

Mount the bearing housing in an upright position in a special fixture or soft-jawed vice as seen in Fig 4.15, making sure that the vice jaws grip the turbine wheel extension nut only, and not the turbine fins.

Unscrew the locknut retaining the compressor

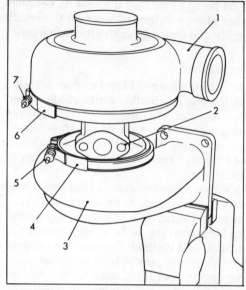

1 Compressor cover	5 Locknut – 'V' clamp
2 Bearing housing	
3 Turbine housing	6 'V' clamp
4 'V' clamp	7 Locknut – 'V' clamp

Fig 4.14 Turbocharger correctly held in a vice

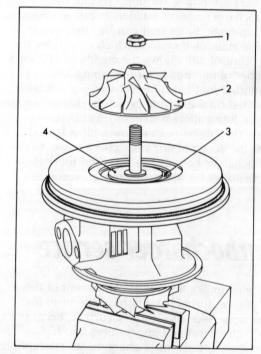

1 Locknut
2 Compressor wheel
3 Circlip
4 Bearing housing

Fig 4.15 Bearing housing correctly held in a vice

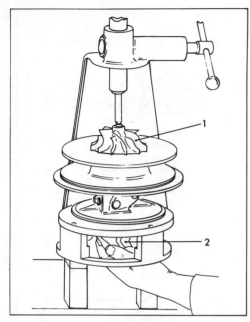

1 Compressor wheel   2 Turbine wheel

Fig 4.16 Pressing the turbine shaft from the compressor wheel

wheel and lift the wheel off the turbine shaft. On some turbochargers the compressor wheel is shrunk onto the shaft and will have to be pressed off as shown in Fig 4.16.

With the compressor wheel removed, the turbine shaft can be removed from the turbine end of the bearing housing.

Next remove the circlip or capscrews from the thrust assembly located at the compressor end of the bearing housing, and, with a piece of wood dowling inserted into the centre hole of the thrust assembly, lever the assembly out of the bearing housing. Remove the remainder of the thrust assembly and the two plain turbine shaft support bearings and circlips. Finally remove all piston ring seals from the turbine shaft and thrust assembly spacer sleeves. The disassembled turbocharger is now ready for cleaning and inspection.

Inspection

Clean all components for inspection using an approved cleaner; caustic solutions will damage aluminium components and must not be used. All parts should be soaked in cleaning fluid until all foreign deposits have been removed. The turbine wheel and turbine housing can be bead blasted to remove carbon deposits provided that the smooth surface of the turbine shaft is

protected. After soaking the components, blow out all passages and compartments with compressed air.

Generally, no parts should show signs of wear, corrosion or damage. A wear evaluation in accordance with the manufacturer's specifications will determine whether parts are replaced or reused. Refer to Fig 4.17 for an exploded view of the layout of the turbocharger parts described below.

Bearing housing—The bearing housing must not show wear marks due to contact with rotating parts. Inspect the bores in which the bearings run for scores, and measure their diameter with a telescopic gauge. The bore diameter is critical, with permissible wear approximately 0.025 mm. If the bores are scored or are worn oversize or oval, a new housing should be fitted or the old housing sleeved.

Turbine shaft bearings—Whenever the turbocharger is overhauled the shaft bearings (21, 14) must be renewed, regardless of their condition.

Thrust bearing assembly—The thrust bearing (10) and thrust rings (12, 9) should be renewed, regardless of their condition. Measure the thrust spacer (11) and spacer sleeve (7) and inspect their surfaces for scoring or heat discoloration. Discard if worn, scored or discoloured.

Rotating assembly—Examine the turbine wheel (16) and shaft (18) for any signs of wear. Inspect the fins of the wheel for cracks, carbon deposits, distortion of shape, erosion wear on the tips and foreign object damage. The shaft must show no signs of wear, scoring or discoloration. Measure the shaft journals for exact size and ovalness and check against the manufacturer's figures. Check the width of the seal ring groove in front of the turbine wheel to ensure that the groove has not worn oversize.

Inspect the compressor wheel blades for tip damage due to foreign objects or rubbing on the compressor housing. The wheel should also be checked for signs of rubbing between the underside of the wheel and the bearing housing. Any foreign object damage or wear marks on either the turbine or compressor wheels will cause them to run out of balance, creating undue vibration and wear within the turbocharger.

Check that the bore of the compressor wheel is of the correct size—it may be either an interference fit or a slide fit onto the turbine shaft, depending on the type of turbocharger.

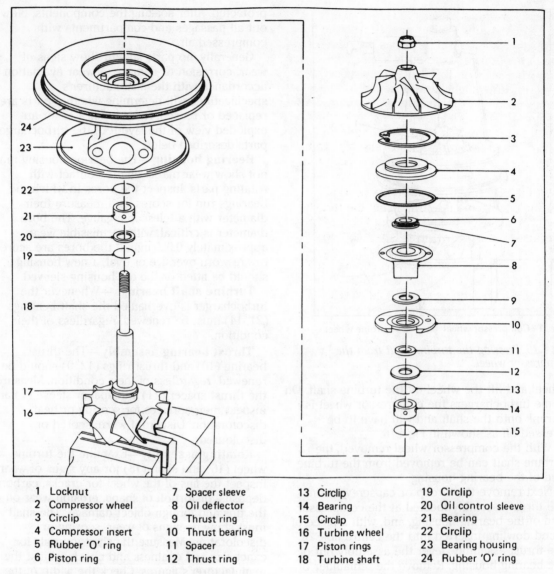

1 Locknut	7 Spacer sleeve	13 Circlip	19 Circlip
2 Compressor wheel	8 Oil deflector	14 Bearing	20 Oil control sleeve
3 Circlip	9 Thrust ring	15 Circlip	21 Bearing
4 Compressor insert	10 Thrust bearing	16 Turbine wheel	22 Circlip
5 Rubber 'O' ring	11 Spacer	17 Piston rings	23 Bearing housing
6 Piston ring	12 Thrust ring	18 Turbine shaft	24 Rubber 'O' ring

Fig 4.17 Exploded view of bearing housing assembly

Turbine and compressor housings—
Inspect the turbine housing for erosion, cracking and rub marks in the vicinity of the turbine wheel. Also, the turbine housing to the exhaust manifold mounting flange is to be checked for surface trueness.

The compressor housing should also be checked for wheel rub marks and cracking.

Reassembly

When reassembling a turbocharger, cleanliness is vital to a long service life. All piston-ring-type seals, 'O' rings, lock tabs, circlips and the compressor wheel retaining nut should be

automatically renewed, together with all other parts that require replacement during the inspection period. Throughout assembly, lubricate all rotating parts with clean engine oil. Install the bearing retaining circlip (19) into the turbine end of the housing, taking care not to scratch the bearing bore.

Fit the oil control sleeve (20) and the turbine end bearing (21) into the bore. Install the two inner bearing retainer circlips (22, 15), and the compressor end bearing (14) in the bore. With the turbine wheel (16) mounted in a vice as shown in Fig 4.17, fit the piston ring seals (17) into the oil ring grooves. Install the turbine shaft

(18) in the bearing housing (23) from the turbine end. As the shaft is pushed into the housing, resistance will be felt as the chamfered edge of the housing bore butts against the piston ring seals on the shaft. Apply moderate pressure with a slight turning action to the turbine shaft, and the chamfer will compress the piston ring seals and allow them to enter the bore. If the piston ring seals don't enter the housing, rotate the turbine shaft and try again. Never use force to install the turbine shaft into the bearing housing.

With the compressor end of the bearing housing now facing upward, install the thrust ring (12), followed by the spacer (11) and thrust bearing (10), over the turbine shaft and down onto the dowel pins in the bearing housing. Then place the upper thrust ring (9) over the turbine shaft and down onto the thrust assembly, followed by the oil deflector (8), which sits on top of the thrust assembly and is located by the dowels in the bearing housing.

Next, install the piston ring seal (6) onto the spacer sleeve (7) and insert the sleeve into the compressor insert (4), using light finger pressure. Fit a new 'O' ring (5) to the compressor insert, and place the insert over the turbine shaft and down into the bearing housing. Secure the insert with the retaining circlip (3).

Install the compressor wheel (2) on the turbine shaft. If it is an interference fit, it will have to be heated in hot clean engine oil to expand it before sliding it onto the turbine shaft. (Refer to the manual for correct oil temperature.) Secure the wheel with the self-locking nut, tensioned to the required torque.

In order to check for correct running clearances of the rotating assembly, mount a dial indicator onto the compressor end of the bearing housing and measure the axial movement of the turbine shaft (refer to Fig 4.18).

Reposition the dial indicator mounting so the rotating assembly can be checked for radial clearance (refer to Fig 4.18).

Refer to the turbocharger specifications for correct turbine shaft clearances. Excessive clearances must be corrected before proceeding further.

Install a new 'O' ring (24) onto the bearing housing and fit the compressor housing, at the same time aligning the assembly marks on both housings. Refit the 'V' clamp and tighten.

Turn the turbocharger over and install the turbine housing, once again aligning the

Fig 4.18 Checking axial and radial shaft movement

assembly marks. Secure the housing with the 'V' clamp or the bolts and lock tabs (as the case may be).

Finally, cover all openings until the turbocharger is to be installed on the engine.

Performance testing

An accurate method of gauging an engine's performance output is by using test gauges to measure certain aspects of engine operation (see Fig 4.19). A quick and accurate way of checking that the turbocharged engine's output is in accordance with the manufacturer's specifications is to measure the charge-air pressure in the intake manifold when the engine is operated under full load. Correct charge-air pressure is indicative of:

- correct metering and delivery of fuel from the injection pump and injectors
- acceptable compression pressures
- correct injection timing
- efficient turbocharger operation
- unrestricted engine breathing.

Full load can be applied to an engine by loading it appropriately, care being taken to ensure that damage is not caused to either the engine or the equipment being driven.

If the engine is installed in a piece of mobile equipment fitted with a torque converter, full load can be applied by applying the brakes and stalling out the torque converter with the

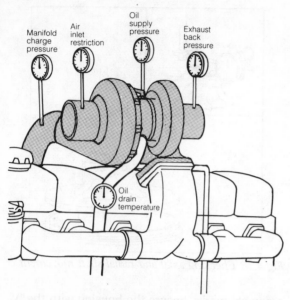

Fig 4.19 Gauge installation for performance testing

transmission engaged in top gear. The engine should be running at full throttle and rated engine speed during any stall test.

Note Do not run the engine under stall conditions for more than 30 seconds at one time as serious overheating of the engine and torque converter will occur.

Blowers

*T*he term 'blower' is used to refer to the air supply pump that supplies the air under pressure to two-stroke engines, the primary purpose being to scavenge burnt gas from the engine cylinder. As a secondary function, blowers usually ensure that the cylinder is completely filled with fresh air by raising the cylinder pressure to above atmospheric. The primary function is achieved by having both the inlet and exhaust ports open together, allowing the fresh air to sweep through the entire cylinder, while the secondary function is performed by closing the exhaust port (or valve) before the inlet port (or valve), thus allowing the pressure to build up in the engine cylinder before the air supply is shut off. Almost all engine manufacturers use Roots blowers for this purpose.

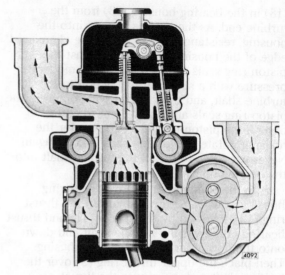

Fig 4.20 Scavenge air flow through two-stroke diesel engine

Construction

The basic Roots blower consists of three major subassemblies: an oval housing, a pair of rotors and associated bearings, gears and seals, and two end covers.

The rotors are geared, one to the other, and turn in opposite directions in the housing, supported in antifriction bearings in the end covers.

Each rotor consists of a steel shaft with (usually) three lobes surrounding it. These are generally twisted along their length and are known as helical rotors (see Fig 4.21). Rotors with two lobes only are also used in some blowers.

Although designed to pump air, the blower rotors are not fitted with seals, but rely on the precise and limited clearances between the rotors themselves and between the rotors and the housing.

Operation

The operation of a blower is similar to that of a gear-type oil pump. The lobes on the rotors fit together like gears in mesh, and turn in opposite directions. As one lobe moves from the valley between the two lobes on the other rotor, it creates a void that is filled with air. This is the inlet action.

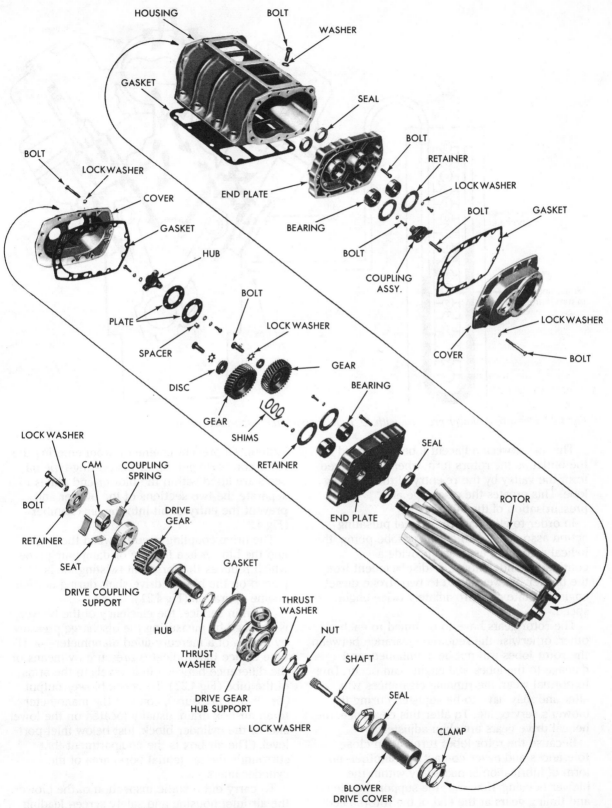

Fig 4.21 Exploded view of blower assembly and drive

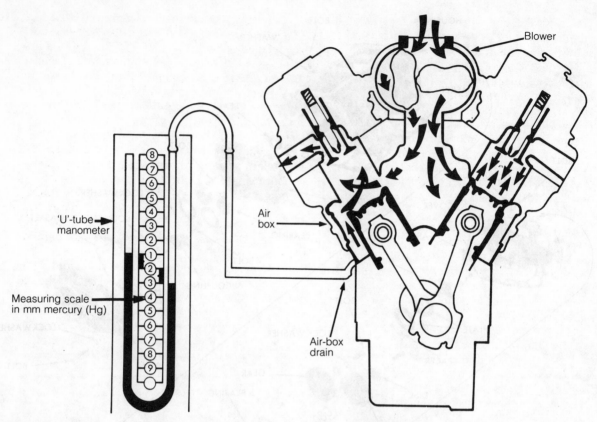

Fig 4.22 *Measuring blower pressure with 'U' tube manometer—schematic view*

The air between adjacent lobes is carried to the outlet as the rotors turn, where it is forced from the valley by the re-entry of the meshing lobe. This creates the discharge and pressurisation of the air.

In order to eliminate the typical pulsating action associated with a gear or lobe pump, the helical rotors are used and provide a continuous and uniform air displacement from the blower. Blowers fitted to two-stroke diesel engines rotate at approximately twice engine speed.

The rotor gears have to be timed to each other, otherwise the required clearance between the rotor lobes will not be maintained, and damage to the lobes and engine can occur. Due to normal wear, the running clearances will alter and may have to be adjusted during the blower's service life. To alter this clearance, the helical drive gears are shim adjusted.

Because the rotor lobes turn within close tolerances and never contact one another, no form of lubrication is necessary within the blower housing. However, the support bearings and timing gears at the end of the rotors need constant lubrication from the engine lubrication system. To prevent engine oil from entering the rotor compartment, lip or piston-ring-type oil seals are fitted within the blower end plates to separate the two sections of the blower and prevent the entry of oil into the air chamber (Fig 4.21).

The drive coupling used between the engine and the blower is a flexible or dampening type, which reduces the torsional twisting loads placed on the blower drive shaft during normal engine operation (Fig 4.21).

When in service, the efficiency of the blower is checked by measuring the discharge pressure by means of a mercury-filled manometer—a 'U'-tube device that indicates pressure by means of the difference between fluid levels in the arms of the tube (Fig 4.22). To check blower output (or air-box pressure), connect the manometer to an air-box drain, usually located on the lower side of the cylinder block, just below inlet-port level. (The air box is the compartment that surrounds the tangential ports area of the cylinder liners.)

To carry out a static inspection of the blower, the air inlet housing and safety screen leading into the blower inlet must be removed. The safety screen is a wire gauze screen located at the blower inlet to prevent the entry of foreign objects.

To detect a worn flexible drive coupling, hold the driving rotor and try to rotate it. The rotor should move, against the flexing of the coupling, from 10–16 mm as measured at the lobe crown. On release, the rotor should spring back at least 6 mm. If the rotors cannot be moved as described above, the drive coupling should be inspected and replaced if necessary. A faulty blower drive coupling can be detected by a rattling noise within the vicinity of the coupling, during engine operation.

The rotors should be examined for evidence of contact by visually checking the edges of the rotor lobe crowns and mating rotor roots for signs of scoring or contact wear marks. At the same time, the drive gear backlash should be checked by mounting a dial indicator on the blower housing with the indicator probe perpendicular to, and in contact with, the side of the lobe. The backlash is measured by moving the rotor in one direction and then the other within the limits of the gear teeth clearance (the second rotor must not move). The allowable backlash is generally 0.1 mm, and if this is exceeded, the blower drive gears will have to be renewed.

During an inspection, oil on the blower rotors indicates leaking rotor shaft oil seals, which may be the result of worn rotor bearings, worn seals or lip-type seals that have been turned inside out due to the closure of the emergency shutdown flaps during high-speed engine operation. The emergency shutdown flap is a shutter mounted on the inlet to the blower, which, when operated, closes off the air supply to the blower (and engine), thereby stopping the engine. The emergency shutdown flap is to be used **only** in an emergency when the normal method of engine shutdown is inoperative.

Finally, the safety screen should be checked for signs of damage and, after the emergency shutdown flap has been refitted, the latch checked to ensure that the flap remains open during engine operation.

Disassembly of the blower

When diassembling, inspecting and reassembling a blower, the appropriate workshop manual should always be used. However, as a guide to procedures, a general description of overhauling a blower fitted to a Detroit Diesel Allison 'V' series 71 engine is detailed below.

After the governor assembly and fuel feed pump have been removed from the blower, the drive gears are ready to be pulled off the rotor shafts. By placing rag between the two rotor lobes to prevent the rotors from turning, unscrew and remove the allen-headed bolts retaining the drive gears. Mount a suitable puller and remove both drive gears together— because they are helical gears, pulling one alone will cause partial rotation of one rotor against the other. Also remove the spacer shims from behind the gears and mark the gears and shims to ensure correct positioning of parts on reassembly.

Remove the bearing retainer bolts and retainers for all four bearings and, with the aid of a puller, remove the rear end plate and bearing assembly. Repeat the procedure for the front end plate. Next, remove the two rotors from the blower housing. With the aid of a press, remove the bearings and oil seals from both end plates.

Inspection of components

Prior to inspection, all the blower components should be washed in a suitable cleaning solution and dried off with compressed air. All the parts of the blower should be examined and measured to determine whether they should be reused.

The rotor lobes should be examined for burrs and scoring, especially on the sealing edges. Witness marks along the full length of the lobe usually indicate worn bearings or excessive backlash in the timing gears. Small imperfections on the lobe or rotor roots can be removed with fine emery tape.

The internal surface area of the blower housing and the blower end housings should be checked for scoring. Any fine score marks can be removed with fine emery tape. Deep score marks will necessitate replacement of the housing. The blower end housings should also be checked for surface flatness and for evidence of bearing rotation in the housing.

Oil seal ring carriers and running surfaces (if fitted) should be examined for wear and scoring. All oil seals, bearings and lock tabs should automatically be renewed during a complete overhaul.

Assembly

The assembly procedure as described below is that for a typical blower with lip-type oil seals sealing the rotor shafts (some blowers use other types of seal).

Install the lip-type oil seals in both of the end plates so that they sit approximately 0.125 mm below the finished surface of each end plate.

Fit the front end plate to the blower housing with the three oil holes on the side of the end plate facing the cylinder block. Because no gaskets are used between the end plates and housing, ensure that the mating surfaces are smooth and clean.

Before installing the rotors, establish where the driving rotor is to be placed in the blower housing relative to the drive shaft coming from the engine. The driving rotor lobe and its associated drive gear are identified by the way they both form a right-hand helix. The driven rotor lobe and drive gear form a left-hand helix.

Match the rotors together so that the master splines (omitted serrations) lie in a horizontal position (as shown in Fig 4.23), and face the left when looking at the rear of the blower (non-drive end). Install the rotors into the blower housing and fit the rear end plate.

Reposition the assembled blower into a vertical position and, with its housing and rotor shafts supported, install the rotor shaft bearings.

After all the bearings have been fitted, reinstall the drive gear spacer shims on their respective shafts.

The drive gears can now be refitted to the shafts, taking note that the left- and right-hand helical gears are matched to their corresponding rotor lobes, and that the master splines on the rotor shafts and drive gears are in alignment with one another prior to the gears being pressed on.

With rag placed between the rotor lobes to prevent them from turning, the drive gears can be pressed onto the shafts by means of a puller bolt screwed into the end of the blower shaft. As with removal, the gears must be installed together, to prevent rotation of one in relation to the other.

With the end plates bolted up and the drive gears installed, the rotors can be timed to each other. During operation, the rotor lobes run with a slight clearance between them. This clearance can be adjusted by moving either one of the helical drive gears on the rotor shaft in

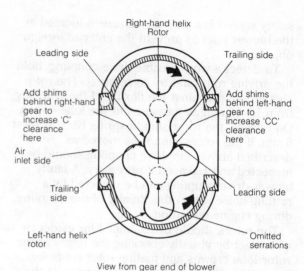

Fig 4.23 Blower lobe clearance adjustment guide

or out relative to the other gear. The positioning of the drive gears is determined by addition or removal of spacer shims from between the gears and bearings.

During the blower timing procedure, if the left-hand helix gear is moved in, the left-hand rotor lobe will turn counterclockwise (CC) and if the right-hand helix gear is moved in, the right-hand rotor lobe will turn clockwise (C).

The running clearance between the rotor lobes should be checked with the aid of a feeler strip as shown in Fig 4.25—the trailing and leading edges as shown in Fig 4.23 must be measured from the inlet and outlet side of the blower.

The measured running clearances can be compared with the specifications (Fig 4.24) and the necessary repositioning of the timing gears carried out. As a reference, the adding or subtracting of 0.075 mm shim material will revolve the rotor approximately 0.025 mm.

When timing of the blower rotors is completed, any accessories mounted on the blower can be refitted and the blower installed onto the engine.

Bypass blowers

Detroit Diesel Allison are now installing a modified form of Roots blower to turbocharged Detroit Diesel two-stroke engines. This 'bypass' blower, as it is known, directs excess and unnecessary boost air back into the intake of

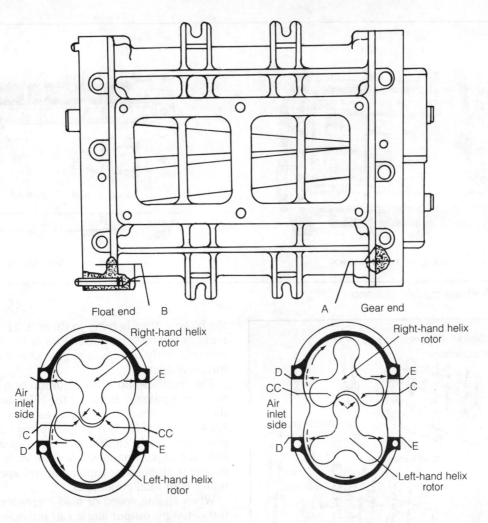

VIEWS FROM GEAR END OF BLOWER

ENGINE	BLOWER	IDENT*	A	B	C	CC	D	E
6 & 12 CYL.	STD.		.007″	.012″	.008″	.006″ TO .010″	.015″	.004″
6 & 12 CYL.	TURBO		.007″	.012″	.008″	.006″ TO .010″	.015″	.004″
8 & 16 CYL.	STD.		.007″	.014″	.010″	.004″ TO .008″	.015″	.004″
8 & 16 CYL.	TURBO		.007″	.014″	.010″	.004″ TO .008″	.015″	.004″
8 & 16 CYL.	TURBO	LT	.007″	.024″	.010″	.004″ TO .008″	.030″	.004″

†

Chart indicating the areas where rotor lobe clearances are to be checked and the allowable clearances in these areas.

*Identification stamped on blower
†Dimensions are in inches as supplied by manufacturer.
   Note   Time rotors to dimensions on chart for clearance between trailing side of right-hand helix rotor and leading side of left-hand helix rotor (CC) from both inlet and outlet side of blower.

Fig 4.24 Rotor clearance data

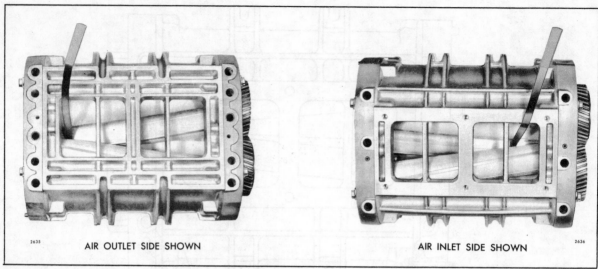

AIR OUTLET SIDE SHOWN   AIR INLET SIDE SHOWN

Fig 4.25 Measuring lobe clearances

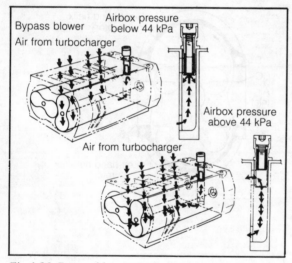

Bypass blower

Air from turbocharger

Airbox pressure below 44 kPa

Airbox pressure above 44 kPa

Air from turbocharger

Fig 4.26 Bypass blower air flow

the blower, thus helping to drive it. This in turn reduces the internal power loss of the engine and ultimately increases engine power and torque output.

The modification is the addition of a bypass valve in a passage between the inlet and discharge sides of the blower, as shown in Fig 4.26. The valve is a simple spring-loaded relief valve, held closed by its load spring to keep the bypass passage blocked while the engine is stopped or running at low speed or under light load conditions.

When engine speed or load increases, the turbocharger output alone can provide adequate air flow for scavenging and charging the engine cylinders, making the use of the blower unnecessary. As air-box pressure increases to 34 kPa, the bypass valve begins to open, and is fully open at 44 kPa, allowing the excess air to flow back into the inlet port of the blower, thus helping to drive it. The blower does less work and so requires less power input, increasing the engine's fuel economy.

To keep the blower bypass valve clean and to maintain its proper function, a small amount of air is allowed to bleed past the valve and through a vent hose into the engine crankcase.

5
The engine indicator and theoretical heat cycles

A heat cycle may be described simply as a sequence of pressure conditions necessary in the cylinder of an IC engine to efficiently convert the chemical energy in the fuel to mechanical energy available at the engine flywheel. A mechanical cycle, either four-stroke or two-stroke, is used to create the necessary pressures in the engine cylinder to fulfil the requirements of the heat cycle.

The cycles are presented pictorially as a graph of the pressure–volume relationship inside the cylinder of the working engine. This graph is termed a heat cycle diagram or indicator diagram, from a device called an engine indicator used to create the graph.

Based on the concept of a perfect engine, engineers developed theoretical heat cycles termed **ideal heat cycles**. In practice, the actual heat cycle of real engines varies considerably from the theoretical.

The engine indicator

B ecause the pressure–volume relationship inside the cylinder is directly related to the performance of the engine, engineers developed a mechanical engine indicator for steam engines and later modified it to suit low-speed internal combustion engines. Indeed, indicator diagrams are regularly taken of low-speed marine engines to give the marine engineer valuable information on condition, performance, timing, etc.

However, the mechanical indicator is not suitable for modern high-speed engines, where electronic indicators are used. These would seldom be seen outside manufacturers' engine-testing laboratories.

The indicator diagram

A heat-cycle diagram, whether theoretical or produced on an engine by an engine indicator, relates cylinder pressures to atmospheric pressure. A line to represent atmospheric pressure—the atmospheric line—is a feature of any indicator diagram.

In Fig 5.1a, the first stage of the development of an indicator diagram is shown. Line *AL* is the atmospheric line and line *AB* represents the partial vacuum created as the piston moves from TDC to BDC on the intake stroke.

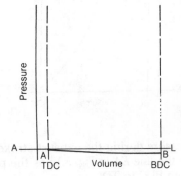

Fig 5.1a

As the piston moves toward TDC on compression stroke, the pressure rise is shown by line *BC* in Fig 5.1b.

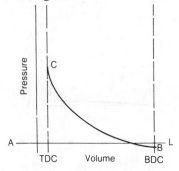

Fig 5.1b

In this example, combustion is considered to begin at TDC and maintain the pressure in the engine cylinder as the piston begins the power stroke (line *CD* in Fig 5.1c).

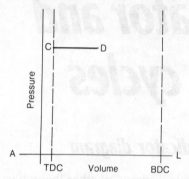

Fig 5.1c

Line *DE* (Fig 5.1d) represents the fall in cylinder pressure from the end of combustion to BDC. At BDC, the cylinder pressure is slightly above atmosphere.

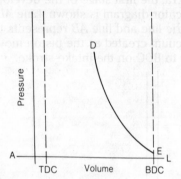

Fig 5.1d

The pressure during the exhaust stroke is shown by the line *EA* in Fig. 5.1e as the piston moves from BDC to TDC.

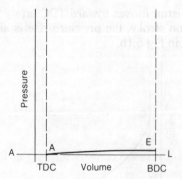

Fig 5.1e

The complete indicator diagram is, of course, a continuous graph, not a series of separate lines. The complete diagram is shown in Fig 5.2, and is clearly the combination of the stages described in relation to Fig 5.1. In practice, the lines representing the pressures on intake and exhaust strokes are usually omitted.

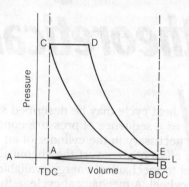

Fig 5.2

Theoretical and actual heat cycles

Engineers have proposed a number of theoretical heat cycles for internal combustion engines, the most noticeable being the constant pressure cycle by Dr Rudolf Diesel and the constant volume cycle, which was first successfully applied to an IC engine by Dr A.N. Otto.

The constant-pressure cycle

The operation of large slow-speed diesel engines, which are used for marine and stationary purposes, is based on the ideal constant-pressure cycle.

In this cycle, air only is compressed in the engine cylinder and when the piston reaches TDC, fuel injection begins. Fuel is injected into the cylinder gradually and ignition takes place when the first fuel droplets come in contact with the hot compressed air in the cylinder. As the fuel charge enters the cylinder, it burns at a controlled rate and so produces a steady expansion of the gases in the cylinder. As this expansion takes place, the piston—which has passed through TDC—moves down the cylinder, thereby increasing the enclosed cylinder volume.

Thus, by controlling the rate of admission of the fuel, the pressure neither rises nor drops as combustion continues, but is constant for a large number of degrees of crankshaft rotation. Combustion of the fuel charge, therefore, is said to take place at constant pressure, the constant pressure during combustion being the same as the pressure at the end of the compression stroke.

As combustion occurs at constant pressure, engines operating on this cycle can use higher compression ratios than those operating on the constant-volume cycle, without damage to engine components.

Fig 5.3a shows the type of pressure–volume diagram that would be taken from the cylinder of an imaginary perfect slow-speed diesel engine operating on the ideal constant pressure heat cycle.

As the piston moves up the cylinder from BDC to TDC, the air in the cylinder is compressed and so its pressure and temperature are increased. This is shown on the pressure–volume (P–V) diagram by the line A–B. At point

B (TDC), maximum cylinder pressure is attained and fuel injection starts. The fuel entering the cylinder is ignited by the heat of the compressed air and combustion begins. Combustion of the fuel charge progresses as the piston is forced back down the cylinder from the TDC position, and so the pressurised gases created by combustion of the fuel are able to expand. The result of this is that combustion pressure in the cylinder remains constant, due to the increased cylinder volume during combustion. This is shown on the P–V diagram by the line B–C.

Once combustion stops, the cylinder pressure falls as the piston is forced back down the cylinder by the expanding gases. This is shown on the P–V diagram by the line C–D. At BDC, the exhaust valve opens and the pressure drops to atmospheric. This is represented by the line D–A.

The P–V diagram shown in Fig 5.3b has been taken from the cylinder of an actual slow-speed diesel engine operating on the constant-pressure cycle. Because such things as valve timing and fuel injection timing have been altered to suit actual engines, the P–V diagram from an actual engine differs from that of a theoretical engine. The mechanical condition of the engine and the quality of the fuel used also affect the shape of the actual P–V diagram. Note that the inlet and exhaust strokes have been omitted from Fig 5.3a and Fig 5.3b.

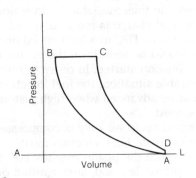

Fig 5.3a

The constant-volume cycle

This cycle was proposed by Beau de Rochas in 1862, and it was successfully applied to an IC engine by Dr A.N. Otto in 1876.

The operation of the petrol engine, which compresses an air–fuel mixture in its cylinders, is based on the constant-volume cycle. In the ideal constant-volume cycle, instantaneous and complete combustion of the fuel charge is assumed to take place when the piston has completed its compression stroke and is stationary at TDC. This means that the complete fuel charge is burnt while the clearance volume on the top of the piston is constant or unaltered. Although a sharp rise in pressure results from the instantaneous combustion of the fuel charge, this pressure is not sustained.

The sudden and rapid rise in pressure that occurs in the constant-volume cycle limits the

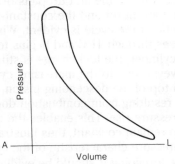

Fig 5.3b

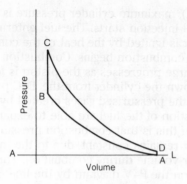

Fig 5.4a

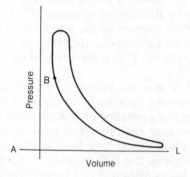

Fig 5.4b

compression ratio that can be used in an engine operating on this cycle. If the compression ratio used is too high, the pressure rise in the cylinder during combustion is excessive and damage to engine components is possible.

Figure 5.4a shows the P–V diagram that would be taken from the cylinder of an imaginary perfect petrol engine operating on this ideal constant-volume heat cycle.

As the piston moves up the cylinder from BDC to TDC, the air–fuel mixture is compressed and its pressure is increased. This is shown by the line A–B on the P–V diagram. At point B (TDC) the air–fuel mixture is ignited by an electric spark and instantaneous combustion occurs. A sharp but unsustained pressure rise results, and this is represented by the line B–C on the P–V diagram. Note carefully that as the pressure rises from B to C as a result of combustion, the volume does not change—it is constant during the combustion period. Due to the sharp pressure rise that occurs at TDC, the piston is forced down the cylinder on its power stroke and the cylinder pressure falls (line C–D).

At the bottom of the power stroke, the exhaust valve opens and the pressure drops to atmospheric. This pressure drop is shown on the diagram as the line D–A.

Note Before the start of the next compression stroke, the piston must execute the exhaust and intake strokes. These are not shown. At the completion of the intake stroke, the pressure will once again be atmospheric and the compression stroke will again begin at A.

A P–V diagram taken from the cylinder of an actual petrol engine is shown in Fig 5.4b. Factors such as valve and ignition timing and engine condition cause this diagram to differ from the ideal diagram shown in Fig 5.4a. Again, the inlet and exhaust strokes have been omitted from the P–V diagrams shown in Fig 5.4.

The mixed cycle

The operation of modern high-speed diesel engines is based on what is known as a mixed cycle. This is actually a combination of the constant-volume cycle and the constant-pressure cycle and has some of the features of each.

Before discussing this cycle, the reader must realise that as the piston speed of a diesel engine is increased, the time available for injecting and burning the fuel charge is reduced. If fuel injection began at TDC in a high-speed diesel, the piston would be well down the cylinder before combustion started. In order to remedy this undesirable situation, the fuel injection timing must be advanced when high piston speeds are used.

In the mixed cycle, air only is compressed in the engine cylinder. Fuel injection must commence before the piston reaches TDC to overcome ignition lag, and, after ignition occurs, combustion continues until just after fuel injection ceases. There is a sharp increase in cylinder pressure as the first droplets of the injected fuel are burnt and the constant-volume characteristic of the cycle is evident. When the piston passes through TDC and begins to move down the cylinder, the fuel charge is still being burnt. However, due to the increasing cylinder volume on top of the descending piston, the gas expansion resulting from combustion does not create a pressure rise. This enables the cylinder pressure to remain constant, thus illustrating the constant-pressure characteristic of the cycle.

From the foregoing it should be evident that the fuel charge is burnt at constant volume

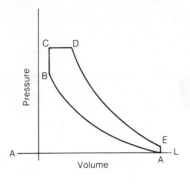

Fig 5.5a

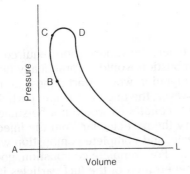

Fig 5.5b

immediately after combustion begins, and at constant pressure after the piston passes TDC and starts its power stroke. As a result of the continued burning of the fuel charge after TDC (constant-pressure combustion), the expanding gases give the piston a steady push down the cylinder on its power stroke while combustion continues. It is this action that enables the diesel engine to pull well at low speeds, and to have a near-constant torque characteristic at all speeds.

Fig 5.5a shows the P–V diagram that could be expected from the cylinder of an imaginary perfect high-speed diesel engine operating on an ideal mixed pressure heat cycle. As the piston moves up the cylinder from BDC to TDC, the air in the cylinder is compressed and so its pressure and temperature increase. This compression stage is shown on the diagram by the curved line *A–B*. Although fuel injection begins some considerable time before the piston reaches TDC (approximately 30°–15°), the fuel charge does not start to burn until just before TDC (approximately 6°–3°) because of ignition lag. As regards the ideal P–V diagram for this heat cycle shown in Fig 5.5a, it is assumed that combustion commences at TDC, so that the constant-volume phase of combustion may be clearly defined.

Fuel injection starts at point *B*, and for the case of the ideal P–V diagram it is assumed that combustion is instantaneous, although in actual practice this is not so. Due to combustion, there is an instantaneous pressure rise, which is shown on the ideal-cycle diagram by the line *B–C*. This phase of combustion is at constant volume.

As the piston moves down the cylinder from TDC, the remainder of the fuel charge is burnt and the pressurised gases created by combustion expand into the increasing cylinder volume on top of the descending piston. As a result, the cylinder pressure remains constant and is represented by the line *C–D* on the P–V diagram. This second phase of combustion is at constant pressure.

Once combustion ceases, the cylinder pressure slowly falls as the piston is forced further down the cylinder by the expanding gases. This is shown on the P–V diagram by the line *D–E*. At BDC, the pressure drops to atmospheric (line *E–A*).

The shape of the actual P–V diagram that would be recorded from the cylinder of a high-speed diesel engine operating on the mixed cycle is shown in Fig 5.5b. Note from this diagram that injection begins before TDC, and that combustion of the fuel injected during the ignition lag period is not instantaneous as in the ideal cycle. Since this combustion takes some time, the piston moves from a few degrees before TDC to a few degrees after TDC during the subsequent pressure rise. This results in the curved line *B–C* in Fig 5.5b.

Once combustion of the fuel injected during the ignition lag period has ceased, the fuel burns as it is injected. The piston moves down the cylinder as this combustion continues, and the cylinder pressure is maintained fairly constant (line *C–D* in Fig 5.5b).

The shape of the remainder of the P–V diagram is dependent on engine characteristics. As is usual practice, the inlet and exhaust strokes have been omitted from Fig 5.5a and Fig 5.5b.

6
The combustion process

In a compression ignition engine, the fuel is introduced into the combustion chamber by the injector, which breaks the fuel up into very small droplets, which spread through the compressed air. As soon as the cool fuel comes in contact with the hot air, it takes heat from the air and an envelope of vapour forms around the droplets. The air that has cooled has to regain its heat from the main body of air, and, as it does, the vapour also becomes hotter until it ignites. As soon as this occurs, the heat of combustion supplies the necessary heat to completely vaporise the fuel.

If the air is stationary in the combustion chamber, the burnt gas from the first combustion will surround each burning droplet, 'suffocating' it. If on the other hand the air is continuously moving, the burnt gas will move away from each fuel particle and fresh air will take its place, ensuring a plentiful supply of oxygen for continued combustion.

The time during which the fuel vaporises and ignites is dependent on three factors:

- The difference between the air temperature and the self-ignition temperature of the fuel. If the air temperature is much higher than the fuel self-ignition temperature, the fuel will vaporise and ignite quickly—the greater the difference, the quicker will be the vaporisation, and ignition will occur sooner.
- The pressure in the combustion chamber, since the greater the pressure, the more intimate will be the contact between the cold fuel and the hot air. The closer the contact between these two, the greater will be the rate of heat transfer from one to the other, again giving more rapid vaporisation and ignition.
- The fineness of the fuel particles. If the fuel could be broken up into fine enough

particles, the vaporisation required for combustion would be practically negligible and ignition would start almost immediately. However, the mass of the fuel particles under these conditions would not be sufficient to carry the particles far from the injector nozzle, and complete combustion would not occur. For complete combustion, good depth of penetration of the fuel particles into the combustion chamber is necessary, and the particles must have sufficient mass to carry them deep into the compressed air.

Again, if the particles are very fine, the total surface area presented to the compressed air will be large, and a great amount of fuel will be vaporised almost immediately. Once combustion begins, all the vaporised fuel will be burnt rapidly, and a very quick and high pressure rise will occur in the combustion chamber.

The stages of combustion

Combustion in the diesel engine may be divided into four distinct stages or phases:

1   **Delay period**—This is the period of crankshaft rotation between the start of injection and the first ignition of the fuel charge. During this time, fuel is being injected continuously.
2   **Uncontrolled combustion**—Once ignition of the fuel in the combustion chamber begins, all the fuel that has accumulated during the delay period burns very rapidly, giving a sudden pressure rise. This stage usually

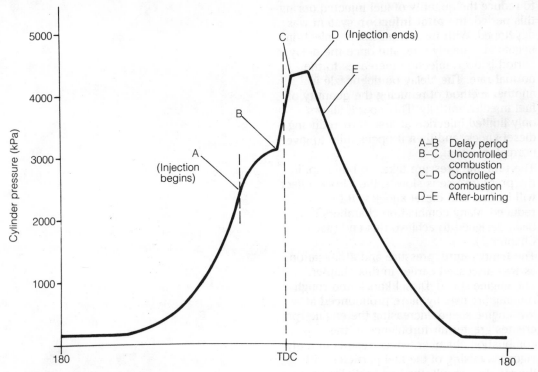

Fig 6.1 The stages of combustion

occurs from a few degrees before TDC to a few degrees after TDC, giving a high cylinder pressure at the beginning of the power stroke. Obviously, the fuel injected during the uncontrolled combustion period burns as soon as it is injected.

3 **Controlled combustion**—When the uncontrolled combustion stops, the fuel burns as it is injected into the combustion chamber, the rate of admission of the fuel giving accurate control of the cylinder pressure. Immediate combustion of the fuel as it is injected during the controlled combustion period is ensured by the heat and pressure generated during the uncontrolled combustion period.

At large throttle openings, the injection and the controlled combustion periods are longer than at small throttle openings, the rate of admission of the fuel being constant for the period during which it lasts. This means that the maximum cylinder pressure (as a result of combustion) for a particular engine is the same regardless of the throttle opening, the extra power being derived from the longer combustion period.

4 **After-burning**—During the controlled combustion period, almost all the fuel burns.

However, some fuel particles fail to find the necessary air for combustion during this stage of combustion and these particles burn after injection has ceased. Again, some of the particles of fuel that settled on the combustion chamber walls during injection evaporate and burn during this final stage—'after-burning'.

Diesel knock

The sudden pressure rise during the uncontrolled combustion period causes a shock wave to spread throughout the combustion chamber. When this wave strikes the metal of the cylinder head or piston crown, a characteristic metallic knock is audible, this being known as **diesel knock.**

The strength of the shock wave, and hence the intensity of the diesel knock, is influenced by the following:

- The amount the pressure rises during the uncontrolled combustion period. This, in turn, is dependent on the quantity of fuel injected during the delay period. In an effort to reduce the quantity of fuel injected during this period, the **pilot injection** system was developed. With this arrangement, injection begins at a steady rate, and once the delay period is over, injection increases to the normal rate. The 'delay pintle' nozzle is another method of reducing the quantity of fuel injected initially. This nozzle allows only limited injection at first, thus reducing diesel knock, and then it opens fully to give normal injection.

- The time the pressure takes to build up. If the pressure rise is slower, the 'shock' effect will be less and diesel knock will be reduced. Many combustion chambers have been designed to achieve this end (see Chapter 7).
- The temperature, pressure and atomisation, as was discussed earlier in this chapter.
- The engine speed. Diesel knock and rough running are usually more pronounced at low engine speed. Increasing the engine rpm creates greater air turbulence in the combustion chamber, which results in more vigorous mixing of the fuel particles with the air. As a result, the fuel particles heat up and vaporise more quickly, reducing the delay period.
- Some fuels are more inclined to encourage diesel knock than others. The tendency for fuels to knock is expressed as the **cetane number** of the fuel (see Chapter 9).

7 Combustion chambers

The combustion chamber is that part of the engine where the fuel charge is burned. In some diesel engines the combustion chamber is formed in the cylinder head, while in others it is formed in the crown of the piston. Regardless of where it is situated, the combustion chamber must be designed so that it is capable of producing maximum turbulence, not only during injection, but also when combustion is in progress.

This high degree of turbulence is necessary for three reasons:

- to ensure complete mixing of the fuel and air
- to provide a continuous supply of air to the burning fuel particles (see Chapter 6)
- to sweep burnt gas from the injection area, so that as more fuel is injected, it meets fresh air.

With regard to the mixing of the fuel and air, it must be remembered that the fuel charge is not injected into the combustion chamber until the piston is between 30 and 15 degrees of crankshaft rotation from TDC. Suppose the speed of the engine is 1900 rpm; the time available for the fuel to be injected, mixed with the air and burnt is approximately 1/500th of a second. It should be obvious that high air turbulence is necessary to mix the air and fuel thoroughly and quickly.

In Chapter 6, the 'delay period' in the combustion cycle was discussed. The duration of this period is largely dependent on the efficient mixing of the fuel and air in the combustion chamber. If mixing is quick and efficient, the delay period will be greatly reduced and this in turn reduces diesel knock. On the other hand, if mixing is slow and inefficient, a longer delay period will result and diesel knock will become more severe.

If the supply of fresh air required to prevent 'suffocation' of the burning fuel particles (and to sweep burnt gas from the injection area) is not continuous, complete combustion cannot take place. If combustion is incomplete, thermal efficiency must drop, with a corresponding increase in fuel consumption and a loss of power.

Complete combustion must also be achieved if a condition known as 'crankcase dilution' is to be avoided. This condition is caused by unburnt fuel passing the rings and mixing with the lubricating oil in the crankcase. As a result of this contamination, the lubricating properties of the oil are greatly reduced and severe engine wear results.

The paramount aim of combustion chamber design, therefore, must be to create very high turbulence so that the atomised fuel and air are thoroughly mixed and complete combustion is achieved.

Due to the amount of research carried out over the years, many types of combustion chamber have been developed. These are generally divided into two categories:

- direct injection combustion chambers, and
- indirect injection combustion chambers.

However, engines are usually classified as either 'direct injection' or 'indirect injection', the words 'combustion chamber' being dropped.

The main types of combustion chamber in use today have been classified as follows:

- open combustion chamber (direct injection)
- precombustion chamber (indirect injection)
- compression swirl combustion chamber (indirect injection)
- air cell or energy cell combustion chamber (indirect injection).

The choice of the combustion chamber design for a particular engine depends largely on its rpm, its principle of operation (two- or four-stroke), the size of the engine, and the purpose for which it is to be used.

The open combustion chamber (direct injection)

The trend in modern high-speed diesel engines is towards the open combustion chamber. This system has developed from a small combustion chamber in the cylinder head to the modern system, which makes use of a flat cylinder head and a specially shaped cavity in the piston crown. It is usual for the piston to protrude slightly above the block surface, giving minimum clearance between the piston crown and the head, thus ensuring good turbulence and combustion **within** the piston cavity. Almost all two-stroke diesels use this system because of the ease of scavenging.

A multi-hole-type injector with a wide spray angle is located directly above the piston cavity, and sprays into this cavity where, since there is very little clearance between the piston crown and the head, almost all the compressed air is concentrated. High injection pressures (175–185 atmospheres approximately, for naturally aspirated engines) are needed to ensure droplet penetration and efficient fuel–air mixing.

Piston cavities and 'squish'

Piston cavities are cast in the piston crown, and take many different forms. The most common type is the toroidal piston cavity—a circular cavity, usually symmetrical about the piston axis, with a small cone projecting upwards from the bottom of the cavity towards the cylinder head. Some manufacturers favour a simple hemispherical cavity while others use a dished piston crown, although this latter is fairly rare. Yet another manufacturer uses a simple deep cylindrical cavity, which is almost flat on the bottom.

It is important to realise that each cavity must be used in conjunction with a particular injector nozzle, and the fitting of a nozzle with a slightly different spray angle can result in holes being burnt in the piston crown.

Regardless of the piston cavity design, the turbulence in the cylinder is created in the same way. As the piston approaches TDC on the compression stroke, the major portion of the air in the cylinder is compressed into the piston cavity. Because of the 'squish' effect, a high velocity is imparted to the air when it moves towards the piston cavity, thus causing a high degree of rotational turbulence in the piston

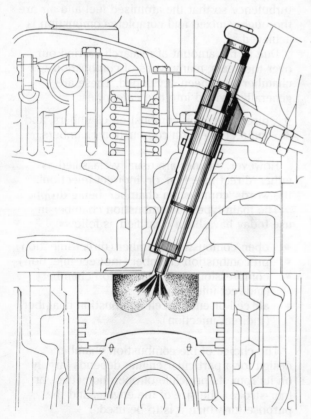

Fig 7.1 Direct injection combustion chamber

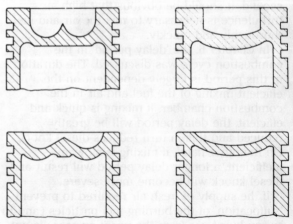

Fig 7.2 Typical piston cavities

cavity itself. The fuel injector is positioned above the piston so that fuel is injected directly into the turbulent air in the piston cavity, promoting efficient fuel–air mixing. It should be mentioned here that some engine manufacturers, such as Perkins, have the toroidal cavity in the piston slightly offset to promote swirl in a predetermined direction.

Squish may be defined as the rapid movement of the air being compressed in the cylinder of an open combustion chamber engine, from the cylinder walls towards the centre of the combustion chamber. It is most effectively created by using either a toroidal cavity piston or a plain cavity piston, the crown of which has only the necessary mechanical clearance from the cylinder head when it is at TDC.

As the piston approaches TDC on the compression stroke, the air trapped between the squish band on the piston periphery and the cylinder head is forced inwards, at high velocity, into the cavity in the piston crown. At or near the centre of the cavity, air meeting air causes mutual deflection towards the bottom of the

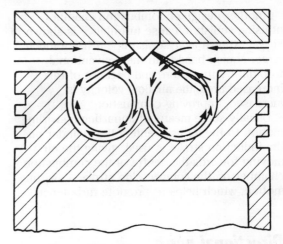

Fig 7.3 The 'squish' effect

piston cavity, beginning the circular turbulence shown in Fig 7.3. The degree of squish produced is largely dependent on the width of the squish band formed on the piston crown, with maximum squish velocity occurring at approximately 8° before TDC.

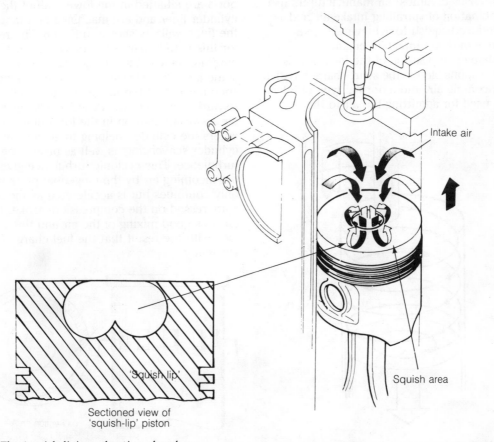

Intake air

Squish area

Sectioned view of
'squish-lip' piston

Fig 7.4 The 'squish lip' combustion chamber

Another design of combustion chamber utilising squish to cause turbulence is shown in Fig 7.4. Known as a 'squish lip', it has a much larger squish band than the conventional toroidal cavity piston, and because of this increased area, the air flow velocity is increased, improving combustion.

Usually some means of imparting a swirling motion to the air as it enters the cylinder on the induction stroke is used, giving the air moving towards the centre of the combustion chamber on the compression stroke a swirling motion, which helps to promote turbulence.

Directional ports

In order to achieve the necessary swirling motion in four-stroke engines using the open combustion chamber system, 'directional ports' are used. These curved induction ports impart a spiralling motion to the incoming air, which not only continues during the compression stroke, but is intensified. Almost all manufacturers use the combination of spiralling intake air and piston-induced squish to achieve effective turbulence in direct injection engines.

The shape and position of the inlet port is directly responsible for the swirl characteristics of the incoming air, and a normal type of inlet valve is used for admitting the air to the cylinder. Some of the larger diesels use two directional ports and, of course, two inlet valves to admit the air charge. Engines designed for automotive work, however, use only one directional port and one or two inlet valves per cylinder.

In earlier designs, masked valves were used to induce a directional swirl into the incoming air charge. The 'mask' consisted of a curved deflecting shield located between the valve seat and the stem. In operation the masked valve had to be prevented from rotating to ensure mask alignment and subsequent correct air deflection. Because of the restriction this design imposed to incoming air flow, manufacturers have discontinued its use in modern diesel engines.

Tangential ports

Tangential inlet ports are used in two-stroke diesel engines to create air turbulence. These ports are situated in the lower half of the cylinder liner and are machined at a tangent to the liner walls as shown in Fig 7.6. The reason for this is to impart a rotary swirl to the air entering the cylinder. Tangential ports used in some large two-stroke diesels are designed in such a way that the incoming air is deflected towards the top of the cylinder. This gives the air a swirling motion in the form of a spiral as it enters the cylinder, helping to achieve efficient cylinder scavenging as well as promoting turbulence. This cyclonic turbulence given to the incoming air by the tangential ports not only continues but is accelerated as the air is compressed on the compression stroke. This ensures good mixing of the air and the injected fuel, with the result that the fuel charge is burnt efficiently.

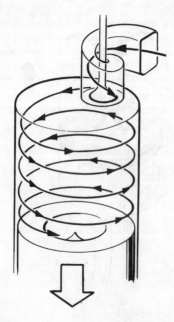

Fig 7.5 Air flow due to directional inlet port

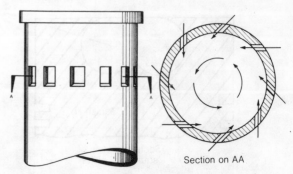

Section on AA

Fig 7.6 Tangential ports

Almost all manufacturers of uniflow scavenged diesel two-strokes incorporate tangential ports in their design, the GM diesel and the Rootes opposed-piston diesel being typical examples.

Advantages claimed for open combustion chambers

- Due to its compact form the surface area of the combustion chamber is small, resulting in a relatively low heat loss to the cooling system, giving high thermal efficiency.
- Because of the low heat loss, the compressed air loses very little heat, giving good cold starting ability without the need for heater plugs (see Chapter 8).
- Again, because of low heat loss, low compression ratios may be employed, while still giving good starting and efficient combustion.

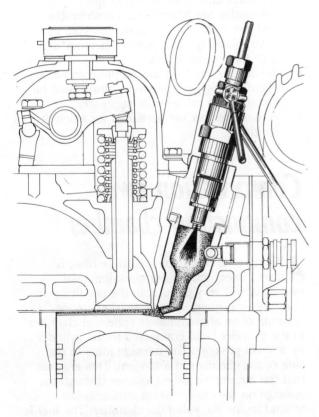

Fig 7.7 Typical precombustion chamber

Disadvantages claimed against open combustion chambers

- Since a number of small holes are used in the injector nozzle, instead of one large one, blocking by carbon deposits is fairly common.
- To ensure penetration of the fuel particles into the compressed air, high injection pressures are necessary. This suggests the need to maintain injection equipment to a higher standard than would be necessary were lower injection pressures required.
- Rough running at low speeds occurs, due to the long delay period that results from rather limited turbulence. In this design, turbulence is largely dependent on the speed of the incoming air, which obviously increases with engine speed.

Precombustion chambers

While more and more manufacturers seem to be changing to direct injection, some still retain the precombustion chamber design, which has some considerable advantages.

A precombustion chamber is a small auxiliary chamber situated in the cylinder head, and connected directly to the main combustion chamber by a small passage. In some designs, the precombustion chamber is separated from the main combustion chamber by a perforated plate made from heat-resistant alloy.

When the piston is at TDC on compression stroke, the major portion of the air charge has been forced through the connecting passage into the precombustion chamber. The remainder of the air charge is contained in the main chamber between the piston crown and the cylinder head. Some engine manufacturers (eg Caterpillar) use a piston with a dish-shaped cavity in the crown, while others use a piston with a flat crown in conjunction with a precombustion chamber. When a dished piston is used, between 35 and 40 per cent of the air is displaced from the cylinder on the compression stroke and is forced into the precombustion chamber, the remainder being left in the main combustion chamber.

When injection takes place (during the compression stroke), the fuel charge is injected into the hot compressed air in the precombustion chamber and ignition occurs. The subsequent burning of the fuel charge produces a substantial pressure rise in the confined space of the precombustion chamber and owing to this increase in pressure, particles of burning and unburnt fuel are forced violently through the narrow connecting passage into the main combustion chamber between the piston crown and the cylinder head. Any unburnt or partly burnt fuel particles mix with the air present, and the combustion process is completed in the main combustion chamber.

Because of the restricted passage connecting the precombustion chamber to the cylinder and the beginning of combustion in that chamber, the sudden pressure rise from the uncontrolled combustion occurs largely within the chamber with minimal effect in the engine cylinder. Thus, high cylinder pressures are eliminated, keeping diesel knock to a minimum and reducing internal stresses in the engine.

The Caterpillar Tractor Company used precombustion chambers in their first diesel engines and are still using them, to a lesser extent, in their heavy duty engines. The design is uncommon in that the precombustion chamber is not cast or machined in the head, but is a separate 'screw-in' unit into which the injector is directly fitted. The engine pistons have a heat-resistant insert fitted into the piston crown directly below the outlet from the precombustion chamber. This protects the alloy piston crown from the burning fuel particles, which are ejected at high velocity from the precombustion chamber, and which would eventually burn right through the piston crown.

Because of the combustion characteristics the degree of atomisation of the fuel charge need not be as great as with direct injection engines, so that low injection pressures and pintle-type injector nozzles are usually employed. Caterpillar engines are a notable exception to the rule, using a single-hole nozzle of the firm's own design, featuring a very large orifice.

Advantages claimed for precombustion chambers

- The pintle-type nozzle, usually used in conjunction with precombustion chambers, has a relatively large orifice, plus a moving pintle, so that blockage due to carbon deposits is practically eliminated.
- The fuel does not have to be as finely atomised as for direct injection engines, and consequently injection pressures are lower.
- Because of high turbulence and efficient mixing of the fuel and air, the fuel quality does not have to be as high as with other types of combustion chamber.
- Engine operation is smooth because maximum cylinder pressure during combustion is low.

Disadvantages claimed against precombustion chambers

- In almost all types, heater plugs must be used when starting the engine in cold weather, because of the heat lost from the compressed air to the relatively large combustion chamber wall area.
- The heat lost from the compressed air to the walls and at the throat of the precombustion chamber is considerable, and generally engines using this system have relatively low thermal efficiency and high fuel consumption.
- Because of the heat loss to the combustion chamber, high compression ratios must be employed to achieve the necessary compressed air temperatures for efficient ignition.

Compression swirl combustion chamber

Although the trend appears to be toward direct injection, the compression swirl system is still very popular in the smaller automotive engines. A basic swirl chamber consists of an approximately spherical chamber in the cylinder head connected to the cylinder by a small passage. The passage joins to the side of the chamber at a tangent. This ensures that the air, entering the chamber through the passage on the compression stroke, swirls around inside the spherical chamber. The fuel is injected at a right angle to the swirling air, ensuring complete mixing. As soon as the pressure rise due to combustion is sufficient, a mixture of burnt, burning and unburnt fuel is ejected violently into the main chamber in the piston crown.

The most popular compression swirl chamber used in modern high-speed diesel engines would appear to be based on Sir Harry Ricardo's 'Comet' design, of which there are many variants. This compression swirl combustion chamber consists of a basic spherical chamber and tangential passage, but in addition the top of the piston has a figure-8 depression in its crown, which acts as part of the combustion chamber when the piston is at TDC, only very slight clearance being allowed between the flat of the piston crown and the head.

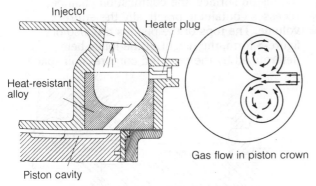

Section through combustion chamber

Fig 7.8 A typical swirl chamber

The swirl chamber itself is made in two parts. As it is formed in the cylinder head, the top half of the chamber is indirectly cooled by the engine's cooling system, but the bottom half is made of special heat-resisting alloy steel, or in some cases stellite, and is not cooled in any way. An air space is used to insulate the bottom half of the combustion chamber from the cylinder head, the only contact being a small area of the mounting flange that touches the cylinder head. This small contact area provides minimal conduction, keeping the bottom half of the combustion chamber hot so that it will heat the incoming air, and so reduce the delay period and improve combustion.

As the piston approaches TDC on the compression stroke, from 50–85 per cent (depending on the particular model of the chamber) of the air is forced through the

tangential passage and compressed in the swirl chamber. Because the connecting passage is tangential to the swirl chamber, the air entering the chamber is given a rapid circular swirling motion. The velocity of the swirling air is dependent on the engine rpm, and so as the engine speed is increased and the air velocity increases, greater turbulence results.

The fuel injector is positioned in such a way that when injection occurs, fuel is sprayed across, and at right angles to, the swirling air. When combustion begins, a pressure rise occurs in the swirl chamber and when the pressure in the chamber becomes greater than that of the incoming air, burning and partly burnt fuel particles are ejected, at high velocity, from the swirl chamber into the air remaining in the cavity in the piston crown. A figure-8 turbulence results, so that air in the cavity and the burning fuel particles are thoroughly mixed, completing combustion of the fuel charge.

Because of the heat lost to the large surface area of the swirl chamber—particularly the upper part—on compression stroke, cold-starting devices are generally needed to allow easy starting in cold weather. However, the Pintaux injector nozzle developed by CAV and Ricardo ensures good starting in swirl chambers under normal conditions without the use of such devices. Under colder conditions, however, a cold-starting aid may still be necessary with this compression swirl chamber even though a Pintaux nozzle has been used.

The compression swirl combustion chamber may easily be confused with the precombustion chamber, since they are both separate chambers in the cylinder head in which the fuel is injected and combustion starts. It is here that the similarity ends. The passage connecting the swirl chamber to the engine cylinder is of much larger diameter than the passage connecting the precombustion chamber to the engine cylinder. Again, the air enters the swirl chamber at a tangent and has the fuel sprayed across it, while air entering the precombustion chamber does so centrally and has the fuel sprayed directly into it.

Low injection pressures are used with compression swirl chambers, since thorough mixing and penetration are accomplished once combustion begins, making fine atomisation unnecessary. For the same reason, pintle injector nozzles are generally used, and these are, of course, self-cleaning. But Pintaux nozzles have one very fine hole in addition to the pintle hole, and this hole becomes choked with carbon very easily.

Advantages claimed for compression swirl combustion chambers

- Because of high turbulence and near-perfect combustion, the odour-producing exhaust gas emissions are minimised.
- Relatively low injection pressures can be used.
- Because of the high turbulence and good fuel–air mixing, the delay period is reduced and diesel knock is practically eliminated.

Disadvantages claimed against compression swirl combustion chambers

- Due to losses in thermal and mechanical efficiency, slightly more fuel is used than in direct injection engines.
- The heat lost by the compressed air to the top of the swirl chamber is considerable, causing cold-weather starting problems.
- Owing to the design of the compression swirl chamber, efficient scavenging of the burnt gas is a problem.

Air cell or energy cell combustion chambers

Over the years many types of air cell have been developed, the most successful of these being the 'Lanova' energy cell. The most common version of this combustion chamber consists of a cell located in the cylinder head directly opposite the fuel injection nozzle. The energy cell is usually a removable unit screwed into the head, and consists of two differently sized and shaped cells in series, connected by a venturi choke. The cells are permanently in

communication with the combustion chamber, through another venturi at the inner end of the inner cell.

The Lanova energy cell

To achieve certain desirable results, the combustion chamber over each cylinder is of figure-8 form. The valves open into the two recesses of the figure-8, between which is the narrower portion, or throat, of the chamber. On one side of the throat is the injection nozzle, which is positioned to spray the fuel directly across the throat towards the small orifice of the energy cell on the opposite side.

The bottom surface of the cylinder head has only gasket clearance over the flat portion of the piston surface, the combustion chamber recesses containing practically the entire volume. The tops of the pistons are also of figure-8 form, the height of the lobes being determined by the required compression space.

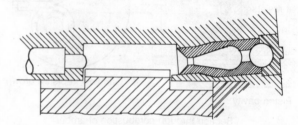

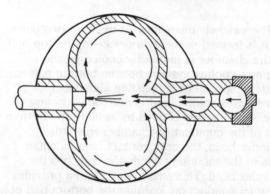

Fig 7.9 The Lanova cell (common type)

In operation, beginning with the induction stroke, the air enters through the inlet valve in one of the two recesses and, because of its offset position with respect to the cylinder axis, is given an initial swirling motion. On the compression stroke, this air is compressed into the confined space of the '8'-shaped combustion chamber, continuing the swirling motion. When the top of the compression stroke is approached, the air above the flat portions of the piston crown is rapidly displaced into the chamber recesses, attaining both high velocity and temperature.

During the compression stroke, a small part of the air is forced into the energy cell, where owing to the restricted cooling, it later attains a pressure higher than that in the combustion chamber. This high pressure, however, is confined within the cell where it cannot cause increased stress on the working parts of the engine but where, on the other hand, it serves a definite purpose in combustion control.

At the proper instant, the fuel is sprayed by the nozzle directly across the chamber throat, the main body of the stream entering the energy cell, those portions near the edges being swept around the circular recesses in opposite directions, thus accelerating the already swirling air therein. As the fuel reaches the energy cell it ignites instantaneously and a rapid combustion takes place. However, since the volume of air within the cell is small, only a small part of the fuel is thus consumed and the balance, the major portion, is blown violently back against the continuing stream from the injection nozzle. It is divided by the form of the throat into two streams of highly atomised fuel and hot air in the process of combustion, to swirl actively in opposite directions in two recesses. In doing so, these streams oppose the direction of rotation of the air already there so that air and fuel are most intimately mixed.

It is the relative volumes of the two cavities of the energy cell and the scientifically developed venturi that provide the control of combustion, a most valuable advantage of the system. They operate to control the rate at which the mixture of fuel and air is fed back to the combustion chamber, so that combustion is consequently controlled in such a way that the pressure rise occurs slightly after TDC, and continues at a moderate rate over a considerable number of degrees of crankshaft rotation, thus giving an expansion without the rapid and stress-inducing rise to excessive pressures, which is characteristic of some combustion systems.

So efficiently does this principle operate that the peak pressures are little higher than occur in many petrol engines, while the more sustained combustion results in exceptionally high brake mean effective pressure. In addition, since the turbulence is being induced by thermal expansion, it is virtually independent of the engine speed and thorough and smooth combustion over a wide range of speeds is attained.

As a high degree of atomisation is not essential, the fuel injector nozzle used in conjunction with the Lanova energy cell is of the pintle type. The fuel injection pressure usually ranges from 110–120 atmospheres.

Fig 7.10 illustrates a particular type of Lanova air cell that has been used in Mack engines. Note that here the main combustion chamber is cylindrical and not the figure-'8' design, and the air cell and injector do not lie across the cylinder diameter.

Although it may seem to have many advantages over other combustion chamber systems, the Lanova system has today almost vanished from the high-speed diesel engine scene.

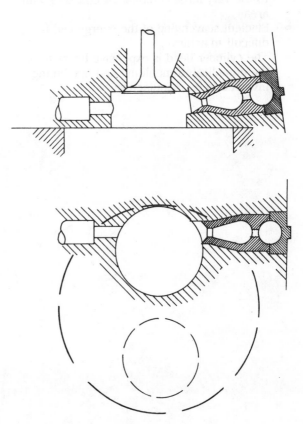

Fig 7.10 The Lanova cell as used on some Mack engines

Advantages claimed for air cell combustion chambers

- There is minimal shock loading on working components due to a high degree of controlled combustion.
- A clean exhaust is possible over a fairly wide range of speeds, because the turbulence in the combustion chamber is induced by thermal expansion and is virtually independent of the engine speed.
- Relatively low fuel injection pressures may be used because a high degree of atomisation is not required.

Disadvantages claimed against air cell combustion chambers

- There are starting difficulties when cold, due to the high loss of compressed air heat to the very large combustion chamber wall area.
- Efficient scavenging of the energy cell is difficult to achieve.
- The cylinder head is expensive because of the complicated moulding and machining involved in its manufacture.

8
Cold starting aids

Because of its compact form, the open combustion chamber system presents only a small surface area to the compressed air, and the amount of heat lost by conduction to the combustion chamber is relatively small. As a result, the temperature of the compressed air at the moment of injection is always considerably higher than the fuel's self-ignition temperature, and the fuel ignites readily.

Unfortunately, the other combustion chamber systems present a large surface area to the compressed air, resulting in a large loss of heat. When the engine (and combustion chamber) is cold, the large difference in temperature between the compressed air and the combustion chamber causes a rapid transfer of heat from the air to the chamber. This heat loss, coupled with the fact that compression only raises the air temperature a certain amount above its initial temperature, makes starting from cold a problem.

Many systems of overcoming this have been developed, the most common being:

- glow plugs
- thermostart devices
- the use of volatile fuel.

Glow plugs

Glow plugs screw into the combustion chamber and supply additional heat to the air during the compression stroke. A heater element on the plug lies flush with the combustion chamber wall, and when prior to starting, a current of 20 to 35 amperes is supplied from the battery for from 5 to 15 seconds, this element glows bright red. If the engine is then cranked, some of the heat from the element is transferred to the compressed air, giving a final air temperature high enough to ensure efficient ignition and combustion.

Although other types have been used, almost all glow plugs used today are of the single-pole

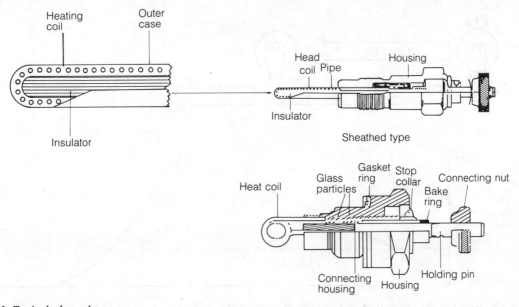

Fig 8.1 Typical glow plugs

type, which provide one insulated terminal post for electrical connection with the circuit being completed to earth through the body of the glow plug (see Fig 8.1).

Single-pole glow plugs are said to be connected in parallel. This means that the current is supplied from the battery to the single terminal of each plug, and the circuit is completed through the cylinder head to earth. In this system full battery voltage is applied to each plug, and one plug can fail without affecting the others. Thus for a vehicle using a 12-volt electrical system, 12-volt single-pole glow plugs can be employed.

Many automotive diesel engines are equipped with a glow plug control system, usually known as a 'quick start' system. The basic quick start system is designed to control the preheating time of the glow plugs by the use of an automatic timer; the more sophisticated models modify that timer operation to take into account the engine water temperature.

When starting a cold engine fitted with a typical system, immediately the starting switch is turned to the 'ON' position, the glow plug warning lamp is illuminated and the main glow plug relay becomes operative. This allows a

high current flow through the glow plugs to earth, causing rapid heating of the glow plugs.

After a set time, the automatic timer switches off the glow plug warning lamp, indicating that the engine is ready to be started, but leaves the glow plugs on. The engine is started by turning the key to the 'START' position (refer to Fig 8.2). Once the engine has started and the start switch returned to the 'ON' position, the main glow plug relay will turn 'OFF', but a second relay will become operative, supplying intermittent current flow for glow plug operation while the engine is running.

The purpose of this 'after glow', as it is sometimes termed, is to improve combustion in a cold engine. By altering the current flow from continuous to intermittent, the glow plugs are protected from extreme heating and subsequent premature failure. This phase of operation is illustrated in Fig 8.3. After-glow operation continues for a period of up to 45 seconds in some systems, and then the glow plugs are switched off automatically.

To reset the system, the start switch must be turned to 'OFF' and then turned to the 'ON' position again. This will restart the cycle.

When the engine is to be restarted after it has

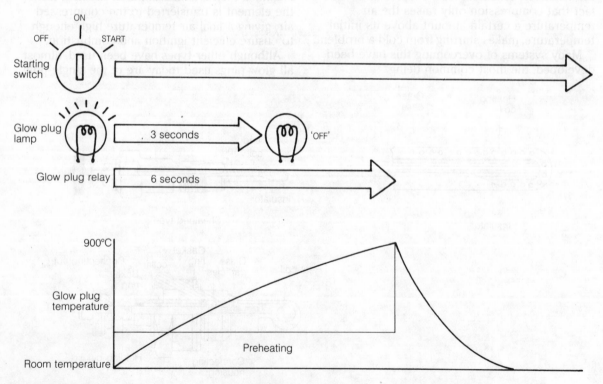

Fig 8.2 Operation of the quick start system with the starting switch turned to the 'on' position

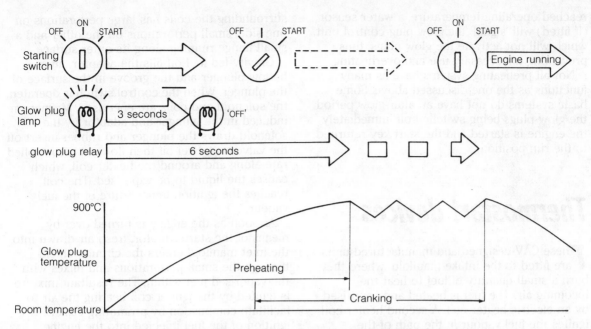

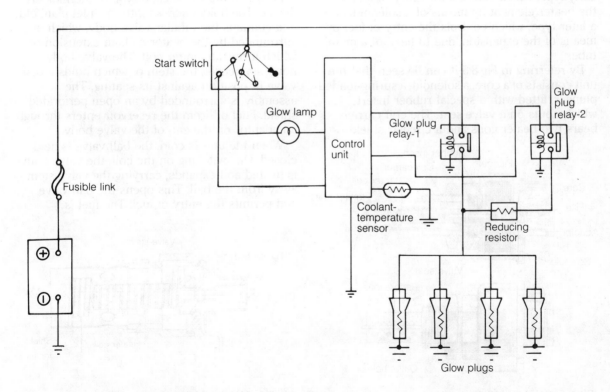

Fig 8.3 Glow plug operation during engine cranking and initial running periods

Fig 8.4 Glow plug control system wiring diagram

reached operating temperature, a water sensor (if fitted) will 'signal' the glow plug control unit, which will not activate the glow plugs, thus preventing unnecessary use and overheating.

Not all preheating systems have as many functions as the one discussed above. Some basic systems do not have an after-glow period, the glow plugs being switched off immediately the engine is started and the start key returned to the run position.

Thermostart devices

These CAV-designed and manufactured units are fitted to the intake manifold, where they burn a small quantity of fuel to heat the incoming air. The fuel is heated and vaporised by an electric heater coil. A second heater coil ignites the fuel vapour in the path of the incoming air, raising the air temperature so that at the end of the compression stroke the air temperature is sufficient to ensure good starting.

There are several variants of the thermostart device, but they fall into two main types – an early type, which controls the entry of fuel to the heater element by means of a solenoid, and a later type, which controls the entry of fuel by means of the expansion, due to heat, of a metal tube.

By referring to Fig 8.5 it can be seen that the unit consists of a core, a solenoid, a spring-loaded plunger, fitted with a special rubber insert, which abuts on a valve seat. The coil carrier bears two heater coils and a circular shield

surrounding the coils has large perforations on one side, small perforations on the other and a small flange running along its outer surface.

Gravity-fed fuel oil fills the adaptor, filter, hollow plunger and the groove in the surface of the plunger. When the control switch is operated, the solenoid and coils are energised. Magnetism induced in the plunger and adaptor by the solenoid draws the plunger and rubber insert off the valve seat. Fuel oil then flows at a controlled rate along and around the heater coil, which causes the liquid to be vaporised. The coil reaches the ignition temperature of the fuel vapour.

As soon as the engine is turned over by means of the starter motor, fresh air drawn into the inlet manifold enters the circular shield through the small perforations and mixes with the vaporised fuel within. The resultant mixture is ignited by the igniter coil heating the air to facilitate combustion by promoting easier ignition of the fuel injected into the engine cylinders.

The flange running along the outer surface of the shield provides a sheltered zone around the outlet holes and protects the flame from the incoming air stream.

Fig 8.6 shows the second type of thermostart device. The holder screws into the inlet manifold, and contains the tubular valve body, which is surrounded by the heater coil, an extension of which forms an igniter coil. The valve body houses a needle, the stem of which holds a ball valve in position against its seating. The assembly is surrounded by an open perforated shield. Fuel oil from the reservoir enters through an adaptor on the end of the valve body.

When the unit is cold, the ball valve is held closed. On switching on the coil, the valve body is heated and expands, carrying the valve stem away from the ball. This opens the ball valve and permits the entry of fuel. The fuel is

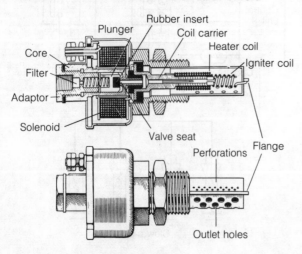

Fig 8.5 CAV 'mark' 1 thermostart

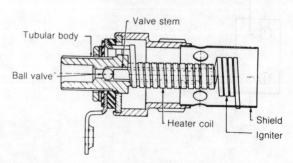

Fig 8.6 CAV 'mark' 3 thermostart

vaporised by the heat of the valve body and
when the engine is cranked and air is drawn
into the manifold, the vapour is ignited by the
coil extension and continues to burn, thus
heating the inlet air.

When the coil is switched off the flow of air
in the manifold cools the valve body rapidly,
causing it to contract. The valve closes.

The cold start aid is a sealed unit and cannot
be dismantled. If the unit ceases to function, it
must be replaced.

Volatile fuel

Instead of using some means of providing extra
heat for the air before, or during, the
compression stroke in order to achieve a
sufficiently high temperature at the completion
of the compression stroke to give efficient
starting, a volatile fuel, which has a much lower
self-ignition temperature than diesel fuel, is
sometimes used. The fuel used is usually some
form of ether, and an ether–air mixture will
ignite at a much lower temperature than diesel
fuel.

The ether fuel may be packed in pressurised
cans so that it can be sprayed into the air
intake as the starter is operated. Or it may be
supplied in a capsule, which fits into the base
chamber of a special carburettor in the inlet
manifold. The base chamber is removed for
fitting the capsule, which is automatically
punctured when the base chamber is refitted.
Puncturing the capsule releases the ether
compound, and air, being drawn through the
inlet manifold into the combustion chamber,
carries a quantity of ether with it. This ether–air
mixture ignites readily on the compression
stroke.

The manifold is fitted with a butterfly valve,
which is nearly closed for starting to give a very
high air speed past the fuel chamber. The high
speed of the air causes a depression, which
draws the ether from the chamber to mix with
the air. As soon as the engine runs evenly the
butterfly is opened fully, so removing its
restricting effect from the air intake.

9
Diesel engine fuels

A number of grades of fuel are available for compression ignition engines, and the use of the correct fuel is extremely important. The fuels available in Australia are listed below.

Automotive distillate

Automotive distillate is a highly refined fuel for use in high-speed diesel engines as fitted to tractors, trucks and motor cars. The distillate is also known as automotive diesel fuel, automotive diesel oil, distillate, diesolene and diesoleum. It is clear, to light straw coloured.

Summer and winter fuels

Because wax solidifies in diesel fuel under very cold conditions and blocks the fuel flow, winter-grade diesel fuel has been introduced; this contains special additives to prevent the wax from solidifying at low ambient temperatures.

Industrial diesel fuel

This is the name given to a less refined fuel suitable for use in slow to medium speed diesels and commercial heating installations. It is a heavier grade of fuel, but is free flowing and does not require preheating. Its colour is light straw to black.

Specifications of diesel fuels

In the refining of fuels certain specifications are laid down for each type, and the fuel's properties must lie within these set limits. The most important properties include:

- ignition quality
- self-ignition temperature
- calorific value
- flashpoint
- pour point
- cloud point
- viscosity
- specific gravity
- carbon residue
- sulphur content
- ash content
- water and sediment content.

Ignition quality

The ignition quality of a fuel may be described as its degree of readiness to burn when injected into the combustion chamber. Fuels with low ignition quality will take longer to ignite than fuels with high ignition quality. Fuels with a low ignition quality, therefore, cause a longer delay period or ignition lag with resultant greater diesel knock.

The **cetane number** of a diesel fuel is the most common method of measuring the fuel's ignition quality. Cetane is a chemical fuel that has the highest known ignition quality, and is given the rating of 100. Alpha-methyl-naphthalene is another chemical fuel, which has a very low ignition quality and is given the number 0.

When testing a fuel to find its ignition quality, a test engine is first run on a sample of the fuel. The knocking tendency and the delay period of the engine are noted.

Then the same engine is run on a mixture of cetane and alpha-methyl-naphthalene. The amount of cetane in this mixture is slowly increased or decreased, until the engine gives the same test results as were obtained when it was run on the fuel sample. For example, if the mixture of cetane and alpha-methyl-naphthalene giving the same test results contained 40 per cent cetane and 60 per cent alpha-methyl-naphthalene, the cetane number of the fuel would be 40.

As an alternative to testing a fuel in an engine laboratory to ascertain its cetane number, an approximate equivalent can be calculated after some simple testing of the fuel. This is known as the calculated cetane number of the fuel.

It is worthwhile noting at this stage that the average high-speed diesel engine requires a fuel with a minumum calculated cetane number of 47.

The effects of the quality of the fuel on the engine may be summarised as follows:

- **Diesel knock**—The use of a fuel of too low an ignition quality results in severe diesel knock, rough running and severe shock loading on pistons and bearings.
- **Engine deposits**—When a low ignition quality fuel is used, deposits in combustion chambers, on rings and piston skirts, become excessive. The use of a suitable fuel keeps these deposits to a minimum.
- **Starting**—The higher the ignition quality of the fuel used, the lower the efficient starting temperature. The use of a fuel of lower ignition quality than is recommended results in harder starting and longer warm-up periods, during which the engine produces white exhaust smoke.
- **Odour and fumes**—If the engine is in good condition, a fuel with a high ignition quality keeps fumes, odour and smoke to a minimum, while a lower grade fuel will aggravate the situation.

Self-ignition temperature

This is the temperature at which the fuel will ignite without the aid of a spark. The lower this temperature, the easier the engine will start and the less diesel knock will occur, or the lower the self-ignition temperature, the higher the ignition quality of the fuel.

Calorific value

This is the quantity of heat released when the fuel is completely burnt. It is measured in joules per kilogram of fuel. Distillate has an approximate calorific value of 44×10^6 joules per kilogram (44 MJ/kg), while lower grade fuels have a slightly higher value. Since IC engines are dependent on heat to produce power, the calorific value of the fuel has a direct bearing on the power output.

Flashpoint (open or closed)

This is the temperature at which the fuel will give off a flammable vapour. To ascertain the flashpoint, a quantity of fuel is heated in a container (open or closed) and a flame is passed through the hot oily vapour. The temperature at which the vapour burns momentarily and then goes out is the flashpoint.

Flashpoint has no bearing on the engine's performance, but is important with regard to safety precautions during storage and handling of the fuel.

Pour point

This is the lowest temperature, under test conditions, at which the fuel will flow under its own weight. It is necessary for a fuel to flow freely at the lowest temperature likely to be encountered, hence pour point is an important factor. However, low pour point is often obtained at the expense of ignition quality, and the use of low pour point fuels, where they are not necessary, should be avoided.

Cloud point

This is the temperature at which the wax in the diesel fuel will begin to crystallise and form a solid, and in so doing, will give the fuel a cloudy appearance. The winter-grade cloud point in southern and eastern Australia is around $-1°C$ to $-2°C$, while that for summer is around $5°C$. The cloud point variation now allows users the benefits of summer and winter grades of diesel fuel.

Viscosity

Viscosity may be defined as the reluctance of a fluid to flow; the more viscous a fluid, the greater its resistance to flowing. From a practical point of view, the viscosity of a fuel is a measure of thickness or thinness.

Viscosity is measured by means of a viscometer, which measures the time taken for a specific quantity of a fluid to pass through a set orifice.

Note   There are three viscosity scales in common use, and each gives a different reading—to compare, one reading may be converted to the other from the formula:

Redwood = 29 × Englar = 0.85 × Saybolt. (The Redwood value is numerically equal to 29 times the Englar value and to 0.85 times the Saybolt value.)

The viscosity of a diesel fuel is very important. If the fuel is not viscous enough, then the lubricating film between the moving parts in the injectors and fuel pump may break down, causing rapid failure of this very expensive equipment. On the other hand, an excessively viscous fuel may not fully charge the pumping element at high speeds, resulting in loss of power. Again, the more viscous a fuel, the less the atomisation but the greater the fuel spray penetration, so that the combustion process may be severely altered if a fuel of the incorrect viscosity is used.

Specific gravity

The specific gravity of a substance is the ratio of the mass of a certain volume of the substance to the mass of an equal volume of water. This specification is not significant from a combustion or performance point of view, but is used for converting weight of fuel to volume of fuel and vice versa.

Carbon residue

The carbon residue value is found by burning a fuel sample, weighing the residue, and expressing this mass as a percentage of the mass of the fuel sample. The conditions under which the test is made are not the same as those existing in the combustion chamber, so the result is not directly related to engine deposits. However, it is an indication of the quantity of slow-burning constituents likely to cause engine deposits.

Sulphur content

Although sulphur is undesirable in fuels, operating conditions have a very considerable bearing on the maximum allowable sulphur content.

When sulphur burns, a very small proportion of it forms sulphur trioxide (SO_3), and if this combines with any water vapour (H_2O) formed during the combustion process, sulphuric acid (H_2SO_4) results. If this sulphuric acid vapour is condensed by any cool surfaces, such as cylinder liners, it will settle on and attack these surfaces.

Ash content

The ash content of a fuel is a measure of the incombustible material in the fuel. The incombustible material exists in two forms:

- hard abrasive solids
- soluble metallic soaps.

Both forms contribute to engine deposits, particularly in high-speed engines, while the solids, which consist of silica, iron oxide and other impurities, are extremely abrasive and contribute to injector and fuel pump wear, as well as piston ring wear.

Water and sediment content

Water in the fuel can wreck the fuel injection equipment very rapidly, and every attempt must be made to keep the water content to a minimum. Unfortunately water collects in fuel tanks due to condensation, and so the only way to draw off clean fuel is to have the fuel pick-up some distance above the bottom of the storage tank, and to regularly drain the water and sludge from the tank.

The same holds for sediment. A certain amount of rust (for example) builds up during storage; this should be allowed to settle and should be drained regularly with the water.

Fuel storage

When it is realised that the clearances in the injection equipment are in the order of 0.0025 mm or a quarter of a hundredth of a millimetre, it is obvious that even small particles of abrasive material are capable of doing severe damage to this equipment. Further, since diesel fuel is a heavy fuel, solids take a considerable time to settle out. For these reasons, diesel fuel containers should not be moved for a few days prior to drawing off fuel.

Storage drums and tanks

Storage tanks should be of as large a capacity as practical and, for preference, cylindrical. Ideally, they should be set up almost horizontal, with a slope of 40 mm per metre of length away from the outlet connection. The outlet connection should be not less than 75 mm above the bottom of the tank, and a sludge drain cock should be fitted to the bottom of the tank at the low end. Regular draining of sludge and water accumulations is essential if clean fuel is to remain so.

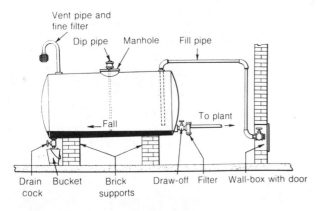

Fig 9.1 Ideal fuel storage tank

Key factors in the proper storage of fuel are as follows:

- Store fuel in steel drums and tanks. Never use galvanised lined drums or tanks; the zinc reacts with the diesel oil and forms a sludge.
- Diesel fuel has a shelf life of 12 months, when correctly stored under cool, dry conditions. Always empty drums or tanks before refilling with new fuel.
- Drain sediment and condensed water from tanks before refilling.
- Avoid carrying over summer fuel for winter use.
- After moving or refilling storage drums, allow them to stand for a number of hours before drawing off fuel.
- Install a filter in the drum or tank outlet hose.

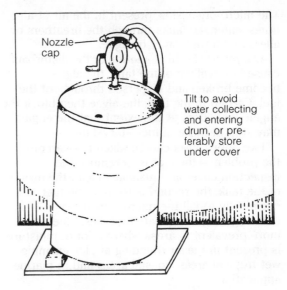

Fig 9.2 Correct method of positioning a fuel drum in current use

Care of storage drums and tanks

- Protect drums and tanks by providing overhead shelter to keep them out of direct sunlight especially during the summer months. Make every effort to minimise variations in temperature, which can cause moisture-laden air to enter the container.
- Store drums off the ground and on their sides with the bungs at the '3 o'clock' and '9 o'clock' positions, as shown in Fig 9.3.

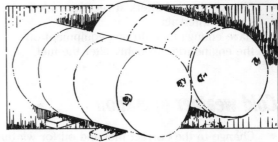

Fig 9.3 Correct positioning of drums in storage

Biological contamination

This fuel contamination problem generally occurs in storage tanks and marine diesel fuel tanks. The contamination begins when micro-organisms multiply at the interface between water on the bottom of the tank and the fuel.

The micro-organisms, present in the air at all times, enter the tanks through the breathers or fillers and form a layer of algae.

As a result of vehicle motion or the disturbance caused by refilling of the tank, the algae become broken and dispersed throughout the fuel. Carried in the fuel, the algae then block the engine fuel filters to the extent that the engine may stop because of fuel 'starvation'.

Two approaches can be taken to overcome the problem—the first is to remove the inspection cover and manually clean the inside of the tank; the second is to treat the fuel chemically to kill the micro-organisms.

Biological contamination of diesel fuel is more prevalent in areas where a lot of moisture is present in the surrounding air, for example wet tropical areas, coastal areas and in marine applications.

under the machine, disastrous results may occur.

—Use 20–25 per cent heating oil mixed into the diesel fuel to reduce waxing. Heating oil has a lower cloud point than diesel fuel.

Care of fuel systems

Machine maintenance

- Refill the fuel tank after use or at the end of the day's operation; this will exclude the moisture-laden air in the tank.
- Drain the sediment and water from the fuel tank and filter once a week.
- Change the fuel filters at recommended service intervals.
- Have enough fuel filtering equipment on the engine to thoroughly filter the fuel.

Cold weather operation

- Change to the correct fuel and oil for winter operation.
- Store the machine under cover overnight.
- Insulate the fuel tank and fuel system components with a suitable cover to protect them from the elements.
- If the engine will not start due to fuel starvation:
    —Renew the filter(s).
    —Warm the fuel system with hot water or a steam cleaner. This will dissolve the wax crystals in the fuel lines. Never light a fire

10 Engine lubrication

During the initial stages of development of internal combustion engines, simple, even crude, lubrication systems were used and were satisfactory. However, internal combustion engines have developed from the low-powered, unreliable engines of the late 1800s to the highly sophisticated high-speed machines they are today. As the speed, power, reliability and performance of these machines have rapidly advanced, so have the lubrication systems advanced to keep pace with changing engine designs, for no engine can perform reliably unless it is efficiently lubricated.

Engine lubricating systems

The first lubrication systems to be used were known as 'splash' systems. This was an apt name, for the engine components were lubricated by oil splashed onto them by the engine connecting rod big end and the crankshaft web ends. In the more sophisticated designs, a dipper on the big-end bearing cap picked up oil from the sump, and a drilling led this oil into the big-end bearing surfaces. The main bearings were lubricated by oil splashed into troughs on the inside of the crankcase by the big ends and crankshaft, and led to the bearings through drillings in the housing. These early systems have long been superseded by the pressure system in which oil is pumped by an engine-driven oil pump to the various engine components.

In a typical system, oil from the sump passes through a strainer into a positive displacement oil pump. Leaving the pump (under pressure), the oil flows past a relief valve, which opens whenever the oil pressure exceeds a set maximum to dump sufficient oil to the sump to limit the pressure. Typically, the oil then flows via galleries to the oil cooler, where it is cooled by the engine radiator water or ambient air, although not all engines are fitted with a cooler. Some engines incorporate an oil bypass valve, which allows the flow of oil to bypass the oil cooler when the oil is below a certain temperature, for example on engine startup.

On leaving the oil cooler, the oil flows through a full-flow oil filter; on some lubrication systems a percentage of this oil flow is directed to a bypass filter, where it is filtered and returned to the sump. From the full-flow filter, the main volume of filtered oil flows to the main oil gallery, which runs the length of the engine block.

The oil flow from this gallery is directed to the following engine components:

- crankshaft
- camshaft and valve mechanism
- timing gears
- underside of the piston
- exhauster or vacuum pump (if fitted)
- turbocharger (if fitted)
- fuel injection pump (in many engines).

Oil for lubricating the crankshaft bearings flows directly to the main bearings through drillings in the engine block. An oil channel cut in the surface of the main bearings carries oil completely around the journal. From this channel, oil passes through drillings in the crankshaft webs to the big-end bearings.

In most designs, the connecting rod small-end bearings (or gudgeon bushes) receive oil from the big-end bearings via axial drillings in the connecting rods. In other cases, the small-end bearings are lubricated by an intermittent spray of oil and by oil vapour. The intermittent oil spray originates at the big-end bearings, where a drilling through the connecting rod aligns with the big-end feed drilling in the crankshaft once in each revolution.

Oil for camshaft bearing lubrication flows through internal drillings in the block, either

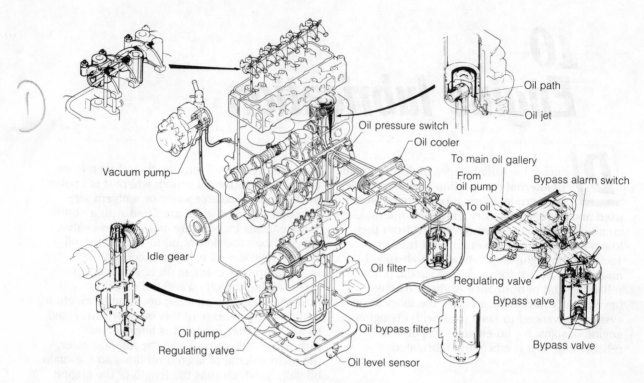

Fig 10.1 Oil flow diagram for a typical automotive diesel engine

from the gallery or from the main bearings. In many designs, a groove in the camshaft journal or in the camshaft bearing provides an oil feed channel around the journal to supply the valve mechanism.

From this supply point, oil flows readily to the valve mechanism, either through an oil feed pipe, or through aligned drillings in the engine block, the cylinder head and the rocker shaft pedestal. Oil seeping through the rocker bearings is splashed about the valve chamber to lubricate valve stems, pushrods, and other moving surfaces. This oil eventually drains down through the pushrod openings, lubricates the tappets and returns to the engine sump.

Oil flow to the timing gears or timing chain (at either the front or rear of the engine) is usually from the main oil gallery. Oil is also directed to a timing chain tensioner (if fitted), the timing and idler gear bearings, and finally is sprayed onto the timing gear teeth.

Many makers utilise the engine oil to cool the pistons, either by actually passing oil through channels in the piston crown, or by directing an oil spray onto the underside of the piston crown.

In a typical system, oil is sprayed upward under the piston through spray nozzles located at the bottom of the cylinder bore. The oil enters special passages in the underside of the piston, circulates under the crown of the piston, and drains out through another passage on the other side of the piston.

Where an engine is fitted with an exhauster (or vacuum pump), it is usually lubricated from the engine lubrication system. The engine oil flows through a supply pipe from the engine lubrication system, lubricates the vanes and bearings of the exhauster, and returns to the engine crankcase through a return line or a drain port via the timing case.

Almost all turbochargers are lubricated from the engine system. Oil flows from the oil filter housing through a supply pipe to the main bearing housing of the turbocharger. There the supporting shaft bearings, the thrust bearings and the oil seals are lubricated and cooled. The oil then drains from the turbocharger via a drain line, back into an open part of the engine crankcase.

In installations where the injection pump is lubricated by the engine lubricating system, the oil is pumped through the pump cambox to lubricate the camshaft lobes, the cam followers and the camshaft bearings. Oil is also directed to the variable timing unit (if fitted) for lubrication. All the return oil leaves the injection pump via the front drive or by a drain pipe to return to the engine sump.

Oil coolers

As engines are developed and produce more power, operating temperatures are becoming higher. One of the functions of the engine oil is to conduct heat away from local hot spots in the engine, and in so doing, the oil becomes hot. The function of an oil cooler is to stop the engine oil from becoming excessively hot under heavy load conditions. The hotter the oil becomes, the greater the danger of lubrication failure, and oil oxidation.

The operation of an oil cooler is similar to that of an engine radiator, particularly an air-cooled unit, where the oil is passed through the core, which is cooled by ambient air flow.

In a typical water-cooled system (see Fig 10.2), the hot oil is circulated through a chamber containing a large number of tubes carrying engine cooling water. Heat passes from the hotter oil to the cooler water, lowering the oil temperature.

Oil filters

During service, engine oil becomes contaminated by oxidation, the products of combustion, abraded metal particles and foreign matter that has entered the engine through the oil filler, engine breather and so on. Oil filters are included in the lubrication circuit to remove any solid contaminants that may cause abrasion or destroy the oil's lubricating quality.

If the oil is passed through a porous material, any solids in the oil larger than the pores will not pass through and will be filtered from the oil. However, the material will eventually become choked, preventing the oil from passing through. The time taken for the filter material to become choked depends on three factors:

- the size of the pores in the material
- the area of filtering material used
- the operating conditions, which directly influence how rapidly the oil is contaminated.

Clearly, the size of the pores in the material must determine the filtering efficiency, but a filter material that is too efficient may remove necessary oil additives from the oil and may become choked far too rapidly to be practical. Many different filter materials have been and are still being used. By far the most common, however, are felt and resin-impregnated paper. Many manufacturers of engines using felt filters recommend that dirty elements be washed in petrol a couple of times before being renewed, thus extending the working life of the filter. This is not recommended with paper filters, but their high efficiency has made them the most popular.

The larger the area of filtering material available, the longer will it take to become

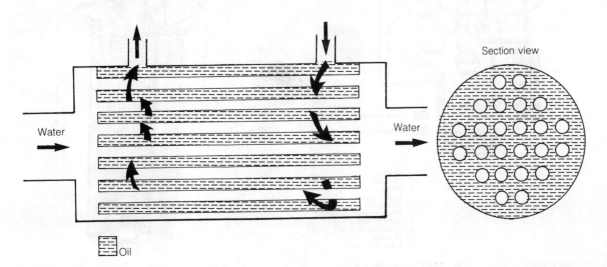

Fig 10.2 Typical engine oil cooler showing oil and water flow paths

choked. Hence it is desirable to make oil filters with the largest possible filtering area, but available space limits the overall size, and therefore manufacturers must design their products to present the greatest possible filter area in the most compact form. Most filters are manufactured from sheet material folded or pleated to satisfy both requirements.

If an engine is operating under severe conditions of dust and load, the oil will become contaminated sooner than under normal conditions. Consequently, the oil filter will choke more rapidly and will require servicing more frequently.

Full-flow oil filters

By far the most extensively used, full-flow oil filters are fitted between the oil pump and the gallery where they filter all the oil delivered to the engine components. Since all the oil flowing to the engine gallery must pass through a filter of this type, the element must not be so fine as to restrict the flow of oil while the filter remains

clean and the oil is not abnormally viscous. To prevent oil starvation should a full-flow oil filter become blocked or choked, a bypass is incorporated in the filter housing to provide an alternative oil circuit to the gallery. This bypass is controlled by a valve that opens when the pressure at the filter outlet is less than the inlet pressure by a predetermined amount, usually in the vicinity of 70 kPa. The bypass valve is held onto its seat by a relatively light spring. Filter inlet pressure is applied to the valve in a direction that tends to open the valve, while on the other side of the valve, filter outlet pressure is applied to assist the spring in keeping the valve closed. Under normal operating conditions the inlet and outlet pressures are very nearly equal and the spring holds the valve on its seat, but when the filter element becomes blocked, the outlet pressure drops. If the pressure drops approximately 70 kPa, the total force applied by the spring and outlet pressure will be overcome by the force resulting from the inlet pressure, and the valve will be forced from its seat,

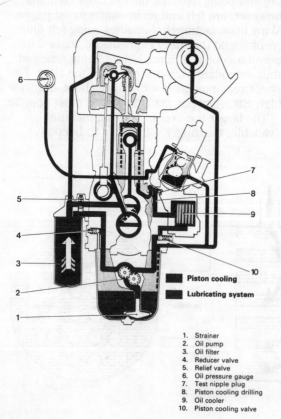

1. Strainer
2. Oil pump
3. Oil filter
4. Reducer valve
5. Relief valve
6. Oil pressure gauge
7. Test nipple plug
8. Piston cooling drilling
9. Oil cooler
10. Piston cooling valve

Fig 10.3 Engine lubrication circuit—full-flow filter system

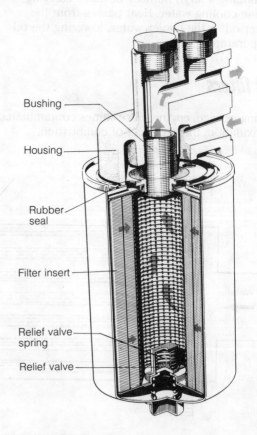

Fig 10.4 A full-flow filter

allowing the oil to bypass the filter. While this allows unfiltered oil to pass to the engine components, it is preferable to insufficient oil.

Bypass oil filters

A bypass filter makes use of a finer element than a full-flow filter, but filters only a proportion of the oil delivered by the oil pump. As its name suggests, the filter is incorporated in a bypass circuit, the filtered oil being returned to the sump. However, because a proportion of the oil is constantly being filtered, all the oil is eventually filtered and kept reasonably clean.

Because of its position in the circuit, a bypass filter must offer sufficient restriction to oil flow to maintain pressure in the lubricating system. If it offered no restriction, almost all the oil would escape back to the sump through the filter and the oil pressure would not build up sufficiently to lubricate the engine components. For this reason, a restricting orifice is usually incorporated in the filter housing outlet.

In many cases a bypass filter may appear to be similar to a full-flow filter, but it will have a finer element and no bypass valve, since a blockage of the filter cannot restrict the flow of oil to the engine components but must increase it.

Centrifugal oil filters

Filters that have no element, but separate heavy solids from the oil by centrifugal force, are becoming increasingly popular with many engine manufacturers. Filters of this type are known as centrifugal filters, and may be driven either by the oil that is being filtered or directly by the engine.

The operation of a typical centrifugal filter may be readily followed by reference to Figs 10.6 and 10.7. Oil from the oil pump enters the filter unit through the oil inlet, passes up the spindle and flows into the rotating canister. When the rotor fills, oil enters the outlet tubes through the gauze strainers and escapes through a nozzle at the lower end of each tube. The oil sprays squirting out of the nozzles spin the rotor (turbine action), the rotor being supported in plain bearings in the top and bottom housings.

The speed at which the rotor spins is governed by the oil pressure, and generally lies between 2000 and 6000 rpm. Any solids in the oil are flung by centrifugal force against the sides of the rotor, where they remain until the unit is dismantled and cleaned during service. Centrifugal filters of this type may be employed as either full-flow or bypass units, the outlet being restricted in bypass applications.

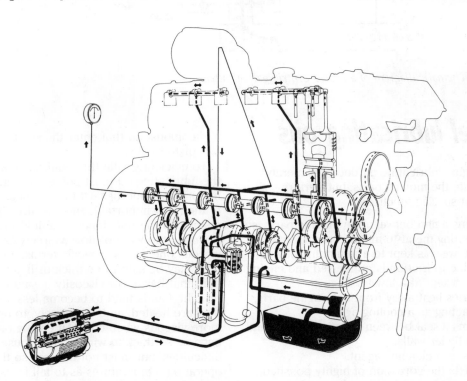

Fig 10.5 Engine lubrication circuit—bypass filter system

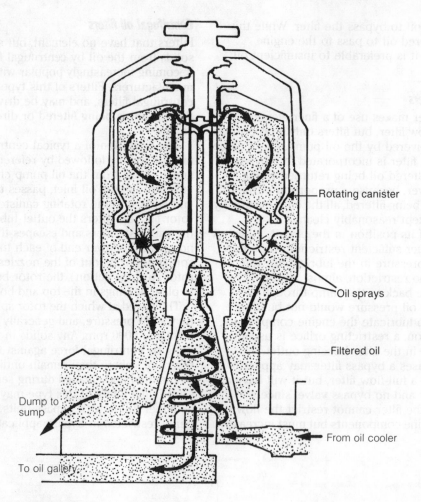

Fig 10.6 Schematic diagram of a bypass centrifugal oil filter

Diesel lubricating oils

The engine lubricating oil does not merely lubricate the moving parts, but performs a number of specific functions:

- It forms a film between the moving parts, so preventing metal-to-metal contact. As a result, wear is kept to a minimum, power loss due to friction is minimised and engine noise is kept at a low level.
- It carries heat away from hot engine parts, thus acting as a cooling agent.
- It forms a seal between the piston rings and the cylinder walls.
- It acts as a cleaning agent.
- It resists the corrosion of highly polished engine surfaces by the acidic products of

combustion that enter the sump past the piston rings.

To efficiently fulfil the requirements listed, the lubricating oil must possess a number of important properties. Of these, viscosity is one of the most important, and has already been defined as the reluctance of a fluid to flow—a viscous fluid will not flow as freely as a less viscous one. In less specific terms, a viscous fluid may be said to be thick or heavy, while a fluid that has a low viscosity is said to be thin, or light. Fluids tend to become less viscous as they are heated, and, conversely, to become more viscous when cold. An engine oil must not become so thick in winter as to cause starting difficulties, but must not become so thin at operating temperatures as to fail in its requirements as a lubricant. The oil should

Fig 10.7 Section view, Glacier GF2 centrifugal oil filter

resist any tendency to become oxidised at high temperatures, since the oxidation of the oil creates deposits and acid products. However, deposits and acids are also formed through the combustion of the fuel, and the oil should have the ability to wash the deposits from engine components and to neutralise the acids. Further, the oil should not foam when agitated in the sump and should be able to withstand the extreme pressures encountered between certain engine components.

Mineral oil, by itself, cannot fulfil all the requirements of an engine lubricating oil, and certain chemicals are added to the basic mineral oil.

Lubricating oil additives

All lubricating oils suitable for use in medium- and high-speed diesel engines contain most, if not all of the following additives. Indeed, many contain more than the major ones listed below:

- detergent/dispersants
- oxidation inhibitors (anti-oxidants)
- corrosion inhibitors (alkaline additives)
- viscosity index improvers
- pour point depressants
- anti-scuff (extreme pressure) additives
- foam inhibitors (anti-foam additives).

Detergent/dispersant additives

The detergent component of these additives is included to wash deposits from the engine components, while the dispersant component is provided to ensure that the solids washed from the engine components by the detergent do not clot and form sludge.

The intense heat in the engine cylinder causes the oil film on the piston rings and cylinder walls to be broken down, producing tarry deposits, which may cause the rings to stick in the ring grooves. In addition, small quantities of unburnt fuel may be broken down into similar substances in the same area, aggravating the situation. It is to prevent these deposits from building up that the detergent additives are included in the lubricating oil. However, if the deposits were simply washed from the pistons and rings, they would accumulate as sludge elsewhere and would eventually restrict the oil passages. Dispersant additives prevent any particles from clotting and forming sludge by maintaining the particles in suspension in the oil. Any such particles are then filtered out of the oil by the oil filter, or drained from the engine when the oil is changed.

Oxidation inhibitors

When mineral oil is heated and agitated in the presence of air, the oil oxidises. The oxidation of oil causes three products to be formed—**gum**, which causes components to stick and/or become coated with a varnish and oil-ways to become blocked, **acids**, which attack bearing surfaces, and **sludges**, which cause thickening of the oil. The ideal conditions for the promotion of oil oxidation exist in the crankcase of a running engine, and to reduce this tendency additives known as oxidation inhibitors or anti-oxidants are included in the oil.

Corrosion inhibitors

Alkaline additives, known as corrosion inhibitors, are included in the engine oil to neutralise any acids created by oil oxidation or combustion. During combustion, sulphur in the fuel combines with oxygen to form oxides of sulphur. The sulphur trioxide so formed then combines with the water vapour created during combustion to form sulphuric acid, while other sulphur compounds combine with water vapour to form less corrosive acids. These acids naturally attack highly polished metal surfaces, particularly if they pass the piston rings and condense on the cooler metal surfaces in the crankcase. Once these acids have entered the sump, they will corrode the running surfaces of the engine whether it is operating or not.

Viscosity index improvers

The rate of change of viscosity with temperature is known as the viscosity index of the oil; the higher the viscosity index, the lower the oil's viscosity change per degree temperature change. Thus an oil with a high viscosity index maintains a fairly stable viscosity over a wide temperature range. To improve this very desirable characteristic of lubricating oils, viscosity index improvers are added.

If a lubricating oil is too thick (too viscous) the amount of power necessary to overcome friction will be extremely high, while if the oil is too thin, the oil film will be easily fractured and lubrication will fail. But the viscosity of oil changes with temperature—the hotter the oil, the less viscous it becomes. If the oil's viscosity changes very much over the engine's operating temperature range, it is likely that difficult starting (owing to the high friction value) will occur when the engine is cold, and subsequently the quality of the lubrication at operating temperature will be poor.

Pour point depressants

At very low temperatures, oil may become so thick that it will no longer flow. When this occurs, it is impossible to start the engine without first heating the oil in the sump until it becomes fluid enough to permit the engine to be turned over and to ensure adequate lubrication. To lower the temperature at which the oil will remain fluid, pour point depressants are added.

Anti-scuff additives

Certain engine components are subjected to extreme pressure of contact during normal running conditions. This high contact pressure is usually combined with a wiping action and this combination readily breaks through the oil film, allowing metal-to-metal contact. To prevent damage, anti-scuff additives are usually incorporated in engine oil. These additives react chemically with the metal surfaces to form very thin films that are slippery and extremely strong. These films prevent metal-to-metal contact during moments when the oil film is broken.

Anti-foam additives

The churning action in the engine crankcase, coupled with a secondary action of some additives, may cause the engine oil to froth and foam. If this occurs, inefficient lubrication and loss of oil from the engine breather, filler and/or dipstick hole result. The foaming tendency is prevented by the addition of anti-foam additives or foam inhibitors.

Diesel engine oil classifications

There are two important aspects to consider in the selection of a lubricating oil. The first is viscosity, which may be considered as the 'body' or thickness of the oil; the second is its additive concentration and content, which determine whether the oil is suitable for use under particular operating conditions.

SAE oil viscosity gradings

Note   This system originated in the United States, where °F are used to designate temperature and test specifications are not given in °C. The following information is given using the original temperature system, but can be converted as follows: 212°F = 100°C; 0°F = −17.8°C.

The viscosity of an oil is indicated by the manufacturer in terms of SAE grade numbers, the higher the number, the more viscous the oil. This number indicates the oil's viscosity at a specific test temperature, namely 210°F. This is a good guide to the viscosity of the oil at operating temperature, but gives no indication of its viscosity at low temperatures, thus there is nothing to indicate that the oil will not become too heavy (too viscous) at low temperatures. However, if an SAE number is followed by a 'W', it indicates that the viscosity is based on a 0°F test temperature, and this gives a good indication of the oil's viscosity under winter starting conditions.

Multigrade oils have been developed to give suitable viscosity under winter cold starting conditions and yet remain viscous enough to ensure efficient lubrication at normal operating temperatures. For example, an oil rated 10W–30 will meet the viscosity requirements of an SAE 30 oil (at 210°F), but will not become any more viscous at 0°F than is specified for an SAE 10W oil.

Note The viscosity specifications for an SAE 10W oil (at 0°F) are not the same as those for an SAE 10 oil (at 210°F). An SAE 10W oil is much more viscous at test temperature than an SAE 10 oil at its test temperature. Further, at 0°F, an oil rated SAE 10W is much more viscous than an SAE 30 oil at 210°F.

Engine service classifications

Diesel engine lubricants are subjected to greater oxidising influences within the engine cylinder than are the oils used in petrol engines due to greater compression pressures and a much more adequate air supply during combustion. When a lubricating oil is oxidised on the cylinder walls, pistons and rings, and this combines with the products of the partial combustion of any of this oil that has combined with fuel, gums and lacquers are formed that cause sticking rings and which build up on the piston skirt. Extreme cases of such build-up may ultimately cause engine seizure and must be prevented. Detergent additives will dissolve such deposits, preventing the above problems. The need to use the correct grade of oil cannot be too greatly emphasised.

If an engine is called on to deliver high continuous power output, then the oxidising tendencies are very high. Operating an engine to deliver fairly high power at a low temperature promotes, in addition to deposits, the formation of acids. The use of fuels with a high sulphur content also causes acid formation, while inefficient combustion, regardless of the cause, leads to engine deposits and dilution of the engine oil.

On the other hand, intermittent operation at the rated load and at normal operating temperature, together with the use of high-grade, low-sulphur-content fuel and efficient combustion, minimise or prevent the formation of excessive deposits, large quantities of corrosive compounds and oil dilution.

Without some standard, manufacturers would each tend to give an oil a classification that may well have no basis for comparison with another oil, or give any indication of its applications. The standard system of classification in Australia is the API (American Petroleum Institute) Engine Service Classification System.

This system provides the standards for the performance characteristics of engine crankcase oils, and for the operating conditions for which they are intended.

There are two classifications within the system—the Commercial 'C' classification (primarily for diesel engines), and the Service 'S' classification (primarily for petrol engines).

Other classifications have also been widely used—the US military system, as in MIL-L-2104B, or the Caterpillar Tractor specifications, such as Series 3.

C—'commercial' classification (primarily for diesel engines)

CA, for light duty diesel engines—This is typically for diesel engines operated in mild to moderate duty with high-quality fuels. Occasionally CA has included petrol engines in mild duty. Oils designed for these conditions provide protection from bearing corrosion and from ring belt deposits in some naturally aspirated diesel engines when run on fuels of such quality as to impose no unusual requirements for wear and deposit protection. Such oils were widely used in the late 1940s and 1950s.

CB, for moderate duty diesel engines—This classification is typically for diesel engines operated in mild to moderate duty, but with lower quality fuels that necessitate more protection from wear and deposits. Occasionally CB has included petrol engines in mild duty. Oils designed for these conditions provide necessary protection from bearing corrosion and from ring belt deposits in some naturally aspirated diesel engines with higher sulphur fuels. Oils designed for these conditions were introduced in 1949.

CC, for moderate duty diesel and petrol engines—CC is typically for certain naturally aspirated, lightly turbocharged or supercharged diesel engines operated in moderate to severe duty, and certain heavy duty petrol engines. Oils designed for these conditions provide protection from high-temperature deposits and bearing corrosion in these diesel engines and also from rust, corrosion and low-temperature deposits in petrol engines. These oils were introduced in 1961.

CD, for severe duty diesel engines—This classification is typically for certain naturally aspirated, turbocharged or supercharged diesel engines where highly effective control of wear and deposits is vital, or when run on fuels of a wide quality range, including those with high sulphur content. Oils designed for these conditions were introduced in 1955 and provide protection from bearing corrosion and from high-temperature deposits in these diesel engines.

Selection procedure

Oil companies are responsible for producing oils that meet the performance requirements of the various API classifications.

Engine manufacturers are responsible for deciding which API classification oil meets the requirements of a particular engine under the conditions of intended use. This API classification will normally be listed in the engine or equipment user's handbook (or on the filler cap), as a guide to indicate the most suitable oil to be used.

The equipment owner or user should then purchase an oil which meets this API classification.

11
Air cleaners

In earthmoving applications, air cleaners have been known to remove over 3.5 kg of dust in an eight-hour shift. Should dust enter the engine at this rate, it is extremely doubtful if even a very large engine could continue to run for the duration of the shift, quite apart from operating efficiently. The entry of dust into an engine causes fast abrasive wear of rings, liners, pistons and valves, while turbochargers are rapidly destroyed by pumping contaminated air. Because the volume of air required by an IC engine is very considerable, a small abrasive dust content rapidly amounts to a very damaging quantity. Hence the air cleaning system must effectively remove the greatest possible amount of foreign particles from the air.

A naturally aspirated engine must have an unrestricted air supply if it is to develop anything like its maximum possible power, and even a turbocharged or a supercharged engine will suffer from considerable loss of power if its air supply is restricted. This power loss is caused by insufficient air being available to allow full combustion of the quantity of fuel being injected, and is evidenced by black exhaust smoke. Thus, the air cleaning system should offer no restriction to airflow.

While the achievement of maximum cleaning efficiency is the most important factor in air cleaner design, closely followed by unrestricted airflow, factors such as durability, service accessibility, service life and cost are also important considerations.

The type of air cleaner fitted to a high-speed diesel engine depends largely on the size of the engine and the environment in which it has to operate. Of the many types available to suit various operating conditions, most manufacturers of modern high-speed diesel engines choose either an oil bath or a dry element type air cleaner. If the engine has to operate under extremely dusty conditions, a preliminary air cleaner (pre-cleaner), usually of the centrifugal type, is generally fitted in addition to the oil bath or dry-type cleaner.

When selecting an air cleaner to suit a particular engine, whether setting up a new installation or improving an existing one, it is of the utmost importance that the air cleaner is suitable for the application. Air cleaner manufacturers have full data available for their various models, including dimensions and airflow capacity. The capacity must be suitable for the engine and so the engine requirements must be known. Although the engine manufacturer can best supply this information, it is possible to calculate the requirements in most instances from the following formulae, bearing in mind that they apply to naturally aspirated engines. For others, refer to the manufacturers' figures.

For naturally aspirated four-stroke engines:

$$\text{airflow} = \frac{\text{displacement} \times \text{rpm} \times \text{VE}}{2000}$$

where airflow is in m^3/min
displacement is in litres
and VE is volumetric efficiency

or

$$\text{airflow} = \frac{\text{displacement} \times \text{rpm} \times \text{VE}}{3456}$$

where airflow is in cu ft/min
displacement is in cu ins
and VE is volumetric efficiency

For naturally aspirated two-stroke engines, double the requirements by halving the denominator.

Note The volumetric efficiency of IC engines varies as follows:

4-stroke side valve petrol engine	70%
4-stroke OHV petrol engine	75%
4-stroke OHV diesel engine	85%
4-stroke OHV turbocharged diesel engine	130% (av)

Dry element type air cleaners

Although oil bath air cleaners have been extensively used, they are rapidly being replaced by dry element type (paper element) air cleaners throughout their entire range of applications. There are a number of reasons for their fall from favour:

- Although the oil bath air cleaners are quoted at 95–98 per cent efficient, dry element air cleaners are rated at 99.5 per cent efficient for the same size dust particles. Hence the oil bath air cleaner may pass from four to ten times as much dust as the dry element type.
- Because the oil bath air cleaner depends on high air velocity to cause dust particles to impinge on the oil surface, its efficiency is affected by engine speed and may fall to 90 per cent at reduced engine speeds. Should it fall to this low figure, the oil bath air cleaner would pass twenty times as much dust as a dry element air cleaner, the efficiency of which does not fall with engine speed.
- The oil bath air cleaner is less efficient at low temperatures due to increased oil viscosity.
- A well designed air cleaning system of the dry element type is less likely to cause restriction to airflow over the entire speed range.
- There is no danger of the airflow picking up air cleaner oil as a result of an overfilled oil container or an extremely steep operating angle. Should oil be carried into a diesel engine from the air cleaner, the engine will run uncontrolledly as it burns this oil and will stop only when the oil is used or when the air intake is closed. Such runaway conditions usually cause extremely high speed and engine damage may well occur.
- An oil bath air cleaner gives no warning when it is in need of service and its efficiency has fallen. A dry element air cleaner, on the other hand, causes air restriction when in need of service, causing loss of power and black exhaust smoke, but suffers no fall off in cleaning efficiency. If a service indicator is fitted to the system—a service indicator is a warning device operated by the pressure differential across the element (although it may be dash mounted), and gives a visible indication when the element is so restricted by dust as to require service—the engine may be operated without risk until the indicator warns that service is required, so extending service periods to the maximum for the prevailing operating conditions, with complete safety.
- In general, the dry element type air cleaner is simpler to service.
- The dry element type air cleaner can be mounted in any position, while the oil bath air cleaner must be vertical.

Although the elements range from small, simple automotive types, perhaps 150 mm diameter and 50 mm thick, to large industrial types, 300 mm diameter and 450 mm long (and larger), the material most often used in their manufacture is resin-impregnated paper, although cloth, fibre, felt or spun glass have also been used. The element may consist simply of a ring of pleated paper carrying a moulded rubbery gasket on its sealing faces, or may well carry a fin (or a series of vanes) on its outer surface, which acts as a pre-cleaner by imparting a swirl to the incoming air, so that large dirt particles are separated by centrifugal force and are deposited in the bottom of the unit in the dust cup. Many industrial air cleaners feature a safety element—a second dry element inside the main or primary one—to protect the engine should the main element be damaged and to prevent any dust disturbed during removal of the primary element for service from entering the engine.

Instead of manufacturing their own air cleaners, most engine and equipment manufacturers buy them from or have them manufactured by specialists in this field. Probably the most extensively used dry element air cleaners would be those manufactured by the Donaldson Company Inc of USA, which are fitted to an extremely large variety of IC-engined machines. Hence we will examine the Donaldson 'Cyclopac' series and the 'Donaclone' series and, in doing so, we should cover the operation and service requirements of those types of air cleaner used extensively on heavy equipment in Australia.

There are two types of Cyclopac currently available—the FW series and the FH series—and both use the same system of air movement

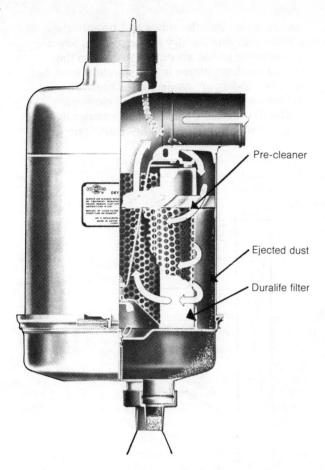

Pre-cleaner

Ejected dust

Duralife filter

Fig 11.1 Donaldson FW Cyclopac air cleaner

to centrifugally remove heavy contaminants.
The FW series (Fig 11.1) does not allow the use
of a safety element, carries the fin and sealing
gasket on the element, and may be mounted
both horizontally and vertically. Incoming air
enters the axial inlet and flows around the
outside of the element, where it passes through
the fin. The fin causes the air to swirl around the
outside of the element before passing through,
and onto the outlet. During the rapid rotation of
the air, heavy dust particles are carried out
against the case and into the dust cup, allowing
relatively clean air only to reach the element.
This feature gives a greatly extended service
period for the element, although the dust cup
must be removed and cleaned at intervals as
dictated by the operating conditions. The
installation of a vacuator valve on the dust cup
will eliminate the need for regular dust cup
service as this valve will automatically eject
dust and water.

The FH series Cyclopac varies from the FW

series in that the outlet is axially positioned
with a side inlet, the fin assembly is built into
the air cleaner body and provision is made for
the installation of a safety element, if required.
However, like the FW series it may be mounted
vertically or horizontally and may be fitted with
a vacuator valve.

Within the Donaclone series of air cleaners
there are three basic types—the SBG series for
medium duty service, the STG series for
medium to heavy duty service and the SRG
series for heavy duty service. All types utilise a
number of Donaclone tubes for primary dust
separation with a dry element for final filtration.

The Donaclone tube (Fig 11.2) consists of a
nylon outer tube with an aluminium inner.
Vanes at the top impart a cyclonic twist to the
air as it enters the unit, causing heavy particles
of dust to be thrown to the outside, from where
they fall to the bottom of the tube and into the
dust cup. Relatively cleaner air is taken up

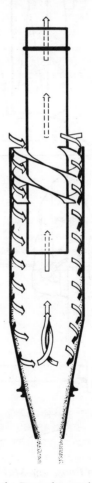

Fig 11.2 Layout of a Donaclone tube

through the centre tube to pass to the paper element. This air leaving the tubes is claimed to be as clean as that obtained from some oil bath air cleaners.

The SBG series Donaclone (Fig 11.3) features the dry element mounted vertically above the lower section containing the Donaclone tubes. Contaminated air enters the unit from one side,

passes down into the tubes and up into the section carrying the dry element, from whence it leaves via a side air outlet. Access to the element is from above in the standard type, but specific models are available for horizontal mounting with end access to the element.

The STG series Donaclone air cleaner is made in a 'T' conformation with the Donaclone tubes

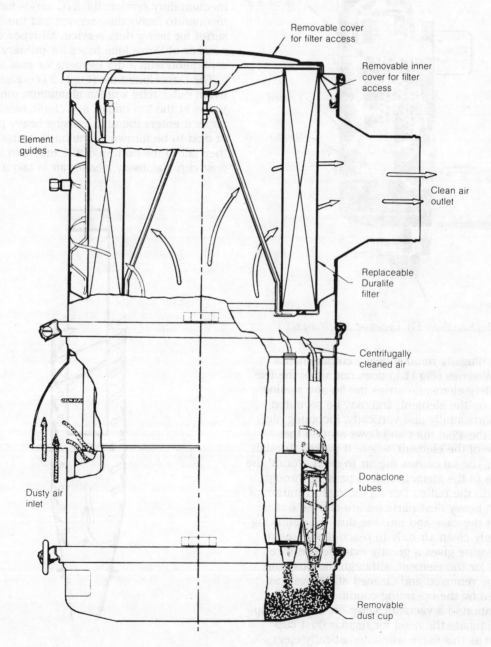

Fig 11.3 Donaldson Donaclone SBG dry-type air cleaner

in the lower section and the dry element mounted horizontally in the upper, where a safety element can be fitted as required. Air entry is through holes in the side of the lower section; from here, it passes down through the outer portion of the tubes, up the centres and into the chamber housing the dry element. By situating the element horizontally above the tubes, ready access is gained when servicing.

SRG series Donaclone air cleaners are designed for the higher airflow requirements of today's high output engines. They are available in three basic models—the smallest utilising single elements (primary and safety), a single dust cup and a single outlet, while the larger models both feature dual elements (in parallel), dual outlets and three dust cups. In standard form, these air cleaners cover the airflow requirement of 56 m^3/min (2000 cu ft per min) to 112 m^3/min (4000 cu ft per min) and are designed to operate under heavy dust conditions for periods of 1000 to 2000 hours without need of service. Special models are available for power and airflows below these.

Dry element service

Commonsense is one of the most important tools required for efficient element service, and while the following recommendations apply to Donaldson air cleaners, servicing of other types will be made possible through application of this tool. However, it should be borne in mind that all elements cannot be treated alike. Some are specially treated and must not be wet, while the manufacturers of others do not approve the use of compressed air for cleaning. Reference to the manufacturer's manual is always the best guide to service.

Dust cups should be regularly emptied, when about two-thirds full, unless a vacuator valve is fitted, in which case a quick check to see that the valve is not choked, inverted or damaged is all that is necessary. When element service is due as indicated by the service indicator (either permanently fitted or periodically attached for element checking), the element should be carefully removed and the housing wiped out with a clean damp cloth. On Cyclopac models the fins should be checked for fouling, while the Donaclone tubes should be checked on Donaclone models. If these tubes are fouled,

light dust may be removed with a stiff fibre brush, but if severe plugging with fibrous material is evident, the lower body section must be removed so that the tubes can be cleaned with compressed air or water at a temperature not exceeding 70°C.

Warning Never clean Donaclone tubes with compressed air unless both primary and safety elements are installed in the air cleaner. Do not steam clean Donaclone tubes.

When the fins or tubes are clean the replacement element may be fitted, care being taken to ensure that all gaskets are in first-class condition and are correctly located. The safety element should be **renewed** every third primary element service, and should not be cleaned and reused.

The primary element can readily be cleaned in one of two ways—by the use of compressed air or by washing in **special** detergent. Compressed air must be employed when the element is to be immediately reinstalled, but this method is not as efficient as washing the element. When using compressed air there are three points to remember—direct the air through the element in the direction opposite to normal airflow, moving the nozzle up and down while rotating the element, keep the nozzle at least 25 mm from the paper, and limit the air pressure to a maximum of 700 kPa.

Element washing instructions should be included with the detergent, but basically the process consists of soaking the element for 15 minutes or more in the solution recommended by the element manufacturer (eg Donaldson D-1400 detergent), rinsing the element until the water is clear (maximum water pressure 275 kPa) and drying with warm flowing air (maximum temperature 70°C) or simply air-drying. To check the element for damage after either method of cleaning, place a bright light inside the element and rotate the element slowly. Any holes in the element will readily be seen by looking in the direction of the light through the element. If any holes, ruptures or damaged gaskets are discovered during inspection, discard the element immediately so that there is no danger of it being subsequently refitted. After an element has been cleaned and is ready for further service, it should be stored under extremely clean conditions in a safe place where it will not be subjected to physical damage.

Oil bath air cleaners

Although oil bath air cleaners may vary in design to suit particular applications, the principle on which they operate remains unchanged.

The type of oil bath air cleaner illustrated in Fig 11.4 is fitted mainly to tractor engines, but may in some instances, be fitted to transport vehicle engines that are required to operate under extremely dusty conditions. A pre-cleaner is usually fitted to the air inlet connection of this type of air cleaner to remove the larger particles of dirt from the air before it enters the main cleaner, thus increasing the interval between successive air cleaner services.

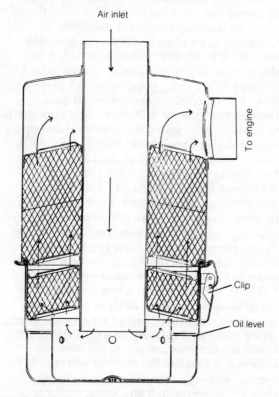

Fig 11.4 Tractor-type air cleaner

By reference to Fig 11.4, it can be seen that air enters the cleaner through the inlet pipe and moves down through the centre to the oil bowl. Here the air impinges on the surface of the oil, and the majority of the dust particles contained in the air are carried into the oil and trapped. The air then turns outwards, travels upwards through the oil-wetted wire element where the remaining dust particles are removed, and passes out through the connection at the top of the air cleaner.

The tractor-type oil bath air cleaner is usually serviced daily, or after 10 hours of engine operation. However, if the engine is operating under extremely dusty conditions, it may be necessary to service the air cleaner twice daily.

To service the air cleaner, first release the clips and remove the oil container from the lower section of the air cleaner. Empty out the dirty oil and scrape all the accumulated sludge from the bottom of the container. Thoroughly wash the oil container in kerosene or distillate. Inspect the wire element and, if necessary, remove the filter body from its installation and wash in either of the solvents mentioned above. Allow the unit to drain before reassembly. Refill the oil container to the correct level with the recommended grade of oil and reassemble to the filter.

Note Care must be taken to ensure that the container is not overfilled, as an overfilled cleaner restricts the airflow to the engine.

Pre-cleaners

Although several types of air cleaning device are used as preliminary (pre-) cleaners, probably the most extensively used are centrifugal in action. The Donaldson PB series (Fig 11.5) is an excellent example of this type of device. It is probably simplest to consider the unit as consisting of three concentric components enclosed under a cover, although the two inner components are in fact manufactured as one piece—the base. In the centre lies the air outlet through which the air with up to 80 per cent of the dust removed passes to the main air cleaner; surrounding this outlet, but integral with it, lies the vaned air inlet—vaned at an angle to cause rotation of the air as it enters the assembly—and round this again lies the transparent dust receptacle.

Contaminated air, entering through the vaned inlet, is caused to move upwards, and to rotate around inside the inner face of the dust receptacle. Heavy dust particles are carried outwards by centrifugal force and pass over the top lip of the dust receptacle inner face to fall into the collection area, while clean air is taken from the centre of the vortex down through the

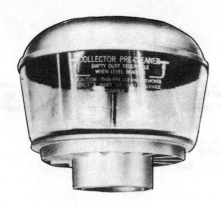

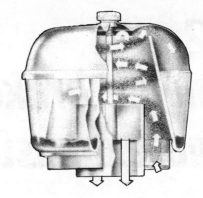

Fig 11.5 Donaldson PB series pre-cleaner

outlet. When the dust level reaches an indicator line on the transparent receptacle, it should be emptied by unscrewing the knurled nut on the top, removing the cover, lifting the receptacle from the base and emptying it.

When an engine is being used under conditions where light, coarse materials such as chaff and leaves may be carried into the air cleaner, a gauze screen may be used as well as or instead of a pre-cleaner to prevent such materials from entering the air cleaner, where they could seriously impair the efficiency of any cyclone-type separator. Such screens are also readily available and are called 'pre-screeners' by Donaldson.

The Donaspin pre-cleaner

The Donaspin pre-cleaner is designed to remove heavy contaminants from the incoming air and automatically eject them from the system via the exhaust pipe. The advantages of this system over the conventional pre-cleaner include higher engine volumetric efficiency as a result of the less restricted air flow, and, most of all, lower maintenance.

Air flows into the Donaspin pre-cleaner via angled slots on the pre-cleaner housing; these slots direct the air in a swirling motion around the inside of the pre-cleaner. Due to centrifugal force, the heavy particles in the air are thrown outwards and collect in the base of the pre-cleaner. This area is connected via a low-pressure scavenge line, to a venturi in the exhaust pipe, and the particles are carried into the exhaust by the air flow resulting from the low pressure created in the venturi.

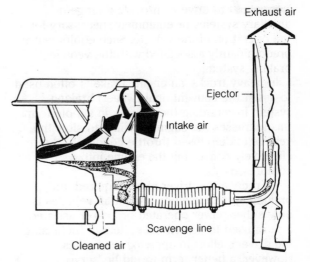

Fig 11.6 Schematic diagram of a Donaspin pre-cleaner

12 Engine brakes, compressors and exhausters

When used in vehicle applications, the diesel engine is often called on to drive or provide a range of ancillary systems or equipment, necessary for the operation of the vehicle. Such equipment is predominantly associated with the vehicle braking system.

In heavy trucks, an engine brake is often used to assist the conventional (wheel) braking system in retarding the vehicle. In simple terms, this increases the engine's ability to slow the vehicle under closed throttle conditions, by effectively converting the engine into a compressor.

To reduce the driver effort required, the braking system itself on almost all vehicles is now either power operated or 'power assisted', a term used to describe a system of increasing the driver's effort in applying the brakes. However, a better term would be 'servo-assisted', and neither power operation nor servo-assistance need be restricted to brakes, but may also be applied to other areas such as steering and gear changing.

Two of the most common methods of obtaining servo-assistance are by utilising either compressed air or a partial vacuum. An adequate supply of compressed air is readily obtained by fitting a compressor to the engine, while the means employed to create the partial vacuum will depend on the type of engine. In a petrol engine the inlet manifold depression may be directly employed, but in the case of a diesel engine the inlet manifold depression is not adequate for the purpose. Consequently a special vacuum pump known as an **exhauster** must be driven by the diesel engine to create the necessary low-pressure condition desired.

Diesel engine brakes

An engine brake is designed to effectively turn the engine into an air compressor that absorbs power while compressing the air charge within the cylinders. In so doing, the engine is converted into a retarder coupled to the vehicle's drive train, which helps slow the vehicle, giving greater safety, with added brake and tyre savings.

Another style of engine braking system is the exhaust brake, which moves a shutter across the exhaust pipe to restrict the exit of exhaust gas, which in turn slows the engine and vehicle.

Jacobs engine brake

In a conventional four-stroke engine, air is compressed during the engine's compression stroke. As each piston passes TDC, this air will act on the piston (like a compressed spring) to drive it back toward BDC. The energy used in compressing the air is thus stored as potential energy, which then drives the piston down again. Because of inefficiencies—heating, leakage, friction—this positive driving effect is somewhat less than the power input.

However, with an engine brake, the exhaust valves are opened just prior to TDC on the compression stroke, thus releasing the compressed air (and, effectively, its potential energy) into the exhaust system. The energy required to return each piston to the bottom of its stroke as well as compressing the air on the compression stroke is gained from the momentum of the vehicle. It is this two-step process of releasing the compressed air from the cylinders, and using the vehicle's

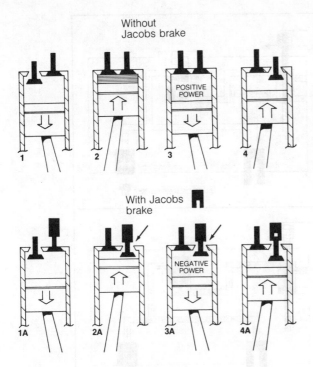

Without
Jacobs brake

POSITIVE
POWER

1   2   3   4

With Jacobs
brake

NEGATIVE
POWER

1A   2A   3A   4A

Fig 12.1 Operating principle of the Jacobs engine brake

momentum to move the pistons that develops the engine brake retarding force.

Operation of the Jacobs engine brake

Under all 'run' conditions, in an engine without a Jacobs engine brake, air is compressed in the engine cylinders during the compression stroke to a high pressure (3100–3800 kPa) and temperature (550°C). Fuel is then injected and combustion occurs, raising the cylinder pressure and temperature to even greater levels than before. As combustion continues, the pressure of the expanding gas in the cylinder forces the piston down, to turn the crankshaft. In other words, the engine produces power.

In an engine equipped with a Jacobs engine brake, whether it be a two-stroke or four-stroke, something very different happens in each engine cylinder during braking, as shown in Fig 12.1. The key component is a slave piston mounted in a housing over the exhaust crosshead or the exhaust valve and hydraulically connected to a master piston in another section of the housing. As can be seen by reference to Fig 12.2, the oil gallery between the master and slave pistons is charged with oil. At the precise engine timing

position, the master piston is driven up, displacing the trapped oil to force the slave piston down, opening the exhaust valve.

The compressed air, with its potential energy, is released into the exhaust system. By venting the compressed air at this point, no positive driving effort is exerted on the piston, thereby dramatically increasing the retarding effect of a normal engine with closed throttle on vehicle movement.

In conclusion, the Jacobs engine brake is a hydraulically operated device, which converts a power-producing diesel engine into a power-absorbing retarding mechanism.

Jacobs engine brake controls

Activation of the Jacobs engine brake is controlled by three main switches, as shown in Fig 12.3:

- An **on–off dash switch** designed to initially activate the engine brake unit, which is left in the 'on' position during vehicle operation. Also, a three-position selector switch mounted on the dash is used to engage the engine brake on selected cylinders only, so altering the intensity of the engine braking, dependent on driving conditions.
- A **clutch switch** mounted beside the clutch pedal with a probe in contact with the pedal itself. The engine brake will only operate when the drive line from the engine to the transmission is coupled. Therefore, when the clutch pedal is depressed to change gears, the clutch switch will operate, preventing the engine brake from operating. However, if there were a malfunction in a switch and the brake did not cease to operate, the engine would stop.
- A **rack travel** or **throttle switch** mounted on the governor housing of the fuel injection pump, which has a probe in contact with the fuel rack. When the vehicle is being operated with the driveline fully engaged and the throttle backed off, the rack travel or throttle switch is activated, due to governor action moving the rack to the 'no fuel' position, thereby setting the engine brake into operation.

In summary, to operate a Jacobs engine brake, the dash on-off switch has to be in the 'on' position, the clutch pedal has to be fully released and the throttle closed.

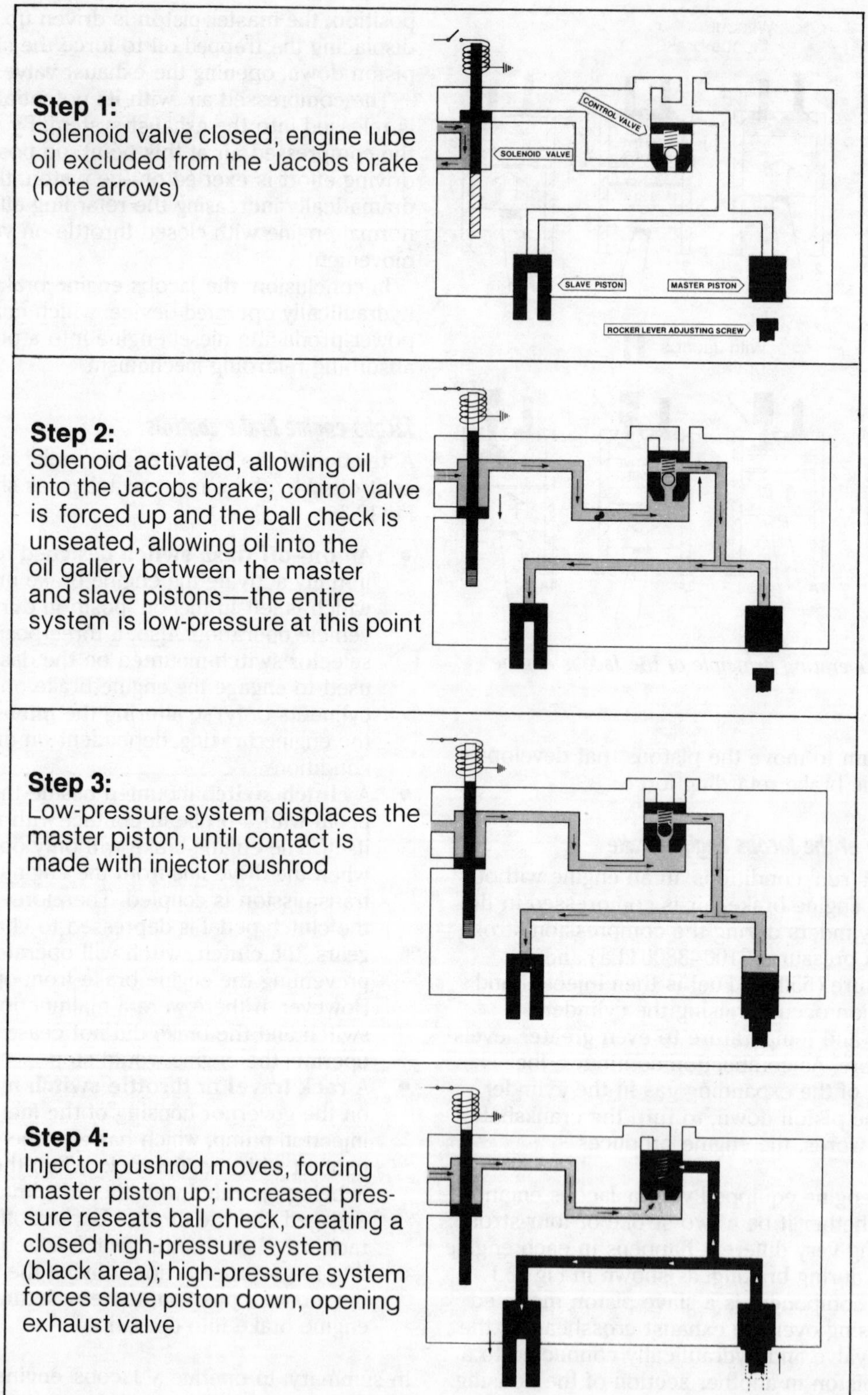

Step 1:
Solenoid valve closed, engine lube oil excluded from the Jacobs brake (note arrows)

CONTROL VALVE
SOLENOID VALVE
SLAVE PISTON   MASTER PISTON
ROCKER LEVER ADJUSTING SCREW

Step 2:
Solenoid activated, allowing oil into the Jacobs brake; control valve is forced up and the ball check is unseated, allowing oil into the oil gallery between the master and slave pistons—the entire system is low-pressure at this point

Step 3:
Low-pressure system displaces the master piston, until contact is made with injector pushrod

Step 4:
Injector pushrod moves, forcing master piston up; increased pressure reseats ball check, creating a closed high-pressure system (black area); high-pressure system forces slave piston down, opening exhaust valve

Fig 12.2 Operation of a Jacobs engine brake hydraulic system

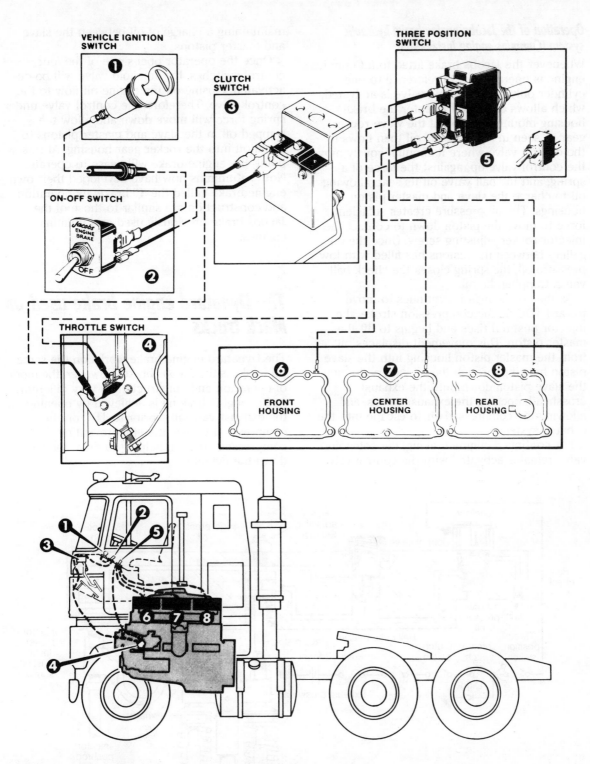

Fig 12.3 Jacobs engine brake control system installation

Operation of the Jacobs engine brake hydraulic system (Cummins engine installation)

Whenever the Jacobs brake fitted to a Cummins engine is operated, (with reference to one cylinder only), the solenoid valve is activated, which allows engine oil to enter the brake housing mounted on top of the valve rocker gear as seen in Fig 12.4. The oil then flows to the control valve where it simultaneously moves the control valve up, against the force of a spring, and the ball valve off its seat, allowing oil to charge the slave and master piston housings. The oil pressure creates sufficient force to move the piston down to contact the injector rocker adjusting screw. Once the oil gallery between the pistons has filled with low-pressure oil, the spring closes the check ball valve, trapping the oil.

As the engine piston continues to move toward TDC on the compression stroke, the injector pushrod rises and begins to lift the master piston. This movement displaces oil from the master piston housing into the slave piston housing through the oil gallery, forcing the slave piston down onto the exhaust crosshead, opening the exhaust valves, and allowing the compressed air to escape into the exhaust system.

For continued engine braking, the solenoid valve remains activated with the control valve maintaining a charge of oil between the slave and master pistons.

Once the operator opens one of the four control switches, the solenoid valve will be de-activated, closing off the engine oil flow to the control valve. Therefore, the control valve under spring force will move down and allow the trapped oil in the slave and master pistons to bleed off into the rocker gear housing. At this point, the engine brake will cease to operate.
Note   Cummins now have developed their own engine brake called the 'C' brake; its operation and construction are similar to those of the Jacobs brake previously used on Cummins engines.

The Dynatard engine brake used on Mack trucks

The Dynatard engine brake works on the same principle as the Jacobs brake. However, the mode of operation and construction is very different. Again, engine braking is achieved by opening the exhaust valve just prior to TDC on the compression stroke, thereby preventing the compressed air from helping to drive the piston down the cylinder.

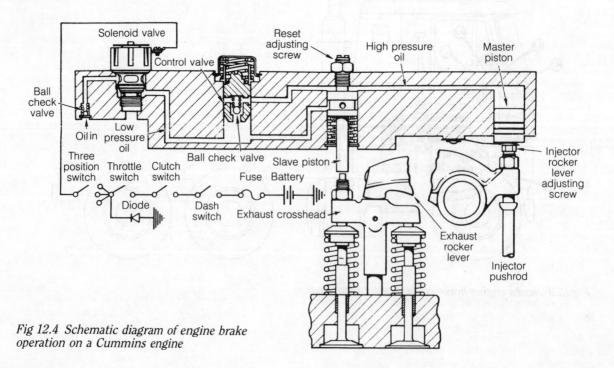

Fig 12.4 Schematic diagram of engine brake operation on a Cummins engine

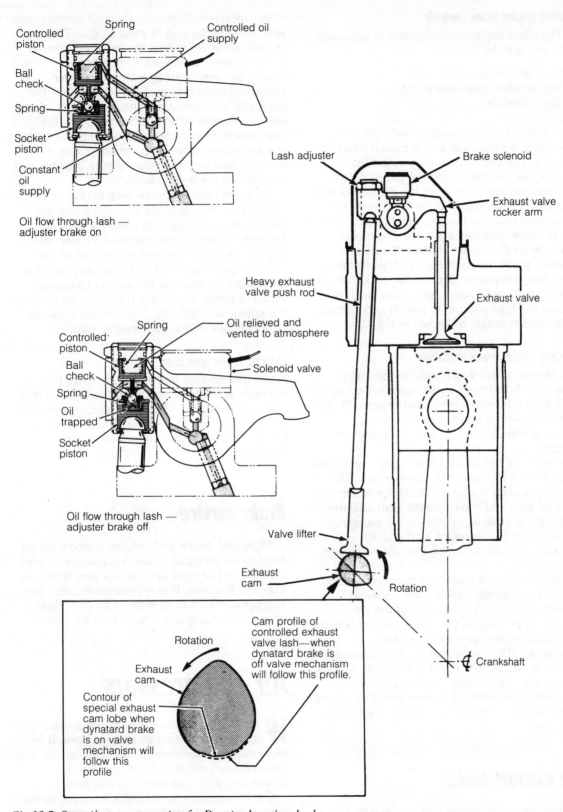

Oil flow through lash — adjuster brake on

Oil flow through lash — adjuster brake off

Fig 12.5 Operating components of a Dynatard engine brake

Dynatard engine brake controls

The Dynatard engine brake system is activated by three controls:

- the cab switch
- the injection pump switch
- the solenoids.

For engine brake operation, both the cab switch and the injection pump switch must be activated. The cab switch is operated by selecting 'on', and the injection pump switch operates automatically when the throttle is fully released as in deceleration, which allows the fuel rack to move to the 'no fuel' position. With both of these switches activated, the electric solenoids on the rocker shafts will operate. However, when the engine speed falls to idle speed, the governor will move the rack from the 'no fuel' position to slight fuel delivery, thus opening the injection pump switch and cutting off the engine brake, as shown in Fig 12.5.

Dynatard engine brake operation

With the dash switch on, and the injection pump switch activated by the accelerator pedal being fully released, the exhaust brake will start to operate, with a number of functions occurring simultaneously, as can be seen in Fig 12.5.

Considering the actions relative to one engine cylinder only, as the piston approaches TDC on the compression stroke, solenoid operation will allow oil flow into the hydraulic lash adjuster, reducing the exhaust valve lash (or clearance) to zero. Prior to the engine brake being activated, normal operating exhaust valve lash is maintained.

With the exhaust valve lash reduced to zero, the valve mechanism will follow a special contour on the cam lobe, causing the cam follower to ride up on the lobe contour, pushing the exhaust valve open as the piston approaches TDC. Consequently, the compressed air charge will escape into the exhaust manifold.

Thus, the engine goes through a power-absorbing cycle without a corresponding power-producing one, resulting in a braking effect by the engine.

The exhaust brake

Another design of engine brake generally known as an exhaust brake features a shutter, which, when operated, closes the exhaust system outlet and restricts the exit of exhaust gas. This type of exhaust brake serves the same purpose as the previously discussed engine brake in that it causes the engine to operate as a power-absorbing compressor. Although it is not the most efficient type, the exhaust brake is widely used in engines of smaller vehicles. In this application, it provides adequate engine braking, is a lot quieter during operation, places a uniform load on the engine during braking, and is probably the least expensive.

A typical exhaust brake system (Fig 12.6) shows the exhaust brake unit fitted into the engine exhaust pipe. It is simply a shutter or a butterfly, which, when operated, closes during engine operation, restricting the exit of the exhaust gas. The result is the production of a large back pressure in the engine cylinders, which opposes the piston travel, hence slowing the engine and vehicle. The exhaust brake is only effective above 1500 engine rpm.

Exhaust brake operation

When the exhaust brake is to be used, the operator activates a switch in the cabin, which causes a compressed air (or vacuum) operated cylinder to close the exhaust brake butterfly. Engine braking begins immediately this happens.

Brake service

All engine brake and exhaust brake systems need to be serviced at various times. However, because adjustment procedures vary from one brake to the next, it is recommended that the appropriate workshop manual be consulted prior to servicing any engine braking system.

Air compressors

While some engine manufacturers make the compressors fitted to their engines, it is more common for them to be made by an independent company (such as Clayton Dewandre) that specialises in this field.

The compressor is normally mounted either on a platform or bracket near the engine timing case, or bolted to the timing case and located

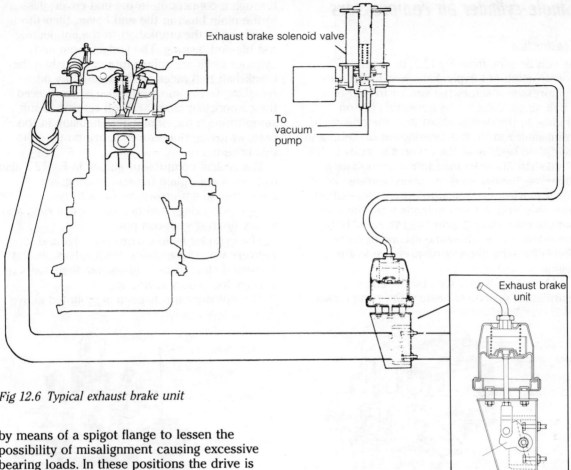

Fig 12.6 Typical exhaust brake unit

by means of a spigot flange to lessen the
possibility of misalignment causing excessive
bearing loads. In these positions the drive is
taken from the timing gears and, where
necessary, is transmitted through the
compressor crankshaft to drive the fuel
injection pump in tandem. In this case, and in
cases where the compressor is platform or
bracket mounted, a short shaft and flexible
couplings are normally used between the
compressor and the fuel injection pump.

Where this arrangement is not possible, the
compressor may be bracket mounted on the
engine, and belt driven from the crankshaft or
auxiliary drive shaft. To accommodate the
compressor in more restricted positions, the
unit may be mounted with the cylinder block in
a horizontal plane, or included at an angle to
clear engine projections. Compressor crankcases
and crankcase end covers, therefore, vary
considerably in design to permit such a variety
of mountings.

The compressor draws air from the engine air
cleaner or separate air filter. The air is then
compressed and fed to the air reservoir. An
unloader valve mounted on the reservoir (or,

alternatively, in the compressor delivery air
line) relieves the compressor of the pumping
load by cutting out when the correct maximum
working pressure is attained in the reservoir.
The unloader valve cuts in again when air has
been used and the pressure in the reservoir falls
to the lower setting of the valve.

Some compressors have an unloading device
incorporated in the cylinder head. This device is
operated by air pressure and is controlled by a
governor valve connected to the reservoir and
the compressor by a small bore pipe.

The compressors used for transport vehicles
may be either twin- or single-cylinder units.
However, the operating principle is the same
regardless of the number of cylinders, and an
understanding of the construction and operation
of a typical single-cylinder unit should provide
the necessary understanding of twin-cylinder
units.

Single-cylinder air compressors

Construction

As can be seen from Fig 12.7, the basic construction of a typical single-cylinder air compressor is somewhat similar to that of a small single-cylinder reciprocating piston engine. In the design illustrated, the one-piece crankshaft runs in ball bearings at one end and in a plain bearing at the other. The ends of the crankshaft are machined to accommodate a drive mechanism, and the plain bearing end cover carries an oil seal to prevent loss of oil from that end. A passage leads from the area between the plain bearing and the seal back to the crankcase to allow the escape of oil and prevent the build-up of oil pressure in the region.

Lubrication is provided by the engine lubrication system. Oil enters the compressor through a connection in the end cover, passes to the plain bush in the end cover, then through an oilway in the crankshaft to the connecting rod big-end bearing. The gudgeon pin and cylinder walls are splash lubricated, while the crankshaft ball races are lubricated by oil returning to the sump after it has overflowed the compressor sump. The oil returns to the sump through the timing gear housing in this case, as the compressor is flange mounted to this housing.

The typical compressor shown in Fig 12.7 also features a split plain big-end bearing, a conventional flat crown piston with two compression rings and one oil control ring, and a fully floating gudgeon pin.

The cylinder head carries the inlet and delivery valve assemblies. Both valves are flat, hardened steel discs held against their seats by springs located by spring guides.

The unloader mechanism is mounted above

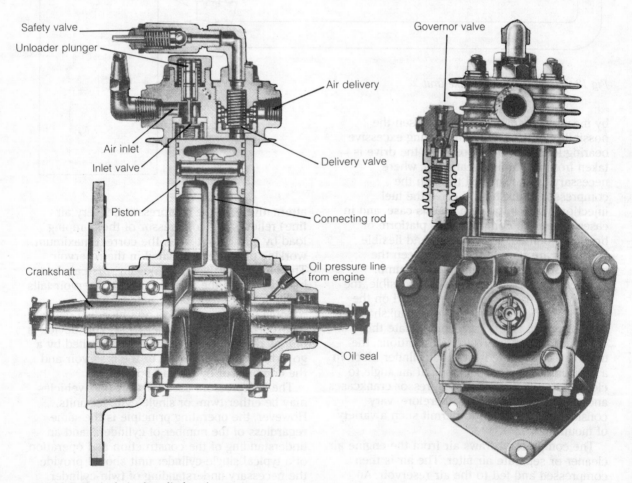

Fig 12.7 Typical single-cylinder air compressor

Labels: Safety valve, Unloader plunger, Air inlet, Inlet valve, Piston, Crankshaft, Governor valve, Air delivery, Delivery valve, Connecting rod, Oil pressure line from engine, Oil seal

the inlet valve, and is controlled by the
governor fitted on the side of the cylinder head.
A safety valve, necessary to prevent excessive
pressure should the governor–unloader system
fail is also mounted on the cylinder head.

Operation

The operation of the compressor may
conveniently be divided into two separate
stages—compressing and unloading.

Compressing—During the downstroke of the
piston, a partial vacuum is created above the
piston, unseating the inlet valve and allowing
the air to be drawn from the air cleaner into the
cylinder above the piston. When the piston
starts the upward stroke, the air pressure under
the valve plus the action of the inlet valve
spring closes the inlet valve. The air above the
piston is further compressed until the pressure
lifts the delivery valve, discharging the
compressed air through the line into the
reservoir. On the piston downstroke the delivery
valve re-seats, preventing the compressed air
from returning to the cylinder, and the cycle is
repeated.

Unloading—When the pressure of the air in
the reservoir reaches the pressure setting of the
governor valve, the spring-loaded ball of the

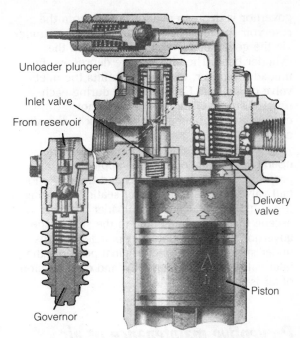

Fig 12.9 Compression of air in a single-cylinder air compressor

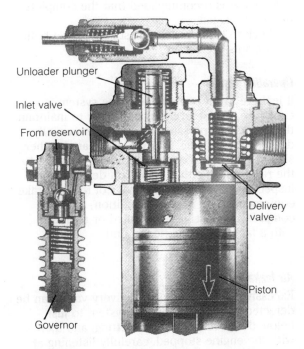

Fig 12.8 Intake of air in a single-cylinder air compressor

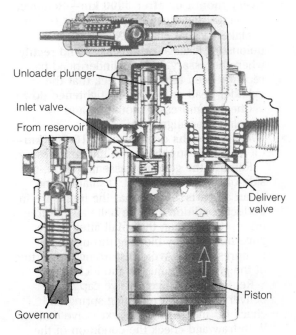

Fig 12.10 Unloading in a single-cylinder air compressor

governor is depressed, allowing air from the reservoir to enter the unloader plunger chamber via the governor valve and a drilling in the cylinder head. The air pressure depresses the unloader plunger, which in turn lifts the inlet valve off its seat. Consequently, during each upward stroke of the piston, air is not compressed, but simply passes back to the air cleaner.

When the brakes or any other of the air-operated vehicle controls are applied, the air pressure in the reservoir falls and the governor ball valve returns to its upper seating, allowing the compressed air in the unloader chamber to escape through a small hole in the side of the governor valve. The unloader plunger returns under spring action to its original position, the inlet valve functions normally, and compression of air is resumed.

Preventive maintenance of air compressors

The following maintenance schedule, by courtesy of Clayton Dewandre, may well be applied to all compressors encountered, in the absence of the correct workshop manual:

- **Every month or after 3000 km**—Remove, dismantle and clean the air cleaner or filter.

    Make a visual check of all joints and unions for leakage or looseness and rectify where necessary. If the cylinder head has recently been removed, check that the cylinder head nuts are fully tightened down.

    Check compressor mounting and couplings for alignment.
- **Every 6 months or after each 16 000 km**—Clean the oil supply line to the compressor.

    Remove the governor assembly from the side of the cylinder head. Dismantle and clean all parts; ensure that the items are in perfect condition and the ball valve is free from pitting. Clean the small filter in the governor body by blowing through with compressed air. Avoid disturbing the setting of the governor adjuster and locknut.

    Remove the delivery valve cap and delivery valve seat retaining spring and check for the presence of excessive carbon. Withdraw and check the condition of the delivery valve. If excessive carbon is found, remove and clean the cylinder head; also

check the compressor discharge line for carbon, and clean or replace the line if necessary.
- **Every 2 years or after each 80 000 km**—Dismantle the compressor, thoroughly clean all parts and inspect for wear and damage. Repair or replace all worn or damaged parts or replace the compressor with a factory-reconditioned unit.

Service checks

The following service checks may well be applied to any compressor:

Inspection

Ensure that the air cleaner or filter is clean and correctly installed.

With the compressor running, check for noisy operation and oil leaks.

Reduce the pressure in the reservoir by operating the brakes and check that governor and unloader mechanisms are functioning at correct pressure. (If possible, the vehicle gauge should be replaced during the test by a master gauge.)

Should the governor unit be found to be defective it is recommended that the complete unit be replaced.

Check to be sure compressor mounting bolts and/or nuts are secure.

Operating tests

If leakage in the remainder of the system is not excessive, failure of the compressor to maintain the normal air pressure in the system usually denotes loss of efficiency due to wear. Another sign of wear is excessive oil passing through to the reservoir. If either condition develops, and inspection shows the remainder of the air brake equipment to be in good condition, the compressor must be overhauled or replaced with a factory-reconditioned unit.

Air leakage tests

Excessive leakage past the delivery valve can be detected by charging the air system to just below the governor cut-out setting, and then, with the engine stopped, carefully listening at the compressor for the sound of escaping air. If this test is satisfactory, fully charge the system and again stop the engine. Check once more for audible leaks, which if present indicate leaking at the unloader valve piston.

Leakage at the delivery valve can be remedied by cleaning, lapping or replacing the valve and/or valve seat. Unloader valve leakage can be remedied by replacing the piston seal or valve piston.

Exhausters

As already mentioned, the exhauster is the engine-driven vacuum pump used on diesel engines to provide the necessary depression for operation of the vacuum-assisted brakes and other vehicle controls. The exhauster is used to evacuate the air from a tank or reservoir, which is, in turn, connected to the control mechanisms of the vehicle systems concerned.

The exhauster may take one of three common forms—a reciprocating piston type, a rotary impellor type, or a vane or eccentric rotor type. The vane-type exhauster, by far the most commonly used type and the most likely to be encountered in the workshop, is the only one to be considered in this book.

Because of various engine and chassis designs, exhausters must be manufactured with housings of different designs to provide different methods of attachment to the engine. Because of this, an exhauster may be designed for spigot, base, flange or bracket mounting. For engine mounting, the exhauster is usually driven from the timing gears with a through drive to the fuel injection pump.

Location by means of a spigot ring concentric with the drive shaft is the most common arrangement owing to less liability to misalignment loads being placed on the bearings. For the same reason, a short shaft with two couplings between the exhauster and the fuel pump is also preferred. It is the practice of some engine manufacturers to mount the exhauster behind the fuel injection pump, a method that reduces torsional loads on the exhauster rotor shaft.

A flange-mounted exhauster is usually bolted directly to the rear of the engine timing gear case and driven directly from the engine timing gears, while a bracket-mounted unit may be bolted to the engine crankcase and be driven by means of a small shaft from the timing gears.

Where it is impossible to mount the exhauster directly onto the engine, a base-mounted belt-driven unit can be used. It is usual to take the belt drive either from the crankshaft pulley, or from the transmission at the input side of the gearbox. Belt-driven exhausters are fitted with larger ball or roller bearings at the drive to take the heavier side loading caused by this type of drive.

On vehicles where the foregoing drives and mounting systems are impractical, the exhauster may be mounted on the gearbox and driven by a spur gear from the input side of the gearbox.

A vane-type exhauster in good condition will create a vacuum to within 25 mm of the prevailing barometric height without difficulty, though prolonged running at such a vacuum tends to create undue heat in the machine. To prevent this occurring, the machine is fitted with a snifter valve, which is set to admit air at 625–650 mm of mercury. This simple valve (adjustable on certain exhausters) consists of a spring-loaded ball fitted in a valve body on top of the exhauster, or alternatively in the exhauster body itself.

While the snifter valve safeguards the unit against overheating, it is essential that only clean air is admitted to the unit when the valve opens. On many exhausters, provision is made to draw air from the engine crankcase through ducting. Alternatively, the valve can be connected to an external source of air, provided

Fig 12.11 A typical vane-type exhauster

a clean location for the inlet is selected. For operation in dusty conditions, it is essential, if externally located, to connect the valve to an air cleaner.

A non-return valve is always fitted in the vacuum system and this valve may be incorporated with the snifter valve in the exhauster body, or it may be fitted to the reservoir itself.

Construction of a typical vane-type exhauster

Fig 12.12 shows a drawing of a typical vane-type exhauster. The rotor and shaft assembly (4) is

supported in the body by ball or roller bearings (6 and 41) housed in the body end covers (15 and 46). On machines having a through drive, a keyway is cut on each end of the shaft to accommodate a gear or coupling. The shaft ends may be tapered or parallel, the gear or coupling being retained in position by a washer, nut and split pin or a cotter pin located in a slot. Seals (9) pressed in the end covers contact hardened steel collars on the shaft to stop air from entering the exhauster body. Double seals are fitted to open end covers (end covers through which the shaft protrudes) to prevent oil leakage due to initial pressure in the exhauster body under starting conditions.

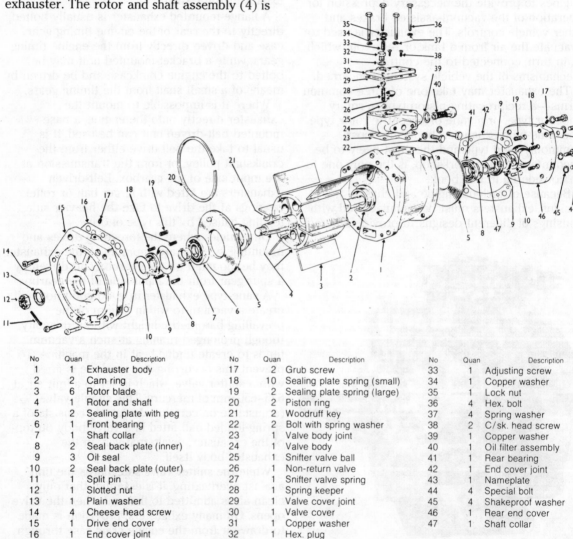

No	Quan	Description	No	Quan	Description	No	Quan	Description
1	1	Exhauster body	17	2	Grub screw	33	1	Adjusting screw
2	1	Cam ring	18	10	Sealing plate spring (small)	34	1	Copper washer
3	6	Rotor blade	19	2	Sealing plate spring (large)	35	1	Lock nut
4	1	Rotor and shaft	20	2	Piston ring	36	4	Hex. bolt
5	2	Sealing plate with peg	21	2	Woodruff key	37	4	Spring washer
6	1	Front bearing	22	2	Bolt with spring washer	38	2	C/sk. head screw
7	1	Shaft collar	23	1	Valve body joint	39	1	Copper washer
8	2	Seal back plate (inner)	24	1	Valve body	40	1	Oil filter assembly
9	3	Oil seal	25	1	Snifter valve ball	41	1	Rear bearing
10	2	Seal back plate (outer)	26	1	Non-return valve	42	1	End cover joint
11	1	Split pin	27	1	Snifter valve spring	43	1	Nameplate
12	1	Slotted nut	28	1	Spring keeper	44	4	Special bolt
13	1	Plain washer	29	1	Valve cover joint	45	4	Shakeproof washer
14	4	Cheese head screw	30	1	Valve cover	46	1	Rear end cover
15	1	Drive end cover	31	1	Copper washer	47	1	Shaft collar
16	1	End cover joint	32	1	Hex. plug			

Fig 12.12 Exploded view of a typical exhauster

In the exhauster shown in Fig 12.12, six equal sized blades are shown, but in some cases four blades may be used. The blades may be of steel or fibre, and fit into slots in the rotor.

The spaces between the blades are sealed by circular plates (5) held in contact with the ends of the rotor by springs (18 and 19) housed in pockets in the end covers. A peg riveted to each sealing plate engages with one of the spring pockets to prevent rotation of the plate. Alternative types of sealing plate are in use, the more usual being located in the body and sealed radially by a spring ring fitted in a groove in the periphery. In less common use, and fitted in the recessed end covers, is a plain type of sealing plate without the radial spring.

On certain exhausters the sealing plates are drilled with an oil transfer hole, which, to ensure its correct situation at assembly, is positioned by means of one of the six springs and spring pockets, which is of larger diameter than the remainder.

Cam rings (2), fitted in the sealing plates and held concentric with the exhauster body, contact the inner edges of the rotor blades and maintain the blades in close contact with the body bore. An intake port in the exhauster body aligns with a passage formed in the valve body, which is pipe-connected to the vacuum tank mounted on the vehicle chassis. The non-return valve is either of the rubber-faced metal type or a hemispherical rubber type as shown in Fig 12.13. According to the design of exhauster, the non-return valve and the snifter valve conform to one of two arrangements, being located either within the valve body attached to the top of the exhauster by set screws or studs or, alternatively, the non-return valve being located at the reservoir and the snifter valve at a convenient place within the exhauster body. In either position, the snifter valve is connected by passages to the intake port on one side and on the other to the atmosphere. The outlet port formed in the base of the exhauster body aligns with the aperture in the drive housing or mounting base and connects with the engine casing.

Exhauster operation

At operating speeds the rotor blades are kept in contact with the bore of the body by centrifugal force, but at engine idling speed, particularly when the oil is cold, the blades have insufficient centrifugal force to keep them in their true position. This is overcome by the action of the cam rings (2), which contact the inside edges of the blades, forcing them to move out radially in their grooves and thus maintain contact with the bore of the body.

When the rotor turns, the spaces between the rotor blades decrease because of the eccentric mounting of the rotor in the exhauster body, and the air between them is compressed (see Fig 12.14). As the point of maximum compression is reached, the air is expelled through the outlet port, together with the lubricating oil, into the engine crankcase. After passing the outlet port the spaces between the rotor blades increase and a depression is thereby created, which is filled by air drawn from the vacuum reservoir. This air is then compressed and expelled.

Where a snifter valve is fitted, the withdrawal of air from the reservoir continues until a partial vacuum of 625–650 mm Hg is obtained, at which point the snifter valve is lifted off its seat to admit air. This action limits the maximum vacuum and prevents unnecessary heating of the exhauster. When the vacuum in

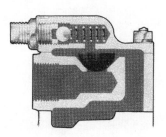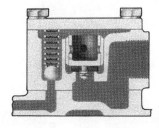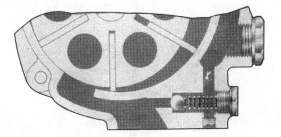

Fig 12.13 Sections showing the adjustable snifter valve and hemispherical rubber non-return valve, the snifter valve and rubber disc-type non-return valve, and the body-located snifter valve

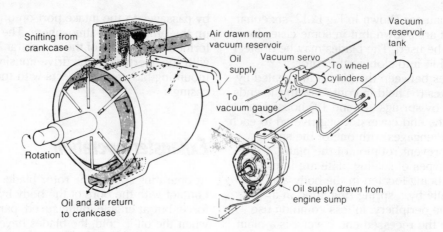

Fig 12.14 Operation of an exhauster and associated components

the reservoir is greater than the vacuum generated by the exhauster due to the slowing down or stopping of the engine, the non-return valve drops on its seat and prevents loss of vacuum from the reservoir.

Exhauster lubrication systems

The exhauster is dependent on an adequate supply of oil for lubrication of the bearings and moving parts and to provide a seal between the rotor blades and the bore of the body. The exhauster draws its own lubricant from the engine sump or, if mounted independently of the engine, from a combined oil separator and reservoir.

Premature failure of an exhauster can often be traced to a fault in the lubrication system, the main causes usually being a partially blocked filter or an indented pipeline, either of which can result in a restricted and reduced oil flow. A defective oil-pipe joint will also affect operational efficiency by admitting air to the system and consequently diminishing or interrupting the oil supply.

Lubrication from the engine sump is the most usual and, because of its simplicity, the cheapest. The oil inlet port is connected by a cross-drilling to the end covers. The vacuum created in the end covers is utilised to draw oil into the exhauster. The oil lubricates all parts and is ejected through the outlet port to the engine sump together with the air evacuated from the reservoir. Adequate crankcase

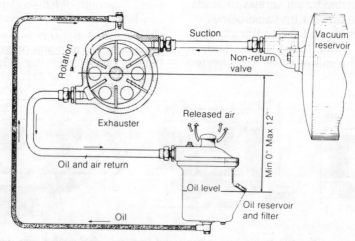

Fig 12.15 Lubrication from the combined oil reservoir and separator

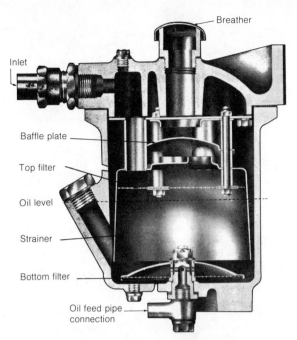

Breather

Inlet

Baffle plate

Top filter

Oil level

Strainer

Bottom filter

Oil feed pipe connection

Fig 12.16 Oil reservoir and separator unit

ventilation must be available to prevent a build-up of pressure. If a filter is not fitted to the exhauster, a sump filter should be incorporated well below the oil level and at a distance of up to a maximum of 300 mm below the centre line of the exhauster drive shaft.

Lubrication from an oil reservoir and separator is used where it is impossible to utilise the engine sump. It is an entirely self-contained lubricating system and, provided the correct oil level is maintained in the reservoir, no maintenance other than the periodic cleaning of the filters is necessary. In the oil reservoir, a series of baffle plates and a filter separate the oil from the air, the oil falling to the reservoir to complete the circulation and the air passing to the atmosphere through the breather at the top.

Periodic inspection and preventive maintenance

The following maintenance schedule (by courtesy of Clayton Dewandre) may well be applied as a maintenance schedule to any

exhausters encountered, where specific instructions are not given:

- **Weekly or every 1500 km**—Check the vacuum lines and fittings. Vacuum leakage may occur through the non-return valve if the valve seat is dirty or pitted.

  Check for oil leaks at the exhauster, particularly at the joint between the end covers and body and around the protruding end(s) of the rotor shaft. If seepage around the rotor shaft is severe, oil will be thrown from the shaft while the exhauster is running. Check the oil supply and discharge lines for leaks at fittings and connections. Rubber connecting hoses, if used, may become hardened due to the hot oil, with cracking and leaking a consequence.

- **Every 8000 km**—Clean the oil filter of the exhauster.

  Check all mounting studs and end cover retaining bolts for tightness.

  If the exhauster is belt-driven, periodically check the belt for wear and tension and the pulleys for alignment. Occasionally remove the belt, open the vacuum port to the atmosphere and, while rotating the pump by hand, check if any binding occurs or if excessive end play exists in the rotor shaft.

- **Every 80 000 km**—The exhauster should be removed for dismantling, and detailed examination of component parts according to the workshop manual.

In applications where the exhauster is lubricated from its own oil reservoir, the oil reservoir and separator unit should be serviced as follows:

- **Weekly or every 1500 km**—Check joints and unions, particularly the oil supply line, for tightness. Check the oil level in the reservoir and replenish if necessary.

- **Every 8000 km**—Check joints and unions for tightness. Remove the plug in the base of the oil reservoir, drain off the oil in the lubricating system and replenish with clean engine oil to the level of the filler plug.

- **Every 30 000 km**—Repeat the 8000 km inspection and remove the separator filters for cleaning as follows:

1 unscrew the securing nuts and remove the top plate

2 withdraw the filter and baffle plate
assembly complete
3 unscrew the nut holding the base filter and
withdraw the filter assembly.

All parts should be thoroughly cleaned and
inspected and the reservoir replenished with
clean engine oil after reassembly.

13
Engine faultfinding and testing

Fault diagnosis is a skill in which the diesel serviceman must become extremely proficient if he is to determine the cause of an engine malfunction and effect repairs quickly and efficiently. He must be able to recognise from the various symptoms shown that certain adjustments are necessary, or that a particular engine component is becoming worn, is about to fail, or has failed. Where the symptoms indicate that the fault could be caused by a number of factors, a process of elimination may be necessary to locate the exact cause of the trouble. However, as this is time-consuming and sometimes expensive, an accurate, on-the-spot diagnosis, based on engine knowledge, experience and commonsense is preferable.

However, without some means of measuring output, it is very difficult to determine if an engine is operating at optimum efficiency, giving rated performance, or perhaps falling slightly below specifications. To ascertain that an engine is performing correctly, particularly after overhaul, and to aid in the tuning operation, many workshops test engines on a dynamometer before delivery to their customers.

Faultfinding

Table 13.1, by courtesy of Perkins Engines Ltd, lists some of the possible causes should an engine fail to operate satisfactorily. As in the original Perkins table, the term 'atomisers' has been retained instead of the alternative 'injectors'.

Table 13.1 Engine faultfinding

Fault and possible cause	Remedy
Engine will not start	
1 No fuel at atomisers:	
• Stop control in 'no fuel' position	Turn control to 'run' position
• Insufficient fuel in tank, air has been drawn into the system	Replenish fuel tank, then 'bleed' system
• Fuel lift pump inoperative	Remove lift pump and rectify or fit replacement pump
• Fuel filters choked or fuel feed pipe blocked	Check fuel feed to fuel pump and filters, rectify as necessary
• Fuel pump not delivering fuel to the atomisers	Remove pump for attention of specialised workshop and fit replacement
Fuel at atomisers:	
• Atomisers require servicing	Service or fit replacement set
• Wrong type of thermostart unit fitted	Check that correct type is fitted
• Thermostart unit inoperative	Visually check unit, fit new unit if unserviceable
• Valve and/or pump timing incorrect	Check and reset if necessary

Fault and possible cause	Remedy

2   Cranking speed too low:

- Battery not in well-charged condition — Fit fully-charged replacement
- Incorrect grade of lubricating oil — Check oil viscosity against approved lists in manual for temperature range
- Poor electrical connections between battery and starter motor — Check and tighten or remake connections if necessary
- Starter motor faulty — Examine and rectify if necessary

3   Poor compression:

With poor compression, starting may just be difficult in normal weather, but in cold weather the engine may just refuse to start altogether, dependent on how much compression there is and the cranking speed. The causes are numerous, and include worn liners, piston rings and leaking valves.

There is no quick remedy for this condition; generally, the engine will have been in service for some time. At least a top overhaul or probably a complete overhaul would be indicated to restore the lost compression, which is so vital for the efficient running of a diesel engine.

Engine starts, runs for a few moments, then stops

- Partially choked fuel feed pipe or filter — Trace and rectify
- Fuel lift pump not giving adequate delivery — Check output of lift pump and rectify or replace as necessary
- Fuel tank vent hole blocked — Check and clear if necessary
- Restriction in induction or exhaust systems — Check and rectify if necessary
- Air leaking into supply or return fuel pipes — Check and trace

Engine misfiring or running erratically

- Atomiser(s) require attention — Isolate offender(s), remove and test; if faulty, service or fit replacement(s)
- Air in fuel system — Check for air in fuel pump; if present prime the fuel system
- Water in fuel pump — Thoroughly check fuel system for signs of water; remove if present, then prime with clean fuel
- Valve and/or pump timing incorrect — Check and reset if necessary
- Valve clearances incorrect — Check and reset if necessary
- Fuel leaking from high-pressure pipe — Observe with engine running and replace pipe if necessary

Engine runs evenly but suffers from loss of power

- Atomisers require servicing — Remove and service or fit a replacement set
- Loss of compression — Refer to previous remarks on poor compression
- Pump not delivering sufficient quantity of fuel to meet engine requirements — Observe throttle linkage for unrestricted travel; if satisfactory, pump should be checked for correct output in specialist workshop
- Air cleaner causing restriction to the flow of air — Check that the correct type is fitted and that it has been serviced in accordance with the instructions given in manual
- Fuel pump timing incorrect — Check and reset if necessary
- Brakes binding, causing excessive load on engine and apparent loss of power — Stop vehicle and check if any brake drum(s) appear over-heated; if drum(s) over-heated, take immediate remedial action

Engine runs but with a smoky exhaust

- Incorrect air–fuel ratio — Check diaphragm and adjustment of air–fuel ratio control
  Check for any restriction to the airflow; if satisfactory, have the fuel pump maximum fuel output checked

Fault and possible cause	Remedy
• Cold starting aid (thermostart) valve leaking	Replace with a serviceable unit
• Valve and/or fuel pump timing incorrect	Check and reset if necessary
• Atomiser(s) require servicing	Remove and service or fit a replacement set
• Excessive oil consumption	Generally consistent with poor compression and long engine life; workshop examination required to give precise details
• Vehicle overloaded	Check that the loading is consistent with the manufacturer's load classification

Engine knocking

• Faulty atomiser (nozzle needle sticking)	Fit replacement atomiser
• Fuel pump timing too far advanced	Check timing and reset if necessary
• Piston striking a valve	Check valve timing, piston topping and valve head depth relative to cylinder head face
• Incorrect fuel	Check that the tank has been filled with diesel fuel and not petrol by mistake
• Worn or damaged bearings etc	Engine overhaul required

Engine overheating

• Coolant level in system too low	Replenish and check if leakage is taking place; if so, rectify at once
• Radiator or system partially blocked	Flush system through thoroughly in accordance with the manufacturer's instructions
• Blockage or restriction due to ice formation	Locate trouble spot and take any action necessary to prevent recurrence
• Fan belt slipping or incorrect type of fan fitted	Check fan belt tension and fan type
• Valve and/or fuel pump timing(s) incorrect	Check and reset if necessary
• Thermostart stuck in the closed position	Check and replace with a new one if found unserviceable

Low oil pressure

• Oil level in sump too low	Replenish to correct level
• Incorrect grade or inferior oil being used	Change to approved grade
• Oil leaking externally from engine	Rectify immediately
• Pressure gauge or oil warning light switch inaccurate	Check either against a master unit
• Oil pump worn or pressure relief valve sticking open	Remove and examine
• Suction pipe to oil pump allowing air to be drawn in	Rectify leak or renew pipe as necessary
• Worn main or big end bearings	Engine overhaul required

High oil pressure

• Incorrect grade of oil being used	Change to approved grade
• Pressure gauge inaccurate	Check against a master unit
• Pressure relief valve sticking closed	Remove and examine

Excessive crankcase pressure

• Partially choked breather pipe	Check pipe for any obstruction
• Worn or sticking piston rings	Engine examination required
• Pipework or tank on vacuum side of exhauster allowing entry of air into the system (only where exhauster is fitted)	Check system for leaks and rectify if necessary

Compression testing

A compression test is given to each cylinder of an engine in order to ascertain its compression pressure. Because a diesel engine relies on compression ignition, it is essential that the compression pressure in each cylinder remains as close as possible to that at which the engine is designed to operate. If the compression pressure in one or more cylinders is reduced by any marked degree the engine will not operate smoothly and efficiently, or may not even start at all. Symptoms such as difficult starting, uneven running, misfiring, excessive fuel consumption or loss of pulling power are usually indicative of low or uneven compression, and when they are present the engine should be given a compression test.

The information gained from a compression test is most helpful in determining the mechanical condition of an engine. For example, low but relatively even compression pressure in all cylinders usually indicates faulty or worn piston rings, or worn pistons and/or cylinder liners, while low but uneven compression pressure readings usually indicate leaking cylinder valves, but may indicate faulty or worn piston rings or worn pistons and/or cylinder liners. Leaking cylinder valves, which result from insufficient tappet clearance, faulty valve operation or incorrectly seated valves, are usually responsible for low compression pressure in a particular cylinder. This, however, is not always the case as broken compression rings or a damaged piston will also cause low compression in a particular cylinder.

Procedure

While the procedure adopted for taking compression tests on both petrol and diesel engines is somewhat similar, care must be taken to ensure that the pressure gauge chosen for a diesel engine is capable of registering pressures up to 5000 kPa. This is necessary as the compression pressure of most high-speed diesel engines is in the vicinity of 3000 kPa—more than twice that of a petrol engine of comparable size. Owing to this high compression pressure it is not advisable, when taking a compression test, to try to hold the gauge in the injector pocket or hole by hand. If a suitable adaptor is

not supplied with the gauge to enable it to be clamped in the injector pocket, one should be made before starting the compression test.

Once a suitable pressure gauge and adaptor have been selected, the compression test should be carried out as follows:

1 Remove no. 1 injector from its pocket in the cylinder head and then cover both the fuel inlet connection of the injector and the disconnected end of the delivery line with clean rag to prevent the entry of dirt or dust.

2 Clean the injector pocket and fit the compression gauge making sure that it is firmly clamped in position to prevent leakage.

3 Either start the engine and increase its speed to a fast idle (eg 600 rpm) or crank the engine over continuously by means of the starter motor. (See the workshop manual for the recommended method.) If the latter method is adopted, the remaining injectors must either be removed from the cylinder head or the appropriate decompression levers set so that compression is not possible in any cylinder other than the one being tested. This allows the engine to be cranked over at a sufficiently high speed for satisfactory testing, while keeping battery drain to a minimum. Should the engine be fitted with a pneumatic governor, it is essential that the venturi butterfly remains in the full open position during the cranking operation, since a restriction in the inlet will cause a low reading.

4 Note the pressure gauge reading and compare it with the recommended compression pressure specified by the engine manufacturer in the workshop manual. In the absence of a workshop manual, an approximate compression of 3100 kPa may be assumed for most high-speed diesel engines tested at idling speed. If the test is taken while the engine is being cranked over by the starter motor, the compression pressure will be decreased by approximately 700 kPa because of reduced volumetric efficiency at lower speeds. The figures quoted above are valid at sea level only and will decrease with altitude. For each 800 m increase in altitude (up to 3000 m), the compression pressure will be reduced by approximately 240 kPa.

5 Test each remaining cylinder in turn, making sure that the pressure gauge readings are accurately noted for further reference.

6 Compare the readings from the different

engine cylinders. If they differ from each other by more than 175 kPa, or if any reading is more than 350 kPa below the recommended compression pressure, it is evident that the engine is in need of attention. While a top overhaul (ie a valve grind) is necessary in most cases, this should not be started before all other possibilities such as loose cylinder head studs, incorrect valve clearances, weak or broken valve springs or incorrect decompressor settings have been eliminated.

Note   Once the engine cylinder head has been removed and a burnt valve discovered, the possibility of worn piston rings and/or cylinder liner is not positively eliminated. Therefore the engine should be examined carefully for evidence of other faults.

Diesel exhaust smoke

*O*ne good aid to faultfinding is the correct interpretation of the exhaust smoke colour. Table 13.2 (overleaf, pp 104–7), by courtesy of CAV, provides a comprehensive guide to the causes of exhaust smoke colours, suggests cures, and lists some pertinent comments.

Diesel engine testing

*T*he most satisfactory method of testing an engine for performance, as well as providing an ideal environment for faultfinding, is to operate the engine on a dynamometer in an engine testing installation. It is also a very convenient way of gaining open access to the engine under varying loads for tuning.

Before proceeding to further discussion on dynamometers and testing procedures, the terminology of testing should be defined to avoid any possible misinterpretations.

Power measurement terminology

Power is the rate at which an engine will do work. Reference to Chapter 1 will provide further basic explanation. The graph in Chapter 1 (Fig 1.1) shows that the maximum power a typical engine can produce is in the upper end of the engine's speed range, as might be expected, since power, by definition, is a 'rate'.

Power is measured in kW (kilowatts) or HP (horse power). For conversion, 1 kW = 1.341 HP or 1 HP = 0.746 kW.

Torque is an indication of an engine's ability to produce a rotational force about its crankshaft. This is also explained more fully in Chapter 1. In addition, the torque curve of a typical engine can be seen in Fig 1.1. It shows that the optimum engine speed to obtain the greatest torque output is less than for maximum power.

The torque output of an engine is rated in newton metres (N.m) or lbs ft. For conversion 1 N.m = 0.737 lbs ft or 1 lb ft = 1.36 N.m.

Brake power is the usable power available at the flywheel of an engine when the engine is placed under load. The brake power of an engine is found by applying a load to the engine output shaft by means of a dynamometer, and measuring, throughout the engine speed range, the maximum torque (in N.m) the engine can sustain without loss of rpm. The brake power of the engine is found by calculation from the torque reading and the engine rpm at which the reading was taken.

Below is an example of calculating the kWbp (kilowatt brake power) of an engine when loaded by a dynamometer. The torque produced under load is 400 N.m at 1800 engine rpm. The calculation formula for the particular dynamometer is:

$$kWbp = \frac{rpm \times N.m}{9549}$$

Therefore

$$kWbp = \frac{1800 \times 400}{9549}$$
$$= 75.4 \, kWbp$$

The 9549 is an internal correction factor, or constant, for the particular dynamometer being used. It is specific to a particular dynamometer based on its performance under test and is supplied by the manufacturer with the machine.

Note   A dynamometer provides a means of applying a variable load to the engine and measuring its torque output and speed. Generally, it is only by using a dynamometer that the kWbp can be obtained. (*Continued, page 106*)

Table 13.2 The use of exhaust-smoke colour in engine faultfinding

Colour of smoke	Symptom	Probable diagnosis
Black or dark grey	Smoke at full load at any engines speed, but particularly at highest and lowest speeds, and power at least normal	Maximum fuel setting of injection pump too high Excess fuel device not tripping automatically to normal after starting
	Smoke at full load particularly at high and medium speeds, engine quieter than normal	Pump timing retarded (or advanced device not correct if fitted)
	Smoke at full load particularly at low and medium speeds, engine noisier than normal	Pump timing too advanced
	Smoke at full load particularly at high and medium speeds, probably with loss of power	Injector nozzle holes (or some of them) wholly or partially blocked
	Smoke at full load at higher speeds only	Air cleaner restricted due to blockage with dirt, or damage
	Intermittent or puffy exhaust smoke, sometimes with white or blue tinge, usually coupled with knocking	Injector nozzle valve struck open intermittently
	Smoke at full loads at high speed, engine running faster than normal when on governor	Governor speed setting considerably above engine maker's maximum
	Smoke at full loads at high speed, engine running slower than normal on governor (vacuum type)	Governor venturi throat partially choked with carbon
	Smoke at most speeds and loads, tending to blue or white when cold and when starting	Nozzle sprays impinging on cylinder head, due to incorrect fitting of injector into cylinder head
	Smoke at higher loads and speeds, not necessarily at maximum	Injector nozzle valve lift excessive, due to repeated valve or seat refacing, without lift correction
	Smoke at all speeds at high loads, mostly low and medium speeds and probably coupled with poor starting	Loss of cylinder compression due to stuck rings, bore wear, valve wear or burning, sticking valves, incorrect valve setting
	Smoke at full load, either at lower or higher speeds only, but in some cases at all speeds	Incorrect nozzle type fitted, or mixed types, or out-of-date type, or type for different duty
	Smoke at full load, mostly at medium and high speeds, probably coupled with low power.	Injection high-pressure pipes of incorrect length or bore, or having badly closed-in bore at ends, or due to sharp bends

Cure	Remarks
Remove pump, have reset to engine maker's maximum flow figure (or less) by authorised service agent if own equipment not available	Some operators may be tempted to reset by trial and error; this may be very misleading, as smoke may be caused by another fault
Have repaired by authorised agent—removal of pump may be necessary	The fault is rare except when caused by deliberate tampering (now illegal)
Correct timing according to engine maker's instructions, taking up pump drive backlash (or rectify advance device if fitted)	Often becomes retarded owing to chain stretch, or backlash not taken up when set; may be critical to 2 crankshaft degrees More likely to apply to indirect injection engines
Replace injectors by reconditioned set on maker's service exchange scheme or clean and recondition with proper equipment	Loss of power will lead to even more smoke if attempt to restore by increasing pump setting
Clean or replace air cleaner element according to type	See note about vacuum governors below
Have injectors examined for sticking valve, broken spring or grossly low opening pressure, or sign of cross-binding in cylinder head; replace as necessary	May be due to neglect of filter maintenance, or water in fuel or bad fitting of injector in head; injector should fit freely in head and should be tightened down evenly and not too tightly
With mechanical or hydraulic governors, reduce governor speed adjustment and seal stops, or better remove pump for attention; with vacuum governors, reset stops on venturi butterfly valve	
Clean carbon from venturi throat	Applies to vacuum governors where engine breather is upstream of venturi
Examine for number of washers between injector and cylinder head—only one required at most (some engines none required—refer instruction book)	Washer often left behind when removing injector and new one fitted on top of old; some injectors fitted with heat shield, which might be incorrectly assembled—refer to instruction book
Can .be rectified by proper equipment during reconditioning	Many improperly trained fuel injection equipment mechanics will not rectify lift
Engine requires top overhaul at least; re-ringing or sleeving, piston renewal if wear indications shown	May be due to unsuitable lubricant or incorrect valve tappet clearance; may cause blue smoke too (if lubricating oil consumption is excessive)
Will be automatically corrected if injectors are reconditioned by an authorised agent, but it is essential to quote exact details of engine type and application	Engine makers sometimes change the nozzle type with new engine marks or for different duties; power may or may not seem satisfactory if the wrong nozzle is fitted
Fit only the engine maker's listed pipe; check ends for closing-in	Pipe bores for vehicle engines are never less than 1½ mm—a 3/64" or no 56 drill should enter freely

Table 13.2 (con't)

Colour of smoke	Symptom	Probable diagnosis
Blue, bluish-grey or greyish-white	Blue or whitish smoke particularly when cold, and at high speeds and light load, but reducing or changing to black when hot and at full load, and with loss of power at least at high speeds	Pump timing retarded (or advance device not correct if fitted)
	Blue or whitish smoke when cold, particularly at light loads, but persisting when hot, probably with knocking	Injector nozzle valve stuck open, or tip broken off nozzle
	Blue smoke at all speeds and loads, hot or cold	Engine oil being passed by piston rings due to sticking or worn bores
	Blue smoke particularly when acclerating from period of idling, tending to clear with running	Engine oil being passed by inlet valve guides due to wear, or valve guide oil shields misplaced
	Blue smoke when running at maximum speed, full or light load	Oil bath air cleaner overfilled
	Light blue smoke at high-speed light loads, or running downhill, usually with acrid odour	Engine running too cold, thermostat stuck or not fitted

Purpose of engine testing

There are many reasons for performing an engine test on a dynamometer, particularly after an engine overhaul. In particular, the tests enable the serviceman to:

- make sure the engine will start and does not run out of control
- check for oil and water leaks
- check for correct oil and water temperature and pressure both under load and without load
- adjust the idle and maximum no-load speed of the engine
- detect unusual noises
- run the engine in, under varying loads
- establish that the engine will produce its rated power and torque as per the manufacturer's specifications
- check the operation of the turbocharger and record the manifold boost pressure
- check for air cleaner inlet restriction
- check for crankcase blowby
- check air-box pressure on two-stroke engines
- check exhaust back-pressure.

If before, during or after testing an engine, it is found that the engine does not perform correctly,

some form of fault diagnosis will be necessary to locate the cause before it can be rectified. Guidance can be obtained from the engine manufacturer's manual, or from the faultfinding schedule listed in Table 13.1.

Engine testing standards

To have any significance, the power and torque output of an engine must be in accordance with set or standard codes. The most common codes in use throughout the world are the:

- BS Standard (British Standard)
- SAE Standard (Society Automotive Engineers)
- DIN Standard (Deutsche Industrie Normen)
- ISO Standard (International Standards Organisation).

The above standards can be either expressed as gross kWbp or net kWbp.

Gross kWbp is the power developed by a bare engine. A bare engine is defined as an engine fitted only with the accessories essential for its operation, such as flywheel, oil pump and fuel pump.

Net kWbp is the power output of a fully equipped engine fitted with all accessories

Cure	Remarks
Reset timing (or rectify advance device if fitted)	Some engines, particularly indirect injection, show this symptom for less retard than gives rise to black smoke, but usually a gross retard is required to give blue smoke when running hot and under load
Examine for sticking valve or broken spring, but suspect handling of injectors out of engine if tip is found broken	Injector nozzle valve sticking or blocked spray holes can lead to this condition if not dealt with quickly
Engine recondition indicated	May be due to unsuitable lubricant; will be associated with high oil consumption
Recondition cylinder head, and make certain that guide oil shields (if any) are in place	May be aggravated with vacuum governors due to depression in inlet manifold when idling; oil consumption may not be noticeably affected
Fill only to the mark or recommended level	May cause an engine uncontrolled runaway in severe cases
Replace thermostat	Low jacket temperatures increase bore wear also

necessary to perform its intended function. It is the usable power from the engine after all accessories are installed, unlike gross power, which, in practical terms, is over-rated and can never be achieved from a fully equipped engine in service.

Intermittent power is the maximum power the engine is capable of producing at a stated speed. The engine should be capable of producing this power for a maximum period of up to 1 hour, after which it should be operated at a reduced power output.

Continuous power is less than intermittent power and is the amount of power an engine can produce continuously for a minimum of 8 hours.

Variations in power output

Engine power output may vary considerably under certain circumstances or conditions of operation. The effect can be quite considerable, and some engine manufacturers give a 5 per cent tolerance in power output rating to cover variations in engine operating conditions.

The major conditions affecting engine performance include variations in:

- ambient air temperature
- barometric pressure
- altitude
- fuel temperature.

Variations in ambient air temperature

As the temperature of the air increases, the air becomes less dense, with a corresponding decrease in combustion efficiency and engine power output, due to the reduction in the amount of oxygen entering the engine cylinder. Conversely, as the temperature of the air decreases, the air increases in density, which improves combustion and increases engine power output.

Variations in barometric pressure

The increase in barometric air pressure has the effect of forcing more air into the cylinders, promoting greater combustion efficiency, with the capacity to burn more fuel, thus slightly increasing the power developed. The opposite effect is caused by a decrease in barometric pressure, producing a corresponding decrease in power.

Variations in altitude

All engines will operate successfully at sea level and to approximately 100 m above without the fuel setting needing to be changed. Naturally aspirated engines operating above the 100 m level will experience a loss of power, because the reduced air density (and subsequent reduced air charge) will lead to less fuel being burned, resulting in a decrease in engine performance. Therefore, adjustments to the fuel system are required to prevent overfuelling and excessive exhaust smoke. However, turbocharged engines automatically compensate for the decreasing air density as the turbocharger rotates faster, forcing more (less dense) air into the engine cylinders.

Variations in fuel temperature

As diesel fuel becomes heated, it becomes less dense and the mass of fuel injected decreases. This causes a corresponding decrease in the power output of the engine. On the other hand, as the fuel becomes cooler and its density increases, a greater mass of fuel is injected to give an increase in power output.

The density of the fuel is measured by its specific gravity (see Chapter 9). One engine manufacturer states that a power loss of 2 per cent per 1°C rise in fuel temperature above 32°C can be expected. Therefore, in a diesel engine, the operational temperature of the diesel fuel is proportional to its available heat energy and subsequent power output.

Dynamometers

A dynamometer is used to simulate engine operational loadings and from this measure the engine's torque reaction to the varying loads. Remember a dynamometer does not measure the direct power output of an engine, but only provides a means of measuring the torque that an engine is capable of producing at the flywheel. After recording the torque output and the speed at which this output was measured, the kWbp of the engine can be calculated using a simple formula supplied with the dynamometer.

Fig 13.1 Go Power DT1000 engine dynamometer

Fig 13.2 A heavy-duty chassis dynamometer

Fig 13.3 A PTO dynamometer attached to the tractor power take off

A number of different types of dynamometers are used to test engine performance:

- the engine dynamometer, which attaches directly to the engine flywheel during testing, as shown in Fig 13.1
- the chassis-type dynamometer, which applies a braking load to the rear wheels of the vehicle (Fig 13.2)
- the power-take-off dynamometer (PTO), which is connected to the PTO shaft of tractors for engine testing (Fig 13.3).

Engine dynamometers can be further classified according to the method by which the load is applied to the engine under test. The load may be applied:

- hydraulically
- electrically.

Hydraulic dynamometers

There are two types in common use—the water brake type and the gear pump type. Dynamometers operating on the water brake design have testing capacities as high as 2700 N.m or 600 kW. However, dynamometers of the gear pump type are designed for use with smaller engines, with a maximum output of approximately 50 N.m or 15 kW.

With the water brake dynamometer, illustrated in Fig 13.4, the braking action or loading is developed by an impeller (driven by the engine), which directs water against a turbine housing, fitted to, or part of, the main dynamometer housing. This turns the dynamometer housing, within limits, in its mountings, so that the turning effort, or torque, of the engine can be measured.

The finned turbine housing is designed to divert the water back against the impeller, so opposing its rotation. It is this turbulence and back flow that causes the braking action on the engine. Engine braking only occurs when water is fed into the dynamometer and the braking action on the engine can be increased by simply increasing the flow of water through the dynamometer.

The second method of providing hydraulic braking action is by means of a gear pump, suspended in mounting bearings to allow partial rotation, and directly driven from the engine crankshaft. Loading of the engine by progressively restricting the outflow of fluid from the pump causes the pump to turn in its mountings, so

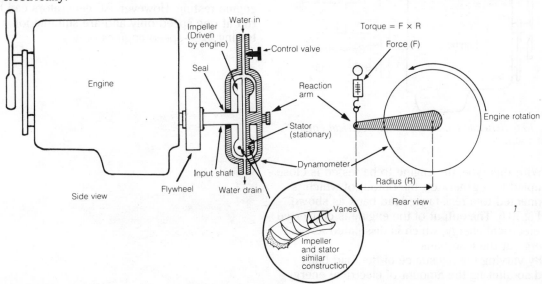

Fig 13.4 Schematic diagram of a water brake engine dynamometer

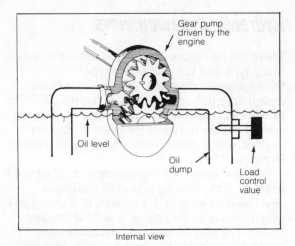 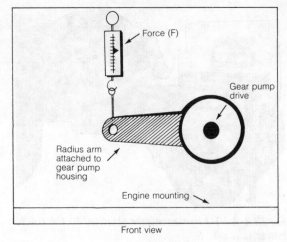

Fig 13.5 *Principle of the gear-pump-type of dynamometer*

that the torque produced can be measured (see Fig 13.5).

Electrical dynamometers

There are several types of electrical dynamometer being used today to test engines. One such type is the direct-reading dynamometer, so named because the brake power of the engine is found directly from the generator's output.

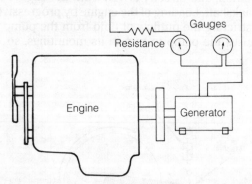

Fig 13.6 *Schematic diagram of an electrical engine dynamometer*

With this type, the engine to be tested is close coupled to a generator, the output of which is connected to a resistance load bank as shown in Fig 13.6. The output of the engine is converted to electrical energy, which is dissipated as heat energy at the load bank.

By varying the resistance of the load bank and so altering the amount of electrical energy

produced, the engine load is readily controlled. However, because the generator is less than 100 per cent efficient, the generator power output (which is readily measured by means of electrical test equipment) is less than the output of the engine under test. To compensate for this, a graph curve is required for determining generator efficiency against load current output.

The formula for finding the brake power (BP) of the engine is:

$$\frac{\text{brake}}{\text{power}} = \frac{\text{volts} \times \text{amps}}{\text{generator per cent efficiency}}$$

$$\text{BP} = \frac{IV}{\eta_g}$$

A DC generator is often used because it can be run at variable speeds, which is desirable for engine testing. However, AC generators can be run at one speed only and are suitable for testing fixed-speed engines only.

14
Fuel supply systems

Two systems are commonly employed to feed fuel from the fuel tank, through the fuel filters, and on to the injection pump of a diesel engine—the **gravity feed system** and the **pressure feed system.**

The gravity feed system can only be used when the fuel tank is mounted at a height above the fuel injection pump sufficient to provide a plentiful supply of fuel at all times. With modern fuel filters, using very fine filtering media, the tank must be well above the injection pump to ensure that an adequate fuel supply is maintained even when fuel flow through the filters is restricted by foreign matter that is filtered from the fuel and chokes the filter element.

Almost all high-speed diesel engines utilise pressure feed systems, even though in stationary applications the fuel tank is mounted above the injection pump. The use of a pressure feed system, with its mechanical feed pump, ensures a constant fuel pressure and guarantees an adequate supply of fuel to the injection pump.

In addition to carrying fuel from the fuel tank, through the filters to the injection pump and on to the injectors, the fuel system must include a means of carrying leak-off fuel (see Chapter 17) from the injectors, fuel (and air) from the self-bleeding filters, and, in many cases, fuel from the injection pump cambox where it accumulates after leaking past the injection pump plungers.

Either one, two or three filters are commonly used between the fuel tank and the fuel injection pump. The fuel tank itself is usually treated to prevent rust, and mounted so that the bottom slopes away from the pick-up pipe towards a drain cock. The pick-up pipe is generally situated above the bottom of the tank, and is usually fitted with a gauze strainer.

The fuel system often used for large marine and stationary engines is worthy of note. The fuel is pumped up from storage tanks below the engine, through a fuel purifier, to the service tank—a fuel tank of sufficient capacity to carry enough fuel for approximately half a day's continuous running. From the service tank the fuel is gravity fed, via the filters, to the injection pump.

Gravity feed fuel systems

The CAV gravity feed fuel system in Fig 14.1 features a hand primer pump, one fuel filter and a distributor-type fuel injection pump. In this system the fuel tank must be located so that the bottom lies at least 300 mm above the injection pump.

The hand primer pump is fitted between the fuel tank and filter in order to facilitate bleeding of the system; because of their design, CAV DPA fuel injection pumps require a pressure of about 20 kPa to force fuel through the pump during bleeding. When an in-line or jerk-type injection pump is used in conjunction with a gravity feed system the primer pump is not necessary, because the pressure due to gravity is enough to bleed the system.

Because the system relies on gravity alone and has the fuel tank above the fuel filter, a self-bleeding type of filter cannot be used, since the fuel cannot flow from the filter back to the tank. However, any fuel that leaks past the pumping elements and injector components flows back to the fuel tank, since pressure soon builds up in the back leakage pipe.

Pressure feed fuel systems

Fig 14.2 shows a pressure feed fuel system, featuring a sedimenter, fuel heater, self-bleeding filter, thermostart (or manifold heater) and a

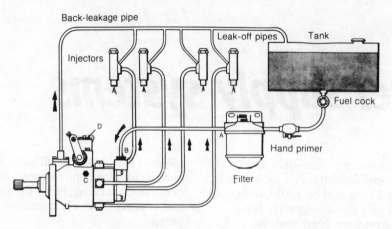

Fig 14.1 Schematic diagram of a CAV gravity fuel feed system for a DPA fuel injection pump

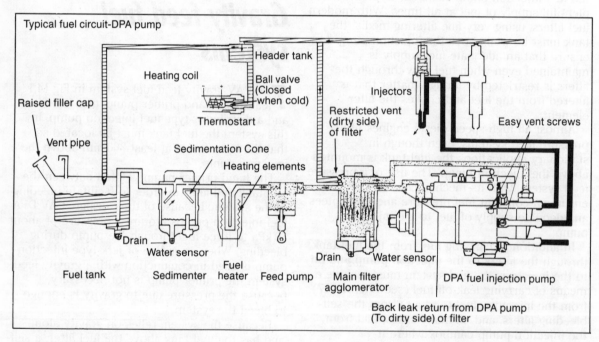

Fig 14.2 Schematic diagram of a CAV pressure feed fuel system for a DPA fuel injection pump

distributor-type fuel injection pump. The engine-driven fuel feed pump draws fuel from the tank through the sediment filter then through the fuel heater (when fitted, as in this example) and pumps it through the self-bleeding filter to the inlet connection of the fuel injection pump. The injection pump raises the pressure of the fuel, which is passed via the injection pipes to the injectors, which spray the fuel into the engine cylinders.

With a self-bleeding filter, a certain amount of fuel is allowed to escape back to the fuel tank,

carrying with it any air that may have entered the filter chamber. In addition to this, some fuel escapes from the injectors (back-leakage) and both quantities of fuel are carried back to the tank via the fuel return line. However, in the system shown, as the fuel returns to the tank, some of it fills a header tank for the manifold heater. (Refer to Chapter 8, Cold Starting Aids for manifold heater operation.) During injection pump operation there is a certain amount of fuel leakage past the pumping components, which builds up in the pump housing. This excess fuel from the distributor-type injection pump is carried back to the second inlet on the fuel filter, where it is refiltered and circulated to the injection pump again. In some systems, however, this fuel may be returned to the fuel tank.

A pressure feed system of a type used with an in-line fuel injection pump is shown in Fig 14.3. The system features a two-stage filter system, a plunger-type fuel feed pump fitted with a sediment filter and mounted on and driven by the fuel injection pump.

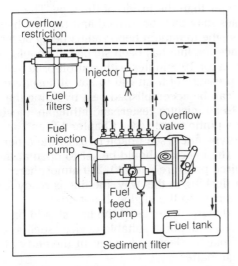

Fig 14.3 Schematic diagram of a pressure feed fuel system for an in-line fuel injection pump

The fuel circuit follows a fairly conventional layout for an in-line injection pump, with continuous fuel circulation through an overflow valve located in the injection pump fuel gallery. The overflow valve allows fuel, excess to engine requirements, to circulate back to the fuel tank to purge the gallery of any gas or air, and to cool the injection pump.

The main fuel filter is fitted with an overflow restriction—an adaptor with a flow-restricting orifice used to connect the highest part of the filter to a return pipe and on to the fuel tank. This allows a small quantity of fuel to return continuously to the fuel tank and, in so doing, bleed off any entrapped air in the main fuel filter housing.

Fuel pipes

*I*n the past, in almost all cases, the fuel pipes that carry the fuel from the fuel tank to the injection pump have been copper. However, copper reacts with certain sulphur compounds in diesel fuel to form greasy deposits, so copper supply pipes are fast becoming obsolete. The modern trend is towards steel pipes and synthetic rubber flexible hoses, or flexible plastic fuel lines.

To ensure an adequate supply of fuel at the injection pump, supply lines of suitable size must be provided. An 8 mm bore supply line is generally adequate for use on a multicylinder engine with injection pump plungers that do not exceed 11 mm, while a 10 mm bore fuel line is necessary when the plunger diameters are between 12 and 15 mm. If the pump plunger diameters are between 15 and 17 mm, either 13 or 14 mm bore pipes are used.

Although all injector leak-off fuel passes through one pipe the rate of leakage is very slight, and 4.5 mm bore tubing is adequate.

Note   When fuel feed pipes are being shaped, sharp bends should be avoided—the bending radius should never be less than 50 mm (2 inches).

The injector lines—the pipes connecting the injection pump to the injectors—have to withstand fuel pressures that may reach 70 MPa. Because of the pressures involved, cold drawn, annealed, seamless steel tubing must be used.

Both the length and the bore diameter of the injector lines affect the injection characteristics, so the lines are kept to minimum length with the minimum bore diameter that will give efficient fuel delivery to the injectors at high speed. To ensure correct timing at each cylinder and uniform injection characteristics, all lines on multicylinder engines should be as near equal length as practicable since a change in the pipe length can alter the injection timing by up to 1° per 100 mm of pipe. The manufacturer's manual should be checked to ascertain the recommended bore diameter when making replacement injector lines, but in the absence of specific information the following chart should provide a satisfactory guide:

Pump plunger diameter (mm)	Injector line bore diameter (mm)
Up to 7.0	1.5 or 2.0
8 to 10.0	2.0 or 3.0
11.0 to 18.0	3.0

Three systems are used to provide a suitable end on injection lines to facilitate attachment to the injection pump and injectors. The oldest method consists of silver brazing conical sleeves or collars to the line, and tightening the collar against the face of the union with the nut.

The second, and most extensively used system, entails cold-forming a conical end or nipple on the end of the line by forcing special dies against the end of the line by hydraulic pressure. This is done in a nipple-forming tool, which utilises a small hydraulic ram completely enclosed in the tool. The nipple-forming tools are supplied with a set of collets to hold each size of pipe likely to be encountered—6, 8 and 10 mm outside diameter—and a separate die or former for each size.

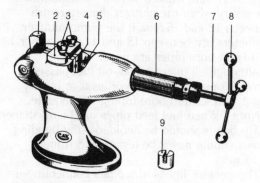

Fig 14.4 Hydraulic nipple-forming tool

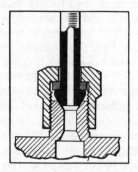

Fig 14.5 Typical pressure-formed nipple and pipe connection

Note   Some manufacturers of fuel injection equipment recommend that the fuel line be annealed after the nipple has been formed by heating to cherry red and cooling slowly.

The third system makes use of a compression-type sleeve, which is clamped onto the pipe when the union nut is tightened, forcing the sleeve into a taper that compresses it. This provides a readily replaceable fitting and is ideal for rapid, on-the-spot repairs.

While actual conditions may make it difficult to achieve, new injection lines should be thoroughly cleaned before being fitted to ensure that no dirt enters the injectors. Ideally, the pipe should be cut to length, the burrs removed, and the pipe then left for 24 hours in kerosene. After soaking, the pipe should be blown out with compressed air and pulled 20 to 30 times over a steel wire about 0.5 mm smaller in diameter than the inside of the pipe. The nipples may then be formed and the line bent to shape, the minimum bend radius being 50 mm.

Once the line has been shaped, it should be flushed for 10 minutes with test oil or kerosene at a pressure of from 140 to 250 atmospheres. This can be accomplished best by fitting the line to an injector tester, and fitting an injector with a pintle nozzle (see Chapter 17) and an edge-type filter (see Chapter 16) to the other end. The injector should be set to open at the required pressure—ie 140–250 atmospheres. After flushing in this way, the line is ready for attachment to the engine. If not fitted immediately, the ends of the line should be covered with some suitable device (such as small plastic sleeves) to prevent the entry of foreign matter.

15
Fuel supply pumps

In all pressure feed systems, a fuel supply pump is required to pump the fuel from the fuel tank, carry it through the fuel system, including the filters, and supply it to the injection pump at constant pressure, and in sufficient quantity, for the engine's requirements. Many types of fuel supply or feed pumps are in extensive use today, but most of them fall into one of the following categories:

- plunger-type feed pumps
- diaphragm-type feed pumps.

The plunger-type feed pump

This type of pump is manufactured by a number of makers of fuel injection equipment, including American Bosch, CAV, Diesel Kiki and Nippon Denso. The pump is mounted on the fuel injection pump and is driven by the pump camshaft. It is self-regulating and will build up to, and maintain, a predetermined pressure.

Construction

The pump consists of:

- an alloy housing
- a spring-loaded, roller-type cam follower, which reciprocates in the housing under the action of a cam
- a spring-loaded plunger, which is a neat fit in the bore of the housing and which is forced away from the injection pump by the cam follower.
- two spring-loaded fibre valves, which control the passage of fuel to and from the pump cylinder or plunger spring chamber.

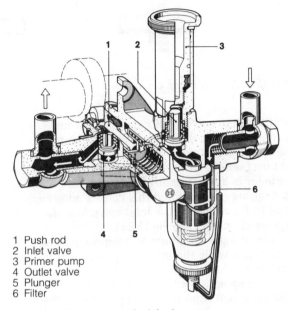

1 Push rod
2 Inlet valve
3 Primer pump
4 Outlet valve
5 Plunger
6 Filter

Fig 15.1 Plunger type fuel feed pump

Operation

When the injection pump camshaft is in such a position that the feed pump cam follower lies against the back of its cam lobe, the pump plunger is forced by the spring towards the camshaft, creating a low-pressure area in the plunger spring chamber. The fuel on the inlet side of the pump, being under atmospheric pressure, opens the inlet valve and fuel enters the chamber (see Fig 15.2a). Then, as the cam lobe drives the plunger against its spring, the fuel is forced by the plunger through the outlet valve and passes around to fill the chamber behind the plunger, which has been left vacant by its movement (see Fig 15.2b).

Further rotation of the injection pump camshaft allows the plunger spring—now under compression—to press the plunger towards the injection pump, thus forcing the fuel lying behind the plunger out into the fuel line leading

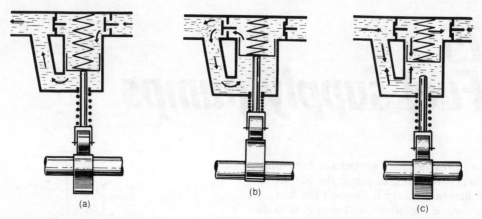

Fig 15.2 Plunger pump operation

to the filters and injection pump. The fuel cannot re-enter the plunger spring chamber since the outlet valve does not allow the fuel to return (see Fig 15.2c). At the same time, the plunger is again creating a low-pressure area in the spring chamber and this causes additional fuel to flow through the inlet valve into the spring chamber (see Fig 15.2a).

The plunger continues to reciprocate and pump fuel while the injection pump uses fuel as fast as the feed pump can supply it. When the feed pump satisfies the engine's requirements, pressure builds up in the supply line to the injection pump and under the plunger, until it is high enough to prevent the plunger from returning with the follower. The supply pressure to the injection pump is controlled by the force exerted by the plunger spring.

As fuel is used, the plunger is forced towards the cam by the spring, causing an intake of fuel into the plunger spring chamber, and the operating cycle starts again. Thus the pump maintains an adequate supply of fuel at a constant pressure, the pressure being controlled by the spring force.

A hand-operated primer pump is usually fitted to this type of feed pump, so that bleeding of the fuel system can be carried out. This consists of a plunger and barrel assembly, which is screwed into the pump housing directly above the inlet valve.

When the plunger is lifted, a low-pressure area is created beneath it, causing the supply pump inlet valve to be forced from its seat, and allowing fuel to flow into the hand priming pump barrel. When the priming pump plunger is pressed downwards, the fuel is forced into the supply pump spring chamber. This causes the

outlet valve to open and the fuel to be forced into the supply line leading to the injection pump.

After use, the hand knob must be firmly screwed onto the top of the plunger barrel. This action seals the top of the plunger barrel to prevent fuel, which can pass the plunger during engine operation, from escaping.

Diaphragm-type feed pumps

There are many variants of the mechanically operated diaphragm-type feed pump, but they are all driven in one of two ways—either by an eccentric on the engine camshaft or by a special cam, or an eccentric on the injection pump camshaft.

The AC-type diaphragm feed pump

This type of pump is identical with the fuel pumps fitted on many petrol engines, and is driven in the same way by an eccentric on the engine camshaft.

Construction

This type of pump consists of an upper and a lower alloy body with a diaphragm (made of fabric impregnated with synthetic rubber) sandwiched between them. The upper body

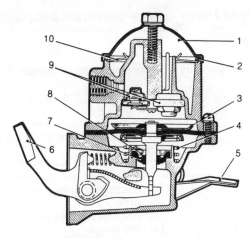

1	Domed cover	6	Rocker arm
2	Gasket	7	Seal
3	Diaphragm assembly	8	Seal retainer
4	Diaphragm return spring	9	Valve assemblies
5	Priming lever	10	Filter gauze

Fig 15.3 AC-type fuel feed pump

carries the inlet and outlet valves, usually spring-loaded fibre discs.

The centre of the diaphragm is clamped between dished circular plates attached to one end of a diaphragm pull rod. A diaphragm return spring is located between the lower circular plate and the lower diaphragm body. A two-piece lever, pivoted on a pin through the lower body, connects at one end to the diaphragm pull rod while the other end bears against an eccentric on the engine camshaft.

A light spring maintains the lever in contact with the eccentric.

Operation

When the engine is turned over with the starter or is running, the eccentric causes the lever to rock on its pivot pin. This draws the pull rod and diaphragm down, compressing the diaphragm return spring and creating a low-pressure area above the diaphragm. The pressure difference—atmospheric at the fuel tank, less in the fuel chamber—causes the fuel to force open the inlet valve and flow into the fuel chamber (Fig 15.4a).

When the lever rocks back, the spring forces the diaphragm upwards, forcing the fuel through the outlet valve (Fig 15.4b).

This cycle continues until the pressure builds up above the diaphragm on the outlet side— when more fuel is pumped than the engine can

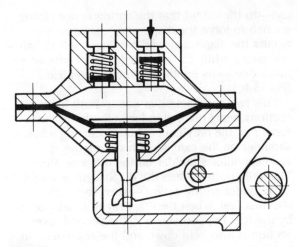

Fig 15.4a Intake

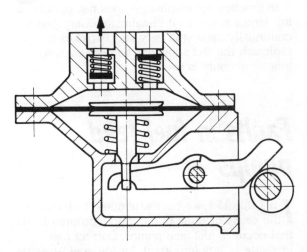

Fig 15.4b Delivery

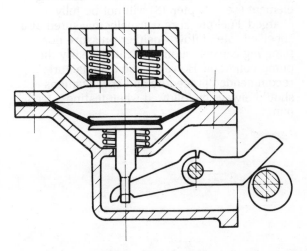

Fig 15.4c 'Freewheeling'

use—to the extent that the spring is not strong enough to force the diaphragm up. When this occurs the diaphragm and one lever will remain stationary, while the other lever will continue to move due to its contact with the eccentric (Fig 15.4c).

This two-lever action is due to both lever sections being pivoted independently on the same pin, the only connection occurring when a shoulder on the cam section strikes the corresponding shoulder on the other section. This only occurs when the diaphragm is able to move under spring action—when the engine requires fuel. When the diaphragm is held down by the fuel trapped above it, the second lever section is also held down and the shoulders do not meet, producing the necessary 'free linkage' action.

In practice the diaphragm does not perform a full stroke nor does it remain stationary, but is continually moving in very short strokes to replenish the fuel supply as soon as even a minute quantity is used.

Faults in fuel feed pumps

In Table 15.1, we have attempted to indicate the cause of some of the more common faults that occur in fuel feed pumps. Correct fuel pressure is very important, for low fuel pressure will not allow the engine to develop full power at large throttle openings, since the pumping element (see Chapter 18) will not be fully charged. Fuel pressure is usually measured at a specified rate of flow—neither pressure nor pumping capacity alone indicates clearly the pump condition. However, the manufacturer's recommended procedure and specifications should always be followed when testing a feed pump.

Table 15.1 Some common faults in fuel feed pumps

Symptom	Cause
Low fuel pressure	Weak diaphragm or plunger spring Worn actuating linkage or eccentric Leaking fuel pump valves Choked, fractured or loose inlet pipes Fractured or torn diaphragm
Fuel output low or nil	Worn actuating linkage or eccentric Fractured or torn diaphragm Leaking valves Choked, fractured or loose inlet pipes
Fuel running from injection pump cambox overflow pipe	Fractured or torn diaphragm

16
Fuel filters

Due to the high pressures and accurate metering necessary when injecting fuel, the pumping components of the injection system are extremely accurately made and finely finished. Research has shown that, while particles of foreign matter of all sizes cause wear to this equipment, most wear is caused by particles from 6 to 12 microns in size (1 micron = 0.001 mm).

Water, even in small quantities, causes corrosion of the highly polished surfaces and contributes to the ultimate failure of both injection pumps and injectors. Therefore in order to obtain the longest possible working life from this expensive equipment, efficient fuel filters are indispensable.

The best fuel filtering system is generally considered to be a progressive one, the first or primary filter removing the large particles of foreign matter, the second or secondary filter removing the finer particles. A third or final stage filter is sometimes fitted in a progressive system, the secondary filter being responsible for the removal of the intermediate particles.

However, there are many systems used widely throughout the world in which two identical elements in series are used, the second providing a safety element should the first become damaged.

Fuel filters range from simple sediment bowls capable of removing water and heavy particles only, to replaceable element filters capable of removing particles down to 2–4 microns.

A sediment bowl is simply a bowl or chamber in the fuel system, and has no filtering element. However, because of its relatively large volume, the fuel passes very slowly through it and heavy particles of foreign matter and water settle to the bottom of the sediment bowl, from where they are periodically drained. Many sediment bowls are made of glass, so that accumulations of foreign matter can be easily seen by the operator, who can then drain or clean the bowl.

Element-type filters are made in a wide variety of types and designs. The self-bleeding filters, referred to in the previous chapter, are element filters, with a small restricting orifice or a spring-loaded pressure relief ball valve in the top of the housing. A pipe leading from this hole carries a small quantity of fuel continuously back to the tank. Any air entering the fuel filter assembly in the fuel rises to the top and passes back to the fuel tank as a bubble in the return line.

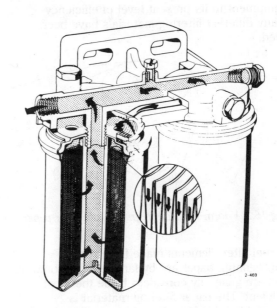

Fig 16.1 Dual filter paper element type

Modern filter elements are generally made to one of two designs. The first design consists of a replaceable cartridge or element fitted inside a filter bowl. The second, and fast becoming the popular, type consists of a canister of pressed steel with the filtering material inside. Replacement of this type consists of unscrewing or unclamping the old element and fitting the new. With this latter type, the outside of the canister is exposed when the element is in service.

Filter materials

The efficiency with which the fuel is filtered is governed by the size of the pores in the filter material through which it must pass, the smaller pores, the more efficient the filtering action. But the material must present sufficient pores to allow free passage of the fuel, or the supply to the injection pump will be insufficient and the engine will lack power; this problem will worsen with use since the filter will become more and more restricted.

The ideal filtering material must therefore efficiently filter the fuel, but must also have sufficient area to allow free and adequate fuel flow for a reasonable period—a filter that requires renewal after only a short period of running is not satisfactory.

During the development of fuel injection equipment to its present level of efficiency, many different filtering materials have been tried.

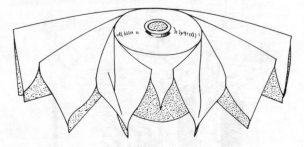

Fig 16.2 V-form resin-impregnated fuel filter paper

Fuel filter elements made from resin-impregnated paper are by far the most common and are generally considerd to be the most efficient. The paper filtering material is invariably folded or pleated to present a very large surface area to the fuel, thus giving maximum life between changes.

Fuel filter service

The need for absolute cleanliness when servicing fuel filters cannot be stressed too greatly.

Where filter bowls and sediment bowls are fitted with a drain plug, this should be removed periodically to allow accumulated sediment and water to drain away.

Before changing a filter element or servicing a filter, it is good practice to thoroughly wash the filter bowl and surrounding engine area with petrol to remove impurities that could fall into the filter components. The filter unit may then be dismantled and the element removed. If the element is a type that can be washed, this must be done in the recommended solvent in a clean container. If the element is to be discarded, ensure that there are no sealing plates adhering to it.

Any parts removed from the filter unit to allow the element to be removed should be washed in clean solvent but should not be wiped with rag. All rubber 'O'-rings and seals should be renewed. The new element may then be fitted and the unit reassembled.

After sevicing the filter, the air bleed screw should be opened and any trapped air bled from the system.

Fuel filter types

Although there are a large number of manufacturers of fuel filtration systems, the majority follow a limited number of basic designs. For this reason, coverage will be restricted to the typical filters manufactured by Robert Bosch and CAV.

Bosch fuel filters

Robert Bosch manufacture a range of fuel filters, the most common of which is a box-type spin-on paper element filter.

The type of filter element used by Bosch is of the 'V'-form spiral design as shown in Fig 16.3. In this type of element, the filter paper is wound round a tube and adjacent layers are glued together, alternately at the top and the bottom. This forms pockets with their openings at the top.

The fuel flows through the 'V'-form spiral element axially from the top to the bottom; the dirt particles that are held back accumulate in the filter pockets. The filtered fuel then flows upwards through the tube in the middle of the filter.

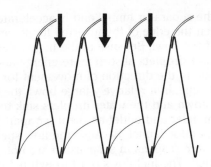

Fig 16.3 Sectioned and expanded 'V'-form spiral element

Fig 16.4 Fuel filter box

Bosch paper filter elements perform with a high degree of efficiency in separating water from the fuel. In addition to a certain amount of molecularly bound water, diesel fuel often contains free water. This free water may enter the fuel through carelessness when filling up the tank, by unsuitable storage in drums or through condensation in the fuel tank.

The molecularly bound water does not cause any adverse effects during operation. However, it is absolutely essential to separate the free water from the fuel for trouble-free operation of the injection system; this is particularly true for distributor-type injection pumps.

The filter paper used by Bosch has not only an exceptionally high filtering efficiency, but it is also a good water separator due to a special impregnation technique, in which a surface-active precipitation agent is deposited on the paper. This chemical treatment enables the water to be retained on the filter paper, but allows the diesel fuel to flow through.

In a spiral element, the diesel fuel passes through the filter while the water accumulates on the dirty side of the paper pockets. Therefore the filter surface area available for the passage of fuel is reduced, the pressure differential across the element increases and finally forces the water through to the clean side. Here it forms larger drops that collect in the lower part of the filter. The spiral filter ceases to function when the water level in the bowl rises above the lower edge of the element.

Bosch fuel filter box

The filter box consists of a metal housing in which a 'V'-form spiral paper element is fitted. At the top of the housing there is a leakproof cover with its edges rolled over. This cover is provided with an M16 × 1.5 mm threaded hole

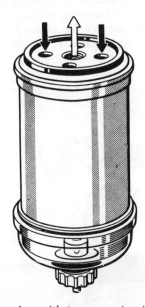

Fig 16.5 Filter box with transparent water trap

in the centre, four outlet holes and a gasket. The threaded hole is used for attachment to the filter head and as the fuel outlet.

The unfiltered fuel enters through the head, flows through the four holes into the filter box, through the filter element and finally exits through the inner tube and the threaded hole.

When the unit no longer allows the passage of sufficient fuel, the filter box must be unscrewed and replaced by a new one. The new filter box should be screwed on by hand until the gasket makes contact with the filter head. It is then tightened a quarter turn. The fuel system should then be bled to remove any air.

Two forms of water trap are used with box-type fuel filters, transparent and non-transparent. In the first case there is an end cover on the bottom of the filter box, which serves for screwing the glass bowl onto the filter with a bolt. In both cases, there is a drain screw with a drain hole in the middle so that the water can be drained out without dirtying the hands. The water level should be checked daily and excess water drained off by turning the drain screw to the left. Replacement of a filter box with a water trap and bleeding of the system is done in the same basic manner as with plain box-type filters.

The Bosch water separator

If the water or dirt content of the diesel fuel is particularly high (tunnelling, desert traffic), it is recommended that a water separator be fitted to increase the service life of the fuel filter. It will not only retain the larger water drops in the fuel, but also the larger rust, metal and dirt particles.

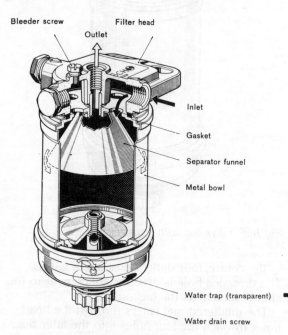

Fig 16.6 Water separator (sectional view) with a filter head for horizontal mounting

At low ambient temperatures the water separator can also hold back paraffinic precipitation without causing the filter to clog.

The water separator is similar in appearance to a box filter with a transparent water trap, but the construction is somewhat different. Instead of the filter element, there is a separator funnel.

The fuel, containing droplets of water, flows

over the separator funnel and is accelerated between the edge of the inverted funnel and the wall of the bowl. Because of the higher density, the water droplets absorb more energy and continue in the direction of movement for a longer time. In a relatively large bowl, the fuel slows down and the water droplets sink to the bottom. Inlet and outlet are kept far apart by the separator funnel. Because the flow speed is low, water droplets in suspension are not entrained. The fuel flows out through the funnel hole and the threaded connection.

The water level can be checked through the transparent water bowl, and the water drained, when required, by the drain screw. To obtain good water separation the water separator must be installed upstream of the fuel feed pump because fuel and water would be homogeneously mixed by the pump due to the pulsating action.

Box filter types

Bosch filter boxes without water traps are available in sizes of 0.4 litres (0.7 imperial pints) and 0.6 litres (1.06 Imperial pints); those with water traps, in the 0.6 litre size only. They are screwed to the filter head, and the assembly is called a box filter. The threads for fuel inlet and outlet connections are in the filter head and the installation holes are in the mounting flange. Filter heads are supplied for vertical or horizontal mounting.

The range of box filters includes single-box filters (as previously discussed), parallel box filters and two-stage box filters.

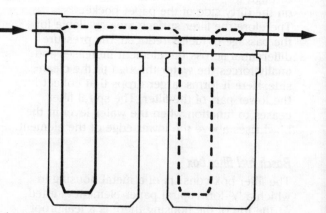

Fig 16.7 Flow principle for fuel filters connected in parallel

The required fuel flow rate determines the size of the filter and the filter configuration. Where there is a large fuel flow requirement, one filter would be insufficient, either because of inadequate flow capacity or limited service life, so a set of filters in parallel are used. (Fig 16.7).

It is interesting to note that two filters in parallel will more than double the service interval of a single unit, since apart from halving the quantity of fuel to be filtered, each element can become more restricted by contaminants, while together they can continue to pass sufficient fuel for the engine.

With a two-stage filter configuration, the fuel flows first through a primary (coarse) filter, which removes the larger particles, and then through a secondary finer filter and on to the fuel injection pump as shown in Fig 16.8.

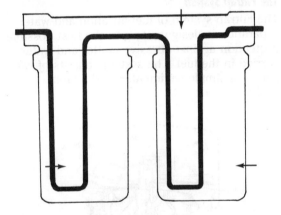

Fig 16.8 Two-stage series filter

Various possibilities in selecting a filter also present themselves in regard to quality. The aim in the development of fuel filters is to obtain, besides a sufficient filter life, the longest possible service life from the injection equipment. Bosch has matched the quality of its filters to the requirements of the widely known Bosch injection pumps. Thus, a filter paper with an average pore size of approximately 9 microns has been developed for standard requirements. It is generally used with in-line injection pumps. A micro filter paper with an average pore size of 4 microns is available for more exacting requirements, for example for distributor-type injection pumps.

CAV fuel filters

CAV Ltd manufacture a large range of fuel filters that feature 'V'-form impregnated paper elements. These may be configured in one of several ways—as single units, as a parallel assembly, or as a series arrangement, depending on the requirements of the engine manufacturer, the rate of fuel flow needed and the desired service life.

FS filter

The FS filter is of the crossflow type. The element is contained within a strengthened steel canister, which forms an integral part of the filter assembly.

With this type of filter the fuel may be passed through in either of two directions. The construction of the filter and the two available directions of fuel flow are shown in Figs 16.9 and 16.10. The unit consists of the following component parts (refer to Fig 16.9):

- a cast aluminium head with inlet and outlet connections and a mounting flange
- a paper filter element enclosed in a strengthened metal canister and
- a cast aluminium base, together with a centre stud.

The three units are held together by the centre bolt, which passes through the head casting and screws into the centre stud.

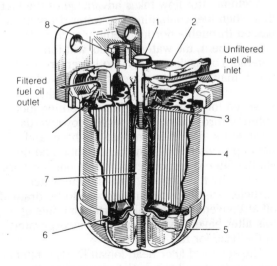

Fig 16.9 FS bowl-less filter, showing filter-flow element

Synthetic rubber sealing rings located at the top and bottom of the filter element canister prevent external fuel leakage. The clean outgoing filtered fuel is sealed off from the incoming unfiltered fuel by a synthetic 'O'-ring held in a groove on the filter head centre boss.

The filter head is supplied with galleries drilled to provide for either filter flow or agglomerator flow. Filter flow is illustrated in Fig 16.9, which shows the incoming unfiltered fuel passing through the centre bolt housing and into the base. It then passes upwards through the filter element and out through the filtered fuel outlet port.

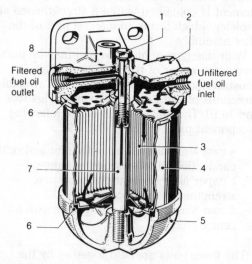

Fig 16.10 FS bowl-less filter showing agglomerator-flow element

Agglomerator flow takes advantage of the fact that when fuel containing fine water droplets is passed through a porous medium such as the filter element, the water droplets will join together (or agglomerate) into larger droplets, which may then be removed from the fuel by sedimentation. Fig 16.10 shows the FS filter arranged for agglomerator flow. Incoming fuel entering through the inlet passes downwards through the filter element into the base, and then upwards through the centre tube and out via the filtered fuel outlet. Abrasive particles are retained by the filter element and the water particles are deposited in the base to be drained off at convenient intervals. Certain models of this filter have a nylon drain plug incorporated in the base for this purpose.

The standard filter head for an FS-type filter has two inlet and two outlet connections to provide alternative piping connections for different types of engine installation. It is suitable for both distributor and in-line fuel injection pumps, but when used with a distributor pump the fuel return from the pump will be to the unfiltered side of the filter. Closing plugs are fitted to the connection ports not used for pump connections.

2FS twin bowl-less filter

This filter is a twin version of the FS-type filter and has a common head casting and mounting flange, with two separate elements. It provides alternative arrangements for fuel flow:

- series flow, for engines working in exceptionally dirty conditions
- parallel flow, for engines requiring double the fuel flow rate of that provided by the single FS filter.

The Filtrap system

The Filtrap system of fuel filtration and water separation is designed to give the best possible protection against abrasive matter and water carried in the fuel. The system consists of two parts—a simple sedimenter, and a filter-agglomerator.

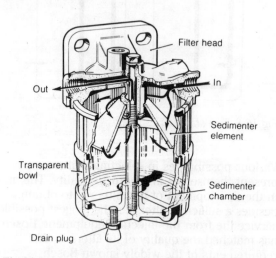

Fig 16.11 Simple sedimenter—Filtrap system

A cutaway view of the simple sedimenter is shown in Fig 16.11. Incoming fuel is fed into the unit above the sedimenter element and passes through a clearance between the sedimenter cone and the sedimenter wall. Sedimentation takes place in the lower transparent bowl, and the fuel, relieved of the large droplets of water and the larger particles of abrasive matter, passes out of the unit via ports in the central portion and galleries in the head. Accumulated water and solid matter are visible in the transparent bowl and may be drained off through the drain plug provided in the base.

This same sedimenter is also available with a water-level detector. Known as a waterscan sedimenter, it operates in exactly the same way as a simple sedimenter, but has the addition of a water-level detector at the base of the filter housing. This detector senses the water level,

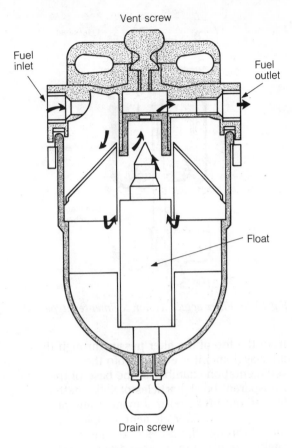

Fig 16.14 Waterstop fuel filter

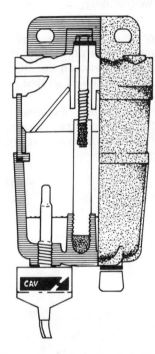

Fig 16.12 Waterscan sedimenter fuel filter

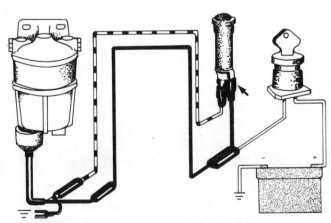

Fig 16.13 Waterscan system wiring diagram

and at a certain point activates a warning light in the instrument panel.

Another type of sedimenter—the waterstop pre-filter (Fig 16.14)—operates on the normal sediment filter principles, but has, in addition, a piston and needle valve, which float on the water in the bottom of the bowl. When the water level in the bowl of the filter rises, the piston and needle valve also rise, and block off the exit of fuel from the filter to the injection pump at a predetermined point, so stopping the engine. On draining the water from the filter, the piston and needle valve drop, opening the passage and allowing fuel to flow again. No bleeding of the fuel system is required after draining water from the waterstop filter.

Fig 16.15 illustrates the filter-agglomerator unit where the smaller water droplets and the smaller abrasive particles remaining after preliminary sedimentation are removed.

Since the desired properties of agglomerator and filter elements are similar, a single filter element is used for both functions. Fuel entering

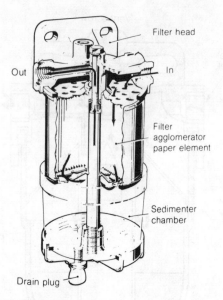

Fig 16.15 Filter-agglomerator sedimenter, type FAS

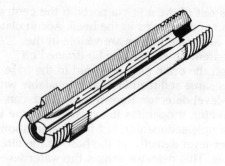

Fig 16.16 Edge-type filter

from the top of the filter passes through the filter-agglomerator element into the sedimentation chamber in the base of the transparent bowl. Particles of solid matter are filtered out in the normal manner and are retained in the filter element. Droplets of water, forced through the pores of the filter, agglomerate and form large drops, which separate from the fuel by sedimentation and accumulate in the base of the housing. The fuel, free from solid matter and water droplets, then passes upwards through the element centre tube to the outlet connection in the unit cover. The accumulated water, visible within the transparent bowl, may be drained off by unscrewing the drain plug provided.

Injector filters

Some injector manufacturers incorporate a filter in the injector inlet connection. In some cases, this filter takes the form of a very fine brass gauze held between the injector inlet adaptor and the housing. Alternatively, an **edge-type** filter may be used.

The edge-type filter consists of a hardened steel rod fitted inside a rather long inlet adaptor. Four grooves are ground along the rod, two from each end, but these do not run the full length of the rod and do not connect. Fuel can run along the two grooves from the inlet end,

but to get to the other grooves and on to the injector it must pass around the outside of the rod. The clearance between the edge and the inlet adaptor is very small, so that no particles of dirt can pass.

Filters of this type are, of course, serviced during injector maintenance. The edges should be inspected for any damage that could allow dirt particles to pass.

17
Injectors and injector nozzles

Over the years, two systems of injecting fuel into the combustion chamber emerged—the air blast fuel injection system and the solid fuel injection system.

The air blast system, the first to be developed, was used on Dr Diesel's first operative engine and was developed and used on marine and large stationary engines until the 1930s, when it fell into disfavour. This system relied on compressed air to blast the fuel into the combustion chamber. When the pressure existing in the cylinder before and during combustion are considered, it becomes obvious that a very efficient compressor is required. Not only a compressor but also an air receiver or reservoir is necessary, making it a relatively bulky system. This, coupled with difficulty in obtaining accurate metering and relatively poor reliability, has resulted in the air blast system becoming virtually obsolete, and it will not be discussed further.

The solid injection system, on the other hand, is compact, meters the fuel accurately, and is very reliable. The system uses a fuel injection pump that raises the fuel pressure to between 12 and 70 MPa. This pressure is sufficient to ensure atomisation of the fuel and to ensure that no difficulty exists in forcing the fuel into the combustion chambers.

In the early stages of the development of solid injection, open injectors were used. In their simplest form these were simply spray nozzles with a single hole leading directly into the combustion chamber. Later developments saw the addition of a non-return valve to prevent combustion gases from entering the injector line.

There is some contention as to where the line of demarcation lies between open and closed nozzles. We choose to regard closed nozzles as those that incorporate a valve that opens, when subjected to fuel pressure, to permit fuel to pass through the spray holes; those that do not include such a valve are considered to be open nozzles.

Typical injectors

Most manufacturers of fuel injection equipment manufacture injectors to the same basic design; injectors following this design may therefore be considered as typical.

Construction

When it is realised that an injector is required to operate at pressures well in excess of 20 MPa, and to inject minute quantities of fuel into a blazing inferno at a rate of from 150 to 1500 injections per minute, some idea must be gained of the quality of materials and manufacture that go into the ordinary fuel injector. With the exception of copper sealing washers, all parts are manufactured from heat-treated alloy steel.

A basic fuel injector consists of seven essential parts:

- A nozzle assembly consisting of two parts, a nozzle body and a needle valve or nozzle valve.

    The two components are lapped together to form an assembly, the needle valve being free to slide, but with minimum clearance, in the bore of the nozzle body. The upper surface of the nozzle body is accurately ground and lapped to form a precision sealing surface where it contacts the body of the injector.

    One or more drillings in the nozzle body lead from the upper lapped surface, which may carry an annular groove, to a pressure chamber (or fuel gallery) above the needle valve seat. The needle valve is stepped or shouldered where it passes through the pressure chamber.

    At the lower end of the nozzle body, below the needle valve seat, lie the passage(s) through

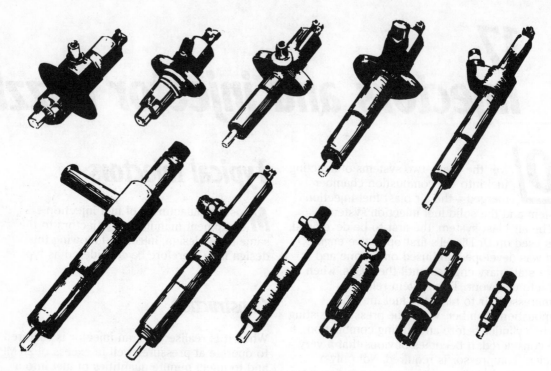

Fig 17.1 A range of typical injectors

which the fuel passes to the engine combustion chamber.
- A steel nozzle holder, complete with mounting flanges for securing the unit in the cylinder head, and with drilled passages for conducting fuel from the inlet connection to the nozzle. The lower end of the holder is provided with an accurately ground and lapped surface, which makes a leak-proof seal with the corresponding lapped surface on the upper end of the nozzle body. A leak-off connection is also provided in the nozzle holder in some cases.
- A steel nozzle nut or nozzle cap nut, which screws onto the lower end of the nozzle holder, securing the nozzle assembly in place.
- A heavy-duty compression spring, which holds the needle valve on its seat.
- A steel spindle located inside the nozzle holder between the spring and the needle valve, transmitting the spring force to the needle valve.
- A spring adjusting mechanism, which is needed for the spring force to be varied. This usually takes the form of a screw and lock nut.

- A cap nut, which is screwed onto the top of the nozzle holder to keep out dirt etc. This nut may be drilled and tapped to accommodate a leak-off connection.

The last six injector parts in the above list are often referred to collectively, when assembled, as the injector body assembly, while the first one, the nozzle assembly, is usually termed the injector nozzle.

Operation

When the metered quantity of fuel enters the nozzle holder through the inlet connection, it passes through the drilled passage(s) to the pressure chamber above the needle valve seat. In some cases, the injector nozzle body drillings may not align with the nozzle holder drilling, but where this can occur, the nozzle body has an annular groove machined in the lapped surface to which the nozzle drillings join. Thus fuel can flow from the holder, around the top of the nozzle, and down the nozzle drillings. When the pressure of the fuel in the pressure chamber

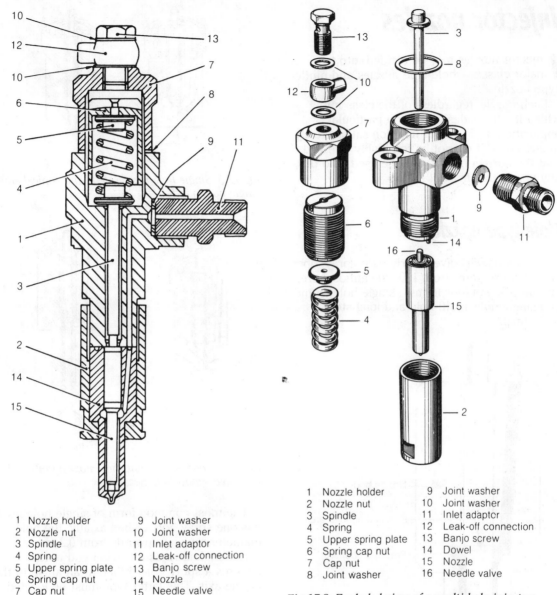

1	Nozzle holder	9	Joint washer
2	Nozzle nut	10	Joint washer
3	Spindle	11	Inlet adaptor
4	Spring	12	Leak-off connection
5	Upper spring plate	13	Banjo screw
6	Spring cap nut	14	Nozzle
7	Cap nut	15	Needle valve
8	Joint washer		

Fig 17.2 Sectional view of a multi-hole injector

1	Nozzle holder	9	Joint washer
2	Nozzle nut	10	Joint washer
3	Spindle	11	Inlet adaptor
4	Spring	12	Leak-off connection
5	Upper spring plate	13	Banjo screw
6	Spring cap nut	14	Dowel
7	Cap nut	15	Nozzle
8	Joint washer	16	Needle valve

Fig 17.3 Exploded view of a multi-hole injector

applies sufficient force to the needle valve shoulder to overcome the spring force, it lifts the valve from its seat, allowing fuel to flow from the nozzle until delivery from the pump ceases, at which point a positive instantaneous cut-off of fuel occurs because the valve is snapped to its seat by the spring force, eliminating the possibility of after-dripping or dribbling.

A certain amount of seepage of fuel between the lapped guide surfaces of the needle valve and nozzle body is necessary for lubrication. This leakage of fuel accumulates around the spindle and in the spring compartment, and the excess is carried away through the leak-off connection provided for that purpose.

Injector nozzles

*I*njector nozzles may be divided into two major classes—hole-type nozzles and pintle-type nozzles.

Each nozzle, regardless of the class into which it falls, is designed for a particular application, and will not function satisfactorily in any other. It cannot be too highly stressed that the correct nozzle must be used, and that any nozzle that fits will not do.

Hole-type nozzles

Hole-type nozzles give a fairly hard spray, open at a high pressure (from 150–200 atmospheres) and take one of four forms—single hole, conical end single hole, multi-hole and long stem multi-hole.

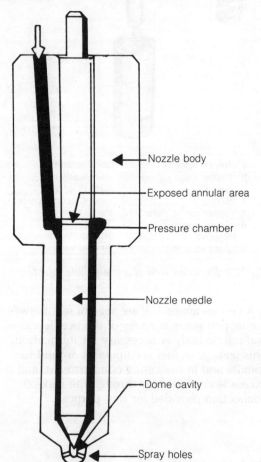

Fig 17.4 *A typical multi-hole injector nozzle*

- Nozzle body
- Exposed annular area
- Pressure chamber
- Nozzle needle
- Dome cavity
- Spray holes

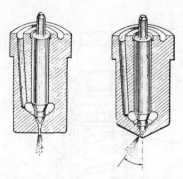

Fig 17.5 *Single hole nozzle (left) and conical end single hole nozzle*

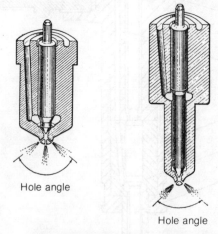

Hole angle

Hole angle

Fig 17.6 *Short-stem multi-hole nozzle (left) and long-stem multi-hole nozzle*

The most common form of single hole nozzle has one spray hole drilled axially, the hole diameter ranging upwards from 0.2 mm. A variation, known as a conical end nozzle, features a spray hole drilled at an angle to the nozzle axis. This latter type must be located in relation to the injector body by dowels in the nozzle holder lower face, so that the fuel sprays in the required direction.

Multi-hole nozzles are widely used and are fitted in direct injection engines. These nozzles utilise a number of spray holes, two or four being the most common. The holes are drilled in such a way as to give the best possible distribution of fuel, and may consist of one axial hole with a number of holes, at an angle to the nozzle axis, surrounding it.

In many applications, the injector is not located vertically above the piston cavity, but lies at an angle to the cylinder axis. The holes on one side of the nozzle must be drilled at a different angle from those on the other side to ensure that the spray does not strike the piston crown on one side and the cylinder head on the other. In applications of this type, the nozzle is again dowelled to the nozzle holder.

In many direct injection engines long-stem nozzles are used. As can be seen from Fig 17.6, the extra length of the long-stem nozzle lies in the greater distance between the fuel gallery and the nozzle tip. The extended small-diameter section of the nozzle makes it suitable for use in direct injection engines, where the space between the valves is limited, and the machining necessary for fitting the thicker, standard type may weaken the head to the extent that it will crack in service. In certain applications there is a tendency for the needle in standard nozzles to seize due to distortion caused by the heat of combustion. This problem is eliminated when long-stem nozzles are employed, since the close-fitting guide section lies a relatively great distance from the combustion chamber, there being considerable clearance between the needle (below the gallery) and the nozzle body.

Pintle nozzles

Pintle nozzles have a pin that protrudes through a single spray hole, and are, to a large extent, self-cleaning. They are not as inclined to become blocked with carbon deposits as hole-type nozzles. The diameter of the pin or pintle is slightly less than the diameter of the hole through which it passes, and the fuel is sprayed from between the pintle and the nozzle body. Because the pintle lies in the centre of the fuel

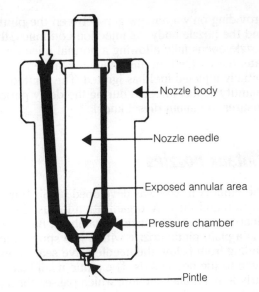

Fig 17.7 A Pintle nozzle

— Nozzle body

— Nozzle needle

— Exposed annular area

— Pressure chamber

— Pintle

spray, the characteristics of the fuel spray are controlled by the shape of this pintle. Pintle nozzles generally give a soft spray, open at low pressure—usually between 85 and 125 atmospheres—and are suitable for use in indirect injection engines.

Plain pintle nozzles are the standard version. The shape of the pintle controls the spray form, and various nozzles are available to provide sprays from a hollow, parallel-sided pencil form to a hollow cone with an angle of 60° or more.

To reduce diesel knock in certain precombustion chamber engines, a special type of pintle nozzle was developed. This delay (or throttling) nozzle allows only a limited quantity of fuel to be injected at first by

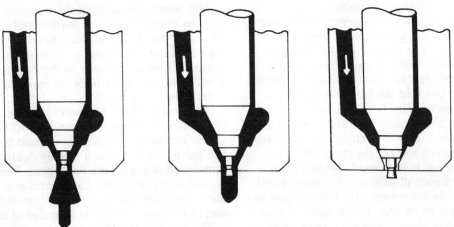

Fig 17.8 Spray characteristics of a delay (or throttling) nozzle

providing only a narrow gap between the pintle and the nozzle body. As injection continues, the nozzle opens fully allowing a normal injection rate, but by the time this stage is reached the initially injected fuel has ignited. The limited quantity of fuel injected during the delay period ensures minimum diesel knock.

Pintaux nozzles

The Pintaux nozzle was developed by Sir Harry Ricardo and the CAV organisation for use in Ricardo Comet combustion chambers. Basically, it is a plain pintle nozzle with a fine spray hole leading from below the needle valve seat at an angle to the nozzle axis. The pintle itself has a fairly long parallel section, which passes through the main nozzle orifice.

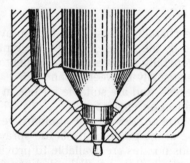

Fig 17.9 Pintaux nozzle

It must be realised that a quantity of fluid can be forced slowly through a small orifice without raising the pressure beyond a certain point. If an attempt is made to force it through at a faster rate the pressure will rise rapidly, but the fuel flow will not increase in proportion.

Now the Pintaux nozzle is designed to spray the fuel through the small hole at cranking speeds, and to spray most of the fuel through the main hole when the engine starts. At low injection speeds the fuel can escape through the small hole so fast that the fuel pressure does not lift the needle valve far off its seat.

This allows the parallel part of the pintle to stay in the main fuel orifice, stopping the passage of fuel. As soon as the engine starts and the speed of injection is increased, the fuel cannot escape from the small hole rapidly enough and the fuel pressure increases. This lifts the needle valve well clear of its seat, the parallel section of the pintle lifts clear of the main hole, and normal injection occurs.

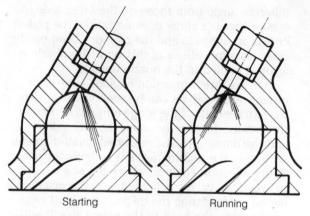

Starting   Running

Fig 17.10 Action of the Pintaux nozzle

It has been definitely established that the area of greatest heat in a Comet combustion chamber on compression stroke lies in the auxiliary chamber at the mouth of the tangential passage. It is towards this point that the fine hole directs the fuel at cranking speed, giving easy starting. However, to ensure good combustion and mixing, the fuel must be directed at right angles to the swirling air and as soon as the engine starts, this is accomplished through the main pintle hole.

Microjector

The Microjector (Fig 17.11) is more compact than the typical injector and has recently been fitted to some high-speed indirect injection automotive diesel engines. The injector screws into the cylinder head in a similar way to the spark plug in a petrol engine.

The nozzle design is of the outward opening type as opposed to the conventional inward opening type. In an outward opening type of injector nozzle, the nozzle valve—similar to a poppet valve as opposed to the conventional needle valve—opens outward towards the combustion chamber.

The design of the Microjector valve is such that, when injecting, a degree of swirl is imparted to the fuel before it emerges around the head of the valve, which forms a closely controlled annular orifice with the valve seat. The resultant high-velocity atomised spray form is a narrow cone, very suitable for efficient burning of the fuel in the precombustion chamber of the engine.

The heart of the Microjector is the nozzle, which is a matched and pre-set assembly that

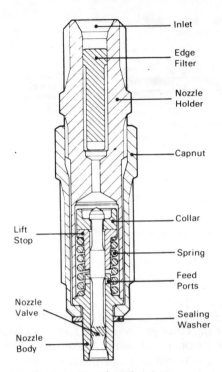

Inlet

Edge
Filter

Nozzle
Holder

Capnut

Collar

Spring

Feed
Ports

Lift
Stop

Nozzle
Valve

Sealing
Washer

Nozzle
Body

Fig 17.11 Section view of a Microjector

cannot be serviced. Therefore, at recommended service intervals the Microjector should be removed for examination and testing. If its functional performance does not meet the set test specifications, the whole Microjector must be replaced. The manufacturer (CAV) recommends that after an operating period of 80 000 kilometres the Microjector be replaced, anyway.

Operation of the Microjector is straightforward. With reference to Fig 17.11, once the unit is filled with fuel, injection occurs as extra fuel is supplied from the injection pump.

Fuel enters the unit at the top, and flows down via the edge-type filter into the spring chamber. It then flows through feed ports into the nozzle body and continues down around the nozzle valve until it reaches the valve head via the swirl helices.

The nozzle valve lifts at a predetermined opening pressure, (approximately 70 atmospheres or 7175 kPa) set by the pre-loaded spring, and continues to move until the lift stop contacts the nozzle body just above the feed ports. As the nozzle valve opens, atomised fuel leaves the tip of the Microjector in a swirling cone-shaped pattern, which continues as fuel is supplied from the injection pump and the nozzle remains open. When the supply of fuel stops

and injection ceases, the nozzle valve is returned to its seat by the spring action and engine combustion pressure.

Some of the advantages claimed for Microjectors are:

- lower nozzle opening pressure
- reduced injection dribble
- no injector back-leakage.

The combination of low opening pressure and reduced nozzle dribble is due to the fact that combustion pressure helps the relatively low spring force to close the needle firmly onto its seat.

The absence of back-leakage is because all the fuel flows in one direction—down along the needle and out into the combustion chamber.

Roosa Master pencil nozzles

While injectors of the Roosa Master pencil nozzle type operate on the same general principle as almost all injectors, they have a number of features that make their construction unique. These features are probably best shown as follows:

- The small body diameter of 9.5 mm allows more room for cylinder valves and large water jackets.
- The thin, light needle valve, is claimed by the manufacturer to increase seat life due to its light impact, and to allow high engine speed due to low reciprocating weight.
- The precision fit of the needle valve is in the main injector body, well away from the heat of combustion.
- Both injector opening pressure and needle lift are adjustable by means of adjusting screws.
- The inlet connection pipe and body length are available in a number of variations to suit different applications.
- The nozzle body, or tip as it is called, cannot be detached from the main body.
- Because of its small size, there is less likelihood of leakage between the injector and cylinder head and a simple clamping device is sufficient.

The operation of the injector follows the standard pattern—fuel flows through the inlet connection and fills the space between the

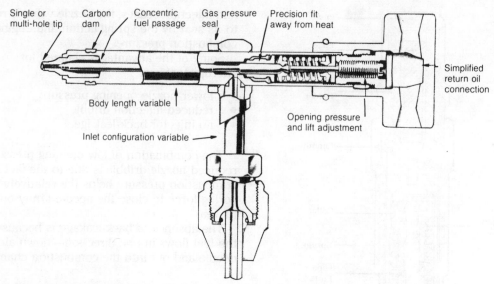

Single or multi-hole tip
Carbon dam
Concentric fuel passage
Gas pressure seal
Precision fit away from heat
Simplified return oil connection
Body length variable
Opening pressure and lift adjustment
Inlet configuration variable

Fig 17.12 The Roosa Master pencil nozzle

precision needle-valve-to-body fit and the seat. As the fuel pressure builds up it acts against the shoulder of the needle valve, lifting it from its seat. Injection then occurs. Back-leakage fuel escapes by passing upwards through the spring chamber and adjustments to the leak-off connection.

Injector service

Injectors will give long periods of service without trouble provided the fuel filters receive regular attention to ensure a clean fuel supply.

Some of the symptoms that indicate inefficient operation are as follows:

- intermittent misfiring on one or more cylinders
- smoky exhaust—black smoke indicating an injector discharging unatomised fuel
- pronounced knock in one or more cylinders
- increased fuel consumption and engine overheating.

The injectors should be removed periodically from the engine (in accordance with the engine manufacturer's instructions) about every 16 000 km for routine testing and examination. Dismantling, cleaning and reconditioning are carried out according to the condition of the injectors.

The injectors are generally tested by means of a lever-operated pump, fitted with a pressure gauge graduated in 'atmospheres'. Clean fuel (or test oil) is pumped through the injector, the pressure, spray pattern, etc indicating the injector condition.

Note   If a test pump is not available, a rough test can be carried out on the engine by removing the injectors then reconnecting them to the fuel supply pipes so that the nozzles are positioned away from the engine. After bleeding the air from the supply pipes, the sprays from the injectors can be examined while the engine is turned by the starter motor or by hand cranking.

Preliminary testing

After removal from the engine, each injector should be checked on an injector tester before any cleaning or dismantling work is carried out to ascertain the general condition of the injector when the injection pressure is applied. This preliminary testing serves as a guide to where any faults lie, or to whether the injector is beyond repair.

There are four standard tests applied to any injector to determine its condition and serviceability. These are:

- pressure setting or pop test
- dry-seat or dribble test

- back-leakage test
- atomisation and spray pattern test.

Pressure setting or pop test

To pressure test an injector, expel the air from the injector tester, fit the injector to the adaptor provided for the purpose, depress the lever very slowly and observe the highest pressure reading obtained before the pressure gauge needle flicks. This is the pressure at which injection begins (although the injection pressure **may** rise to many times this figure when the engine is running), and must correspond with the manufacturers' recommendations.

The injector opening pressure is almost invariably expressed in atmospheres. The term 'atmosphere' means the pressure of the atmosphere, and as this is very nearly 101.3 kPa, a pressure of 100 atmospheres is 10 130 kPa or 10.13 MPa.

If the injector opening pressure is lower than it should be, poor fuel atomisation will result. This will lead to black smoke, hard starting and loss of power.

The effects of high injector opening pressure will vary from combustion chamber to combustion chamber. In some cases the finer atomisation from the high pressure will result in lack of penetration by the fuel spray, while in other combustion chambers a slightly higher pressure may cause the spray to touch some part of the engine, and overheating of this part may follow.

Fig 17.13 Hartridge Poptest—3

If the injector opening pressure is incorrect it should be able to be adjusted by means of the spring tension adjustment—either screw and lock nut or shim. If it cannot be adjusted in this way, this indicates a major fault and the injector will have to be dismantled for further checking.

Note When a new injector spring has been fitted, the injector opening pressure must be set slightly higher than normal to allow for bedding-in, eg CAV injector opening pressures should be increased by $5\pm2\frac{1}{2}$ atmospheres.

Dry-seat or dribble test

Injector manufacturers' recommendations as regards dry-seat testing vary, but the basic test may be performed as follows. Fit the injector to the injector tester, wipe the nozzle dry, and raise the pressure to approximately 10 atmospheres below the injector opening pressure. After a specified time, during which the pressure is maintained, the nozzle tip should remain dry.

CAV injectors are checked by holding a piece of blotting paper against the dry nozzle, raising the pressure to 10 atmospheres below injection pressure, and noting the stain produced on the blotting paper. It should be not more than 12 mm in diameter after 60 seconds, the pressure being maintained constant during the test.

However there is an exception to the above procedure and that is with pencil injectors which are allowed to dribble during a dry seat or dribble test. Depending on the type of injector the dribble rate ranges from 3 to 20 drops within a maximum time frame of 15 seconds.

The dry-seat test shows whether fuel is leaking past the needle valve seat or not, when the needle is held on its seat by the spring. An injector that dribbles may cause one or more of the following faults:

- Engine overheating and/or piston seizure. The fuel entering the combustion chamber will be both excessive and poorly atomised in part. The possible consequences include extra fuel being burnt after combustion should have finished, incomplete combustion leaving unburnt fuel to destroy cylinder lubrication, and oxidation of unburnt fuel in and above the ring area causing gum and lacquer formation.
- Severe engine wear, due to dilution of the lubricating oil by unburnt fuel escaping past the rings.
- Carbon deposits on the nozzle tip.

If the nozzle does not remain dry the injector will have to be dismantled and cleaned, but it is most unlikely that this will rectify the fault and the nozzle will probably have to be reconditioned. It should be noted at this stage, however, that a loose nozzle nut or distorted faces where the nozzle body and nozzle holder mate will allow the leakage of fuel, which will collect on the nozzle tip. This condition is often confused with a leaking needle valve seat.

Back-leakage test

The generally accepted procedure when testing an injector for back-leakage is to pump the injector tester to near opening pressure, then isolate the pump if possible. The pressure shown on the gauge will slowly fall, due to fuel leakage past the needle valve. The rate at which the pressure falls indicates the clearance between the needle valve and the nozzle body.

Most manufacturers give maximum and minimum amounts of time for the pressure drop through a specified range, eg for some CAV injectors, the time for the pressure to fall from 150 to 100 atmospheres should be not less than 15 seconds and not more than 35 seconds.

Note Some manufacturers recommend that the back-leakage test be made at pressures in excess of normal opening pressures. In these cases, the injector pressure setting must be altered before the tests are made.

There are a number of faults that may produce the symptoms of excessive back-leakage. These include:

- dirt between the nozzle body and the nozzle holder, allowing fuel to escape
- a loose nozzle cap nut, again allowing leakage
- loose fuel pipe connections
- high temperatures, causing thinning of the test oil.

The back-leakage test indicates the amount of fuel leaking between the needle valve stem and the nozzle body. Some leakage at this point is necessary for lubrication purposes, but when an excessive amount of fuel is lost at this point the quantity of fuel injected is less than it should be. If the time taken for the specified pressure drop is too long, then the clearance between the needle valve and the nozzle body is not sufficient and the nozzle will probably seize.

This test is very important at the preliminary stage, since it indicates whether or not the nozzle is worth reconditioning. Once the needle-to-body clearance is excessive, the nozzle cannot be repaired.

Atomisation and spray pattern test

The atomisation and spray pattern test is simply a visual check in most instances. The usual procedure is to fit the injector to the injector tester, isolate the pressure gauge and pump the handle at about 60 strokes per minute. The spray should be without distortion and without visible streaks of unvaporised fuel. A multi-hole nozzle should produce sprays of equal length.

It is good practice to check the angle of the sprays when a new multi-hole nozzle has been fitted by allowing the spray to strike a piece of paper and checking this against the manufacturer's recommendations. There have been cases where all the holes have not been drilled right through the nozzle body and cases where one of the holes has been drilled at the wrong angle.

Note Usually, an audible 'grunt' is made when the fuel issues from the nozzle. This is the result of the needle being lifted from its seat, snapping down again and being lifted once more. Thus the needle 'chatters' on its seat. Although this chatter is often considered an indication of a good nozzle, this is not always so. A delay pintle nozzle in top condition may operate without the noise. Again, quite often nozzles with leaking seats may chatter very well, as will nozzles with excessive back-leakage. Thus chattering may only indicate that the needle is free in the body.

If the injector does not atomise the fuel completely, incomplete combustion, causing black smoke, a loss in engine power and poor fuel economy, will result. Increased diesel knock, due to the longer delay period that follows poor atomisation, will also become evident.

Poor atomisation and/or distorted spray pattern usually result from semi-blocked nozzle holes, provided the opening pressure is correct. In such a case the injector should be dismantled, the nozzle cleaned, the unit reassembled, and tested again.

If the injector operates satisfactorily on preliminary testing it may not have to be dismantled. The nozzle should be cleaned with a brass wire brush to remove any carbon deposit. The exterior of the injector should then be cleaned with petrol and dried off with compressed air. After cleaning the exterior, the injector should be checked again for operation on the test pump.

Dismantling a typical injector

If the injector is faulty on preliminary testing or is due for injector service, it must be dismantled for cleaning the interior, examination, and further testing.

For dismantling and reassembling, the injector should be held in a jig fixture instead of a bench vice to avoid damaging the components.

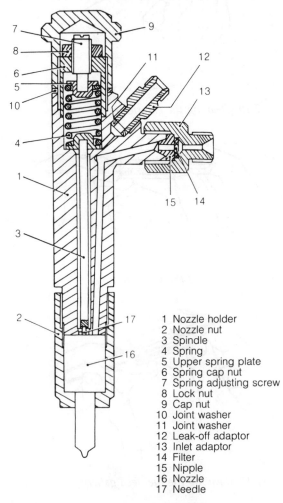

Fig 17.14 Injector with spring adjusting screw

1 Nozzle holder
2 Nozzle nut
3 Spindle
4 Spring
5 Upper spring plate
6 Spring cap nut
7 Spring adjusting screw
8 Lock nut
9 Cap nut
10 Joint washer
11 Joint washer
12 Leak-off adaptor
13 Inlet adaptor
14 Filter
15 Nipple
16 Nozzle
17 Needle

For any type of injector, the usual practice is to begin dismantling at the top to release the spring force before removing the nozzle. To dismantle the injector shown in Fig 17.14:

1 Remove the cap nut (9) and joint washer (10).
2 Slacken the lock nut (8) and adjusting screw (7) to release the spring force.

3 Remove the nozzle nut (2) and nozzle (16) complete with needle (17).
4 Remove the spring cap nut (6), upper spring plate (5), spring (4) and spindle assembly (3).
5 If a filter is fitted in the inlet adaptor (13), remove the inlet adaptor, filter (14) and nipple (15).
6 Place the nozzle holder (1) and the dismantled parts in a container. When several injectors need to be dismantled at the same time, it is important that there is a container for each injector to avoid any interchanging of parts.

Cleaning the dismantled injector

Clean all the parts in petrol and dry off with compressed air. Do not use rags or fluffy cloths. Special tools to facilitate cleaning are available.

A nozzle-cleaning kit consisting of the following tools should be used for removing the carbon deposits:

- a brass wire brush for the exterior of the nozzle
- a brass scraper for the gallery
- one or more scrapers for the cavity in the nozzle tip (the scraper stems are of different diameters, to suit the nozzles in which they are used)
- one or more scrapers for the nozzle seat; (the stems of these scrapers are also made in different diameters to suit the various nozzles in which they are used, and the cutting end of the tool is made to the same angle as the nozzle seat)
- steel pricker wires for the spray holes; (these wires are supplied in different diameters to suit the nozzles in which they are used)
- a pin vice or pin chuck to carry the pricker wires
- one or more brass needle scrapers; (these are designed to accommodate needles of different diameters)
- a number of brass cleaners for pintle nozzles; (these fit into holders, with stems of different diameters to suit various nozzles).

To clean a multi-hole nozzle, proceed as follows:

Remove all traces of carbon from the exterior of the nozzle with the brass wire brush (Fig 17.15).

Clean carbon from the feed hole (between the annular groove and gallery) with a piece of wire or a drill of the correct size (Fig 17.16).

Clean the gallery with the correct scraper (Fig 17.17).

Remove any carbon from the dome in the end of the nozzle with the correct scraper. Select a scraper the stem of which fits neatly in the nozzle body (Fig 17.18).

Note   Pintle nozzles do not have a dome end. Special pintle hole cleaners are supplied for cleaning nozzles of this type.

Clean all carbon from the nozzle seat with the brass scraper. Again ensure that the stem fits neatly in the body, and check that the scraper tip angle coincides with the angle of the seat (Fig 17.19).

Check the size of the spray holes in the manufacturer's manual, and select the appropriate pricker wire. The wire should be fitted in the pin vice so that just sufficient protrudes to clean the full depth of the hole. Avoid having an excessive length of wire protruding, as it is very easy to break part off in the hole. Grinding a chamfer (about 45°) on the end of the pricker wire to form a cutting edge facilitates the cleaning operation.

Hold the nozzle body firmly, and move the pricker wire carefully in and out of the spray holes, while rotating the pin vice slightly between the thumb and forefinger (Fig 17.20)

A hard carbon deposit in the spray holes may be softened by the following method:

1   Dissolve 50 grams of caustic soda in 0.5 litres of water, and add 12.5 grams of detergent.
2   Place the nozzle in the liquid and boil for a minimum period of 1 hour and not more than 1½ hours.
3   The concentration of caustic soda must not exceed 15 per cent and water should be added to replace that lost by evaporation. Should the concentration of caustic soda exceed 15 per cent, the bore and the joint face on the nozzle may be corroded and thus render the nozzle unserviceable.
4   Remove the nozzle and wash it in running water. Remove surplus by draining or by using compressed air.

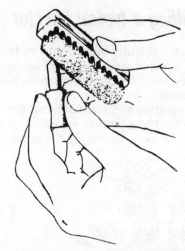

Fig 17.15 Cleaning the exterior of the nozzle

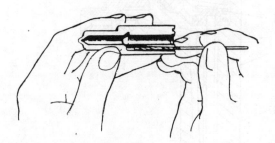

Fig 17.16 Cleaning the fuel feed hole

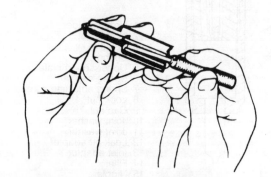

Fig 17.17 Cleaning the fuel feed gallery

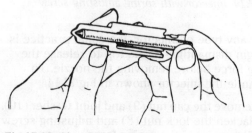

Fig 17.18 Cleaning the dome

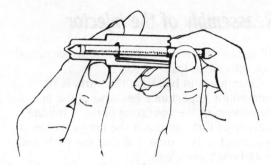

Fig 17.19 Cleaning the nozzle seat

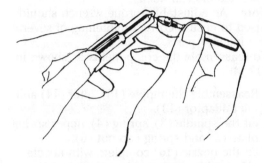

Fig 17.20 Cleaning the spray hole

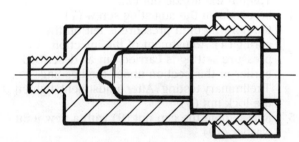

Fig 17.21 Nozzle in reverse flushing adaptor

Flush all carbon particles from the nozzle body by fitting it in a reverse flushing adaptor and fitting the adaptor to the injector test pump. When the pump is operated, the fuel is forced through the spray holes and nozzle body in the reverse direction to normal fuel flow, thus washing foreign particles out through the top of the nozzle body (Fig 17.21).

Clean any carbon from the needle with the brass wire brush and finish with the brass needle scraper (Figs 17.22 and 17.23). Instead of a scraper, CAV recommend the use of a piece of soft wood for polishing the needle.

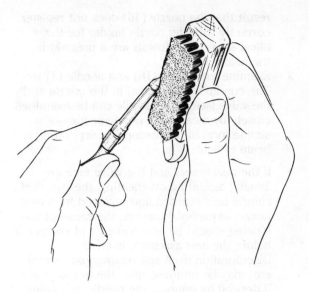

Fig 17.22 Cleaning the needle with a brass wire brush

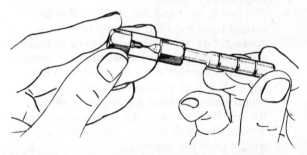

Fig 17.23 Cleaning the needle with a brass scraper

Inspection of the injector components

Referring to Fig 17.14:

1 Examine the condition of the joint faces of the nozzle holder (1) and the nozzle (16). Discoloration or dark stains between the annular groove and the edge of the nozzle holder indicate that the surfaces are distorted and must be 'trued-up'. Should this be necessary, the correct procedure to follow is outlined in this chapter in the section entitled 'Nozzle reconditioning'.

2 Examine for distortion of the dowels (if fitted) in the nozzle holder. The dowels may be distorted (partially sheared) by excessive tightening of the nozzle nut (2) with the

result that the nozzle (16) does not register correctly with the nozzle holder for the offset spray. The dowels are a press fit in the nozzle holder.

3 Examine the nozzle (16) and needle (17). The condition of the seat in the nozzle and the valve face of the needle can be examined closely by means of a nozzle-scope—a simple form of microscope incorporating a beam of light:

- If the nozzle seat and the valve face are 'blued', indicating overheating, the two parts should be discarded and replaced by a new nozzle assembly. However, the cause of the blueing should be ascertained and corrected before the new assembly is fitted. Discoloration from any other cause (stains etc) may be removed from the needle valve if desired by gripping the needle in a lapping chuck and rotating it, while firmly gripping the discoloured area of the needle between two pieces of pine.

- The nozzle seat and the needle valve face should have fine matt finished lines of contact. Pits, scratches and scores in these areas, or wide contact areas are sure signs that the assembly must be reconditioned before it will give satisfactory service.

- When clean (without any traces of oil on the needle stem or in the nozzle bore), the needle should drop freely under its own weight onto the nozzle seat.

- As another check for binding, the needle should be pressed by hand against the nozzle seat (nozzle in normal vertical position). The needle should fall freely when the nozzle is inverted and the hand is removed.

4 Examine the spindle (3). If the recessed end of the spindle is flattened, a new spindle complete with spring cup must be fitted.
5 Examine the inlet adaptor filter (14). Renew the filter if damaged.
6 Examine the spring (4). It may be corroded due to condensation of moisture on the spring as a result of temperature changes in service. It should be cleaned.
7 Examine the nozzle nut (2). Carbon deposits can lodge inside the nut, particularly in the hole for the nozzle. Remove the carbon with a suitable tool.

Reassembly of the injector

All the injectors parts should be reassembled wet with clean fuel.

Cleanliness in handling the parts is of paramount importance because of the fine clearances of the operating parts. A minute particle of dirt can upset the performance of any injector. It is essential that the tools and work bench are clean.

If the injector is being reassembled in anything but dust-free conditions, it should be done under the surface of clean fuel.

Note An adjustable torque wrench should be used to ensure correct tightening of several of the injector parts.

To reassemble the components as shown in Fig 17.14:

1 Reassemble the nipple (15), filter (14) and inlet adaptor (13).
2 Fit the spindle (3), spring (4), upper spring plate (5) and spring cap nut (6).
3 Fit the nozzle (16) complete with needle (17) to the nozzle holder (1) and ensure that the dowels (if fitted) locate the nozzle. Tighten the nozzle nut (2).
4 Screw down the adjusting screw (7) temporarily until pressure is felt on the spring (4). Adjustment of the injection pressure setting is carried out as described earlier in this section under the heading 'Preliminary testing'. After adjustment, tighten the lock nut (8).
5 Reassemble the cap nut (9) with a new joint washer (10).

After reassembly, the four basic injector tests should again be applied to the injector. The pressure setting or pop pressure, however, must be correct, since it is set during reassembly.

If the injector dry-seat test does not come up to specifications, and all other possible points of leakage have been eliminated, the nozzle will have to be reconditioned.

If the original nozzle is being tested, the rate of back-leakage should be the same as in the preliminary test. If excessive back-leakage during preliminary testing indicated the need for a new nozzle, then the rate of back-leakage of the new nozzle should be correct. In either case, then, the rate of back-leakage will probably be correct but should be checked.

Nozzle reconditioning

*S*hould inspection of the injector components indicate the need for reconditioning the nozzle, the following procedures should be followed.

To true nozzle body and nozzle holder joint faces

Lapping the joint face of the nozzle or nozzle holder is usually done on a power-driven machine or by hand on a cast-iron surface plate.

In the latter method, the lapping compound should be smeared sparingly on the surface plate. The nozzle or nozzle holder (with dowels removed) should be held evenly and firmly between finger and thumb of both hands, and moved lightly in a figure-8 motion on the plate.

Note   The operation of lapping a nozzle joint face should be carried out before lapping the nozzle seat. The impact of the needle shoulder on the joint face of a nozzle holder is usually indicated by a mark around the centre hole. Removal of this mark by lapping is not necessary.

To recondition valve face and seat

The two parts of an injector most affected by the stress of operation are the nozzle and the needle valve. The continual hammering effect due to the opening and closing of the needle

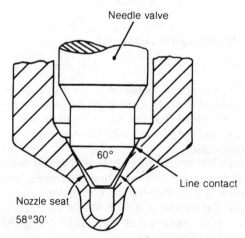

Fig 17.24 Typical needle and seat details

valve ultimately affects the needle seat and the nozzle seat.

The reconditioning of an injector nozzle involves the restoration of the nozzle seat surface and the face of the needle valve. The needle face is almost invariably refaced by a grinding process: the nozzle seat may be ground using specialised equipment, or resurfaced by a lapping process.

It is important to realise that the angle of the nozzle seat is different from the angle of the needle face and, when reconditioning is carried out, it is necessary to maintain these angles. The purpose of having this interference angle or differential angle is to maintain a line contact between the two. Line contact ensures three desirable factors:

- A sharp cut-off of fuel, since there can be no fuel trapped between the valve faces as the needle snaps closed.
- Extremely high pressure of contact, because the seat is narrow and therefore has a very small area. Thus, in terms of the force applied and the area of contact, a small seat area ensures a higher contact pressure than a large seat area.
- Little likelihood of any minute particles being trapped between the needle and its seat because of the very narrow contact.

After long service, leakage occurs when the needle seat and the nozzle seat become pitted and scored. The line contact may become wider or the seat may become grooved where the needle seats against it. If the wear and pitting are excessive, reconditioning should not be attempted.

The nozzle seat and the needle face are reconditioned separately. The nozzle seat and the needle face must never be lapped together to rectify seat leakage because such an operation would create the very conditions (wide contact and seating impression) that have to be corrected.

The needle is made of a harder steel than that of the nozzle, while the nozzle is case-hardened to a depth of about 0.4 mm. The hardened surface of the nozzle seat can be removed if reconditioning is carried out several times. It may be assumed, broadly, that about 0.05–0.075 mm of metal is removed each time a nozzle seat is rectified. Under normal conditions of wear, therefore, the nozzle should be reconditioned only five times at the most, after which it should be scrapped.

Lapping nozzle seats

To recondition nozzle seats by the lapping process, two items of equipment are required—a lapping machine and the 'lap'. A lap may be likened to a long nozzle needle with a hardened body where it fits into the nozzle body. The tip of the lap (equivalent to the face of the needle) may be made of one of a number of materials, cast iron being the most popular. When a small quantity of lapping compound—an extremely fine grinding paste—is applied to the tip, it is retained in the pores of the cast iron. The lap becomes, in effect, a fine grinding tool. Usually, two grades of lapping compound are used, a coarse one for 'roughing' and a fine one for 'finishing'.

Laps are made to suit the different types of nozzle with different bore diameters, and for each type of nozzle, laps are made with stem diameter increases of 0.01 mm. The correct lap should be selected to recondition each nozzle seat, and should be a very fine clearance fit in the nozzle bore. Once the lap has been selected, it should be fitted into a nozzle grinding and lapping machine and the tip ground carefully to the correct angle for the nozzle body seat. If there is any doubt as to the trueness of the grinding wheel, this should first be dressed.

In machines of this type the lap is turned, while it is being ground, by a rubber belt or wheel (depending on the make of machine). The lap body fits against a 'V'-block and is held and turned in this 'V' by the drive mechanism. Thus a centreless grinding process is used, and this ensures concentricity.

Very fine cuts should be taken when grinding the lap, which should be moved back and forth across the full face of the wheel.

As soon as a true, fine finish on the lap end is obtained, the nozzle seat may be lapped. A very good lapping procedure (by courtesy of CAV is outlined below:

1   Mount the lap in the collet of the machine lapping head. Smear the stem of the lap with tallow.
2   Place a small bead of lapping compound such as chromic oxide (about 1 or 2 mm diameter) on the lap face. Too much lapping compound should not be applied.

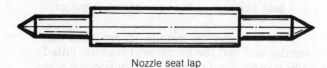

Nozzle seat lap

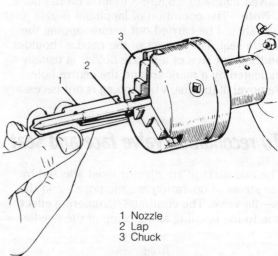

1 Nozzle
2 Lap
3 Chuck

Fig 17.26 Lapping the nozzle seat

3   Hold the nozzle between finger and thumb and slide it on the revolving lap. Care must be exercised that the nozzle bore is not damaged by contact with the lapping compound.
4   Move the nozzle quickly to and fro, at the some time partially rotating the nozzle back and forth, so that the lap comes into contact with the nozzle seat. The nozzle seat should not be in contact with the lap for more than 5 seconds at a time.

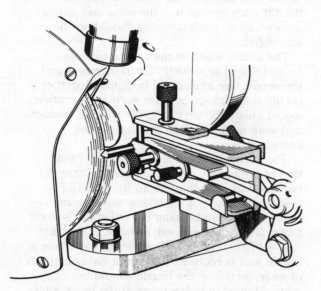

Fig 17.25 Grinding the nozzle seat lap

5 After about 30 seconds, remove the nozzle and wipe the lap. Apply lapping compound and repeat the procedure.

6 After one or two lappings, depending on the condition of the nozzle seat, wash out the nozzle with cleaning fluid by reverse flushing. Then examine the seat in a nozzlescope. The correct finish should appear as a clean unbroken surface.

Machine grinding nozzle seats

For faster and more precise reconditioning of nozzle seats than can be obtained by hand lapping, automatic nozzle re-seating machines are now being used. One such machine is the Hartridge Injectomatic shown in Fig 17.27.

A grinding tool similiar to a nozzle needle but with a grinding stone attached to the tip is used to grind the nozzle seat. After a correctly fitting grinding tool has been selected, it is placed in a needle grinding machine and faced to the

correct seat angle. The grinding tip of the tool is then dipped in cutting oil and the tool placed inside the nozzle.

The nozzle is mounted on the adaption plate of the reseating machine and the machine is set into operation for a period of from 30 seconds to 2 minutes, depending on the seat condition and the type of grinding stone used. Experience is the best guide to selection of the grinding time.

Valve reconditioning

The needle valve face itself may be reconditioned in one of two ways. The first entails fitting the needle into the lapping and grinding machine and carefully grinding just enough metal from the face to eliminate pits and marks. Of course, since the needle face angle is invariably different from the nozzle body seat angle, the machine will have to be reset to grind the needle correctly. Once the face has been ground, it should be

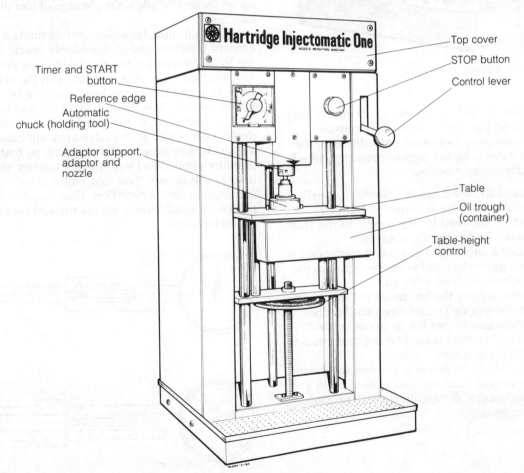

Fig 17.27 'Hartridge Injectomatic One' nozzle reseating machine

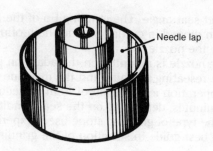

Fig 17.28 Needle lap

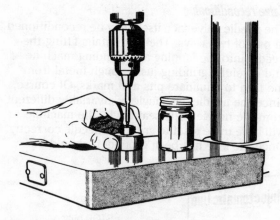

Fig 17.29 Lapping the needle

examined through a microscope to ensure that any pitting has been completely removed.

An alternative method of reconditioning the needle valve is by the lapping sequence, which is clearly outlined below.

1 Mount the needle in the chuck of a vertical spindle machine as shown in Fig. 17.29.
   Note The need for the needle to run true cannot be too greatly stressed.
2 Apply a small bead of lapping compound to the lapping face of the needle lap. Run the machine at about 460 rpm.
3 While holding the lap, move the revolving needle quickly up and down so that the needle face comes lightly in contact with the lap; contact should be not more than 5 seconds at a time.
4 After about 30 seconds, remove and clean the needle. Examine the needle face in a nozzlescope. Repeat the lapping procedure if necessary.

Needle lift

Once the needle valve face and the nozzle body seat have been reconditioned, no further work should be necessary—if the job is carefully done, a perfect seal will result. The two components should be thoroughly cleaned and assembled while wet with fuel or test oil. However, the removal of metal from the seat area lets the needle move deeper into the body, while the removal of metal from the joint face of the nozzle body results in the needle sitting relatively higher in the body. Thus both these reconditioning procedures affect the **needle lift**, which is described in the next paragraph.

The amount the needle can lift off its seat during injection is limited by the amount the needle can move before the shoulder on its upper end strikes the joint face of the main injector body. Thus needle lift may be defined as the maximum distance that the needle may lift off its seat to allow the passage of fuel during injection.

Needle lift must lie within certain limits if efficient injection and/or reasonable nozzle life are to be obtained. If the needle lift is below specifications, the clearance between the needle valve face and the nozzle body seat will be insufficient to allow the full charge of fuel to pass, unrestricted, to the injector hole or holes. Restriction of fuel at the valve area will cause a considerable drop in fuel pressure, so that the pressure applied to force the fuel through the hole(s) will be less than that required for atomisation and penetration. Thus insufficient needle lift considerably affects the fuel spray characteristics.

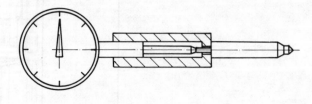

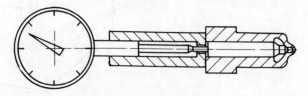

Fig 17.30 Checking needle lift

If, on the other hand, the needle lift is greater than is required for efficient injection (the area through the open valve seat should be approximately twice the orifice area), the needle valve will travel through a greater distance than is necessary. When the valve closes under the influence of the injector spring, any extra distance of travel results in an increase in the impact of the needle on the seat. Thus excessive needle lift greatly reduces the life of a nozzle assembly, due to the hammering of the needle on the seat.

The needle lift may be found very easily by using a dial gauge and adaptor.

First, take the needle and place the spigot in the adaptor hole. Hold the needle so that the shoulder is firmly and squarely against the face of the adaptor. Note the dial gauge reading. Then refit the needle to the nozzle body, making sure it seats properly. Hold the nozzle assembly to the adaptor, so that the joint face of the nozzle body is firmly and squarely against the adaptor face. Note the dial gauge reading.

Now the difference between the two readings is the amount of needle lift from when it is on its seat to when its shoulder is against the joint face of the main injector body. Thus it is a measurement of the amount the needle can lift from its seat—the needle lift.

During service, no problems will occur with insufficient needle lift, but on many occasions excesive needle lift will develop after a nozzle has been reconditioned a number of times. When this occurs, it can be rectified by the removal of metal from the joint face of the nozzle body. Some lapping and grinding machines are equipped with a special chuck that allows the joint face to be accurately ground. If this is done, the face should be carefully lapped on the lapping plate to obtain the required finish. Care should be taken to remove no more metal than is required, and the needle lift should be carefully rechecked after the face has been lapped.

Reconditioning pintle nozzles

Reconditioning pintle nozzles is carried out in the same way as for other types. However, the following points should be noted before attempting to reclaim any type of pintle nozzle:

- The seat is lapped in the normal way, but the removal of excessive metal from the

seat not only affects the needle lift, but also the position of the pintle during injection, relative to the pintle hole. This can result in an entirely different spray pattern.
- The removal of excessive metal from the needle face also affects the position of the pintle relative to the pintle hole, and the spray can again be affected.
- The clearance between the pintle and the nozzle hole is extremely critical, and care must be taken not to damage the hole in any way during reconditioning.

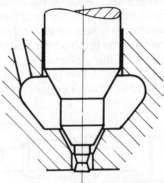

Good nozzle — open position

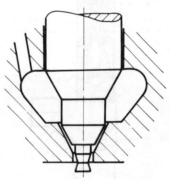

Excessively ground nozzle seat

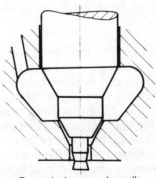

Excessively ground needle

Fig 17.31 Faults when reconditioning pintle nozzles

- Care must be taken not to damage the pintle during needle valve face repairs. If the face is reground, the wheel will probably have to be specially faced to accommodate the pintle. If the face is lapped, the correct laps—with a hole to accommodate the pintle—must be employed.

Testing Pintaux nozzles

In order to check the required injection rate and delivery of this type of nozzle, it is necessary to fit the injector to a special test rig, which is, in turn, connected to the injector tester. The rig consists of an additional nozzle holder fitted with a special nozzle. A cap nut to suit A size (12 mm) or B size (14 mm) connections is available.

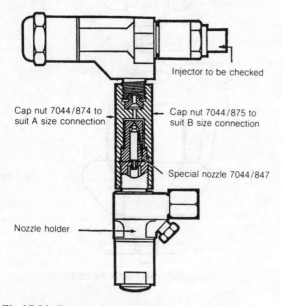

Injector to be checked

Cap nut 7044/874 to suit A size connection

Cap nut 7044/875 to suit B size connection

Special nozzle 7044/847

Nozzle holder

Fig 17.32 Test rig for checking Pintaux nozzles

The cap nut enables the injector, complete with the Pintaux nozzle under test, to be screwed onto the end as shown in Fig 17.32. By setting the opening pressure of the special nozzle to a value where its closing pressure is higher than the opening pressure of the Pintaux nozzle, a sufficiently high rate of injection can be obtained on hand test to determine the quality and form of atomisation of both auxiliary and main sprays.

To test a Pintaux nozzle, the following procedure, (by courtesy of CAV), should be followed:

1   **Seat tightness**—Assemble the Pintaux nozzle to a suitable holder, ie a type with which the particular nozzle is used, and set to an opening pressure of 100 atmospheres.
    Connect the injector to the injector tester and atomise several times to expel air from the system. Wipe the face of the nozzle dry, pump up the pressure to 90 atmospheres, and hold for 10 seconds. Wipe a finger across the face of the nozzle and inspect for wetness. Reject the nozzle if wet.

2   **Auxiliary spray**—Fit the test rig to the nozzle setting outfit and set the opening pressure to 220 atmospheres. Screw the injector to be tested onto the end of the special cap nut. Atomise several times to expel air from the system and then observe the atomisation from the auxiliary hole. This should be well formed and free from splits or distortions. A slight centre core may be disregarded.
    The hand lever of the nozzle setting outfit should be operated at a minimum frequency of 60 strokes per minute during this test.

3   **Main spray**—Operate the hand lever at 140 strokes per minute and observe the main spray. This should be well atomised and free from large splits or distortions. A slight centre core may be disregarded.

Servicing pencil nozzles

Because of their compact construction, pencil nozzles are not able to stand rough treatment and two spanners must always be used when disconnecting and adjusting these injectors. To undo the inlet pipe, for example, the two spanners should be used in one hand—one spanner on the union nut, the other on the hexagon provided on the inlet connection. This will ensure that the injector is not distorted.

These injectors should be subjected to the four standard tests during service.

The injector opening pressure may be adjusted without dismantling the injector, but a leaking seat or excessive back-leakage indicates that the injector requires dismantling.

Dismantling is obviously done from the top, and a special injector service kit is available to facilitate service work. If the seat leaks, it may be cleaned with the aid of an old needle and a little metal polish. However, no grinding or lapping of the seat or needle valve face is recommended. Excessive back-leakage may be rectified by fitting an oversize needle valve.

Once reassembled, the injector opening pressure may be adjusted by moving the outer threaded adjustment.

The needle lift is set by screwing the inner adjustment inwards until it is felt to contact the needle valve. The correct clearance is obtained by unscrewing the adjustment a specified amount from the point where the contact is felt.

Testing and correcting faulty nozzles

The chart shown in Table 17.1 by courtesy of CAV, lists the common nozzle faults and gives an indication of the necessary service procedure.

Table 17.1 Service procedure for common nozzle faults

Fault	Possible cause	Remedy
Nozzle does not buzz when injecting	1 Needle valve too tight, binding, or valve seating leaky	Clean nozzle. Examine cap nut, if necessary replace nozzle and needle valve
	2 Nozzle cap nut distorted	(**Note** Delay-type nozzles and poppet nozzles do not usually buzz at slow plunger velocities given by testing outfit.)
Excessive leak-off	1 Needle valve slack	Replace nozzle and needle valve
	2 Foreign matter present between pressure faces of nozzle and nozzle holder	Clean
	3 Nozzle cap nut not tight	Tighten cap nut after inspecting joint faces
Nozzle blueing	Faulty installing, tightening or cooling	Replace nozzle and needle valve Check cooling system
Nozzle opening pressure too high or too low	1 Compression screw shifted	Adjust for prescribed pressure
	2 Needle valve seized up, corroded	Replace nozzle and needle valve
	3 Needle valve seized up, dirty, sticky	Clean nozzle
	4 Nozzle openings clogged with dirt or carbon	Clean nozzle
Nozzle pressure too low	Nozzle spring broken	Replace spring and readjust pressure
Nozzle drip	Nozzle leak due to carbon deposit; sticking needle valve	Clean nozzle. If this does not clear fault, replace nozzle and needle valve
Form of spray distorted	1 Excessive carbon deposit on tip of needle valve	Clean nozzle
	2 Injection holes partially blocked	Clean nozzle
	3 Nozzle needle valve damaged (pintle type only)	Replace nozzle and pintle valve

Special injectors

By far the largest percentage of injectors in use today follow the typical pattern already discussed. However, some engine manufacturers, including Leyland, Gardner and Caterpillar, have produced, or are still producing, special injectors of their own design. In addition, Stanadyne/ Hartford division—manufacturers of Roosa Master Fuel Injection Equipment—manufacture a 'pencil' nozzle, while CAV have developed a 'Microjector' (both of these were discussed earlier in this chapter), and Lucas Bryce manufacture a unit injector.

Caterpillar injectors

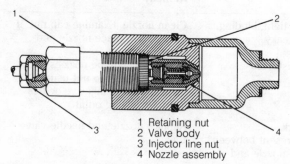

1 Retaining nut
2 Valve body
3 Injector line nut
4 Nozzle assembly

Fig 17.33 Fuel injection valve cross-section

Owing to their precombustion chamber design, Caterpillar engines do not need the same degree of fuel atomisation as do most other engines. Because of this, Caterpillar injectors operate at lower than normal pressures and have one large-diameter spray hole.

The construction of modern Caterpillar injectors is completely different from all others. The injector assembly consists of three parts—a retaining nut, a valve body and a capsule-type nozzle assembly.

The nozzle assembly contains the valve and seat, the spring and a fine gauze filter. The valve body is, in effect, an adaptor connecting the fuel line to the nozzle assembly, which screws onto it. The nut secures the valve body and the nozzle assembly in the precombustion chamber, the tapered face of the valve seating against a corresponding taper in the precombustion chamber housing.

Two drillings in the valve body carry the fuel from the injector pipe to the nozzle assembly. An axial drilling leads from the injector pipe union to a diametrical drilling, which emerges between the thread and the shoulder against which the nozzle abuts and seals. The fuel flows around the thread into the nozzle assembly. When the fuel pressure becomes sufficient, it forces the valve downwards from its seat, compressing the spring and allowing fuel to pass through the single spray hole. As soon as the pressure drops, the needle is snapped back on its seat by the spring.

Servicing Caterpillar injectors

As is pointed out in Caterpillar manuals, the most likely causes for faulty injection performance are:

- air in the fuel system
- low fuel supply
- water in the fuel
- clogged fuel filter
- insufficient fuel transfer pump pressure.

If these conditions are checked and corrected as necessary and the engine continues to perform poorly, the fuel injection equipment should be checked.

Black exhaust smoke, severe knocking and misfiring may indicate either a faulty nozzle assembly or a faulty fuel pump. If one of the fuel lines becomes hot, the fault almost certainly lines in the nozzle assembly, the valve of which has stuck open. However, other faults are not as easy to locate, and the cylinder, the injection equipment of which is defective, should first be located. This is readily done by running the engine at a speed that makes the defect most pronounced, and loosening, then tightening, each fuel line nut at the fuel pump, thereby cutting out injection to each cylinder in turn. If one cylinder is found where loosening the line makes no difference to the engine's operation or causes the black exhaust smoke to cease, the fault must lie in either that nozzle or that fuel pump.

Once the cylinder has been located, the nozzle may be removed for checking by unscrewing the fuel line and retaining nut, withdrawing the valve body and nozzle assembly from the precombustion chamber, and then unscrewing the nozzle assembly from the valve body. The nozzle orifice should be cleaned with a drill of the correct size held in a pin vice or chuck, extreme care being taken to ensure that no metal is reamed from the sides of the orifice.

Should there be any doubt as to the condition of the nozzle assembly, it should be returned to the manufacturer's agent for checking. However, an indication of the nozzle's condition may be gained by fitting it to a spare valve body and assembling this to an old precombustion chamber that has had the combustion section machined off. This assembly can then be fitted to a standard injector tester and the nozzle's performance noted. A stuck valve or leaking seat can be identified in this way as well as a fault in the fuel spray, which should be symmetrical about the injector axis. A finely atomised spray should not be expected.

When fitting a new nozzle assembly or refitting the original to the valve body, it should be screwed on finger tight only. The condition of the nozzle seat in the precombustion chamber should be checked, and the retaining nut tightened to the correct torque. If the nut is not tightened sufficiently, it will allow the nozzle to leak and in some instances may cause the nozzle case to bulge or split. Excessive tightness will also cause damage to the nozzle.

Gardner injectors

One of the oldest manufacturers of diesel engines in England, L. Gardner and Sons Ltd, manufacture their own injectors to a unique design. Termed 'fuel sprayers' by the Gardner company, these injectors all follow the same basic design although the Gardner range of engines is comprehensive. However, not all Gardner injectors are interchangeable, and no attempt should be made to swap injectors from engines of different models.

Gardner injectors vary from standard injectors in the following areas:

- nozzle assembly
- opening pressure adjustment
- needle lift adjustment.

The injector needle is located by its fit in the main injector body and protrudes into the nozzle body, which is held in place by a cap nut. However, the valve and seat arrangement is similar to that of a standard injector, the seat angle being more acute than the valve face angle.

The injector opening pressure is controlled by the force exerted by the injector spring, and is adjustable, within certain prescribed limits, by fitting shims between the top of the injector spring and the end plug shoulder.

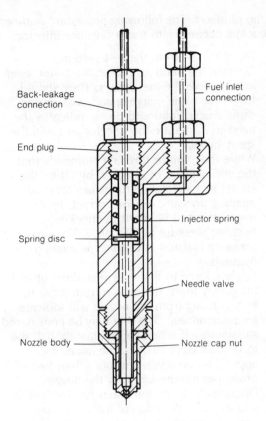

Fig 17.34 Section view of the Gardner injector

The needle lift is controlled by the distance between the spring disc and the spigot extension of the end plug. Adjustment is obtained by machining metal from either the spigot end or the face of the end plug.

Servicing Gardner injectors

The manufacturer recommends that Gardner injectors be regularly tested in position on the engine. This can readily be accomplished by operating the hand priming levers fitted to each pumping element on all Gardner injection pumps. As each priming lever is operated (with the engine stopped), the corresponding injector needle should audibly vibrate. Failure to do so indicates a faulty injector, probably an imperfect valve seat or a binding needle valve.

Should the preceding test or the hours of operation indicate that the injectors must be tested further, they should be removed from the engine and fitted either to one of the injector pipes that has been swung away from the engine, or to a standard injector tester that has a plunger of the same diameter as the injection pump plungers. The following procedure outlines the steps necessary to test a Gardner injector:

1 Bleed the air from the test system.
2 Operate the hand primer or the tester lever and observe the fuel sprays. They should travel the same distance and possess the same shape. Failure to do so indicates the need to dismantle the injector to clean the spray holes.
3 While the manufacturer recommends that the spring force be checked by fitting the spring in a special jig to ensure that the opening pressure will be correct, there would seem to be merit in checking the opening pressure at this stage. The opening pressure is listed in Gardner workshop manuals.
4 Apply a force to the lever just short of that necessary to lift the needle from its seat. Fuel running from the nozzle will indicate an unsound seat. The seat may be considered satisfactory if, when approximately half the force necessary to open the injector is applied to the lever, not more than two drops per minute fall from the nozzle.
5 Operate the lever and listen for the needle vibration. While different injectors may produce differing noises, all injectors should produce an audible noise that may broadly be described as a squeak. Failure to vibrate indicates the need for dismantling the injector for further service. Possible causes of this fault include a leaking valve seat, a worn and consequently wide valve seat, misalignment of the valve and nozzle causing friction, and in rare cases a leak past the large diameter of the valve.
6 Apply a force to the operating lever just short of that which will lift the needle from its seat. Provided that the pumping element is in good order and the needle valve and seat are satisfactory, the lever should have a 'solid' feel when operated. Failure to achieve this condition is an indication that a fair quantity of fuel is escaping past the stem of the needle, and the quantity of fuel escaping from the leak-off connection should be noted. A slight leak is desirable, and a fairly considerable leak may be allowed since, as the manufacturer points out, it will have little effect on the engine performance. Should the feel of the operating lever and the quantity of fuel escaping indicate excessive back-leakage, a new needle valve

should be fitted and it is recommended that the injector be returned to the manufacturer for this service.

If the injector has to be dismantled, the end plug or valve stop should be removed first, followed by the spring and spring disc. The cap nut may then be unscrewed and the nozzle body and needle valve removed. In some instances, the nozzle body may be stuck firmly in the cap nut due to carbon, and may be removed without damage by driving it from the nut using a special Gardner drift.

The bore of the nozzle body may require cleaning, and this may be accomplished by means of a piece of cane or wood shaped to fit neatly inside the bore and against the seat, together with a small quantity of metal polish. The metal polish should be smeared on the wooden tool, which may then be rotated until the inside of the nozzle body is clean.

The manufacturer suggests that a leaking seat may be corrected by lightly lapping the needle and seat together. This may be done by mounting the main injector body horizontally in a vice or jig. The upper end of the needle valve is threaded and a lapping tool or threaded rod may be screwed into this. The needle valve should be fitted to the body while on the end of the lapping tool, and a minute quantity of very fine lapping paste added to the tip as it protrudes from the body. The nozzle should then be held in one hand against the end of the injector body with the needle valve lapping tool gripped in the other hand. While applying a very light load to keep the needle valve in contact with its seat, the needle and body should be rotated in opposite directions. This process should be kept to a minimum, since it widens the needle valve seat and a tight seat may become impossible to obtain.

If the line contact is lost, then lapping and grinding of the nozzle components becomes necessary. This work is not recommended and injectors should be forwarded to the Gardner works for this service.

After thoroughly cleaning the assembly and servicing the nozzle, the unit may be reassembled in the following order:

1 Grip the injector horizontally in a vice or jig.
2 Fit the needle valve, with the lapping tool attached, to the injector body.
3 Loosely fit the nozzle and cap nut.
4 Slowly tighten the nozzle cap nut with one

hand while gently, but firmly, tapping the needle valve against the seat with the other. This action should ensure alignment of the nozzle body with the needle valve. The cap nut should be tightened with a spanner and the freedom of the needle rechecked. Any tendency for the needle to stick is an indication of misalignment, and indicates that the nozzle must be released and aligned again.

5 Unscrew the lapping tool.
6 Refit the spring disc.
7 Using a micrometer depth gauge, measure the distance from the top of the injector body to the spring disc. Then measure the distance from the end plug face (where it seats against the injector body) to the end of the extension that limits the needle lift. The second measurement should be less than the first by a distance equal to the required needle lift. The measurement may be corrected by machining metal from either the face of the end plug or from the end plug spigot extension.
8 Refit the spring. If the spring force has been checked as recommended by Gardner, then the required shims will be known and should be fitted. However, if the pressure is to be set by checking it on an injector tester, then the spring only should be fitted.
9 Refit the end plug.
10 If the pressure is to be set, check the opening pressure on the gauge. If it is much more than 10 atmospheres below the recommended opening pressure, then shimming should not be attempted but a new spring should be fitted. However, if the opening pressure indicates that shimming may be carried out, the necessary Gardner shims should be fitted **between the top of the spring and the end plug shoulder**.
11 Once the opening pressure is correct, the injector should be finally tested again before being replaced in the engine.

It is well to note that Gardner injectors do not seat on a copper gasket, but directly against a tapered seat in the cylinder head. To ensure that this seat is kept in good condition, two tools are available—a cleaning tool for reaming out carbon, and a correcting cutter for trueing the seat. When installing the injector, care must be taken not to overtighten the securing bolts. A small 'T'-spanner is provided by the

manufacturer, and, using this, the nuts should be done up finger tight only. Alternatively a tension wrench should be used.

Unit injectors

As an alternative to the conventional injector and separate fuel injection pump, unit injectors have grown in popularity in recent times due to their compact design and improved injection characteristics. The unit injector combines an individual fuel injection pump and fuel injector in a single compact unit that is fitted directly into the engine cylinder head, where it is operated from a lobe on the engine camshaft in a similar way to the engine valves. Apart from the widely used Detroit Diesel Allison system, unit injectors are now being fitted to some Caterpillar engines, while Lucas Bryce manufacture unit injectors for several engine manufacturers.

The operation of a basic unit injector is described in Chapter 21 under Detroit Diesel Allison fuel systems. Because the operating principles are the same with all unit injectors, only the manufacturers' design and installation variations will be discussed.

Unit injectors are currently fitted to a number of Caterpillar twelve- and sixteen-cylinder engines and will soon be fitted to their automotive truck engines. GEC diesel have also fitted unit injectors to their six-and eight-cylinder Dorman engines.

The unit injector used by Caterpillar (in their twelve- and sixteen-cylinder engines) is comparable in design and operation to the unit injector used in the Detroit Diesel Allison 149 series engines. The Dorman diesel uses a unit injector developed by Lucas Bryce and, although different in construction, its basic design and operation is the same as the Detroit Diesel Allison unit.

Unit injection offers a number of advantages over the conventional separate pump and injector system. The injector itself contains a much smaller volume of fuel trapped under high pressure than conventional systems, thereby allowing higher mean effective injection pressures to be employed. This increases the injection rate and atomisation of the fuel, and is claimed to give improved performance and reduce fuel consumption.

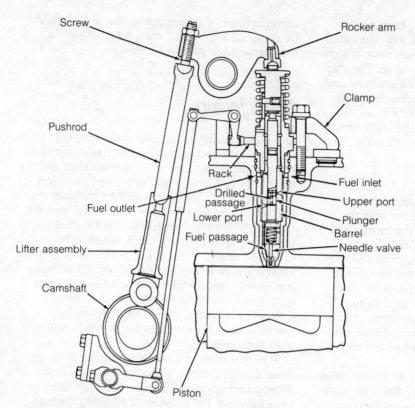

Fig 17.35 Unit injector installation in a Caterpillar diesel engine

Another advantage of the unit injector is the
elimination of the high pressure flow losses
through the injector lines and the problems
associated with fuel compressibility that occur
in separate pump and injector systems
(Chapter 14).

Servicing of unit injectors

Due to installation variations between various
unit injectors, the timing, calibration, and
maximum fuel adjustments vary from one
engine installation to another. Therefore, prior
to carrying out unit injector adjustments, the
workshop manual should be consulted.

18
Jerk-type injection pumps

T he solid fuel injection system, as mentioned in Chapter 17, consists of two (usually separate) components— the fuel injector and the fuel injection pump. While these two units are united to form a single camshaft-operated component in some specific cases, the common practice is to use a separate fuel injection pump, which is connected to the injectors by steel injector pipes.

The function of the injector in such a system was clearly shown in Chapter 17; the fuel injection pump must perform three separate functions, which involve very high pressures and accurate control. These functions are as follows:

- **To raise the pressure of the fuel high enough to be injected into the engine combustion chamber before and during combustion**. Fuel is supplied, either by gravity or by feed pump, to the fuel injection pump at a pressure that usually lies between 20 and 105 kPa. The fuel injection pump must raise the fuel pressure to between 12 and 70 MPa, depending on the injector type, engine speed etc. This very high pressure is required for two reasons—to force the fuel into the combustion chamber of the engine against the gas pressure and to thoroughly atomise the fuel as it leaves the injector nozzle and passes into the combustion chamber.

- **To accurately meter the quantity of fuel passing to the injector over the entire fuel requirement range**. The quantity of fuel required per cylinder for each firing is very small—a droplet of fuel about the size of a matchhead is sufficient for an average sized engine operating under normal load conditions. The fuel pump must measure out the **exact** quantity of fuel for each engine cylinder and pass it on to the cylinder via the injector pipe and injector to be burnt. It is vital that the cylinders each receive the same amount of fuel per injection if the engine is to run smoothly.

- **To accurately time the fuel injection for each cylinder**. For efficient combustion of the fuel charge, injection must begin at a specific point in the engine cycle before the piston reaches TDC on its compression stroke, continue for a certain time, and then cut off sharply. The fuel injection pump must start to supply fuel to each engine cylinder at a specific point in the engine cycle. The point of commencement of injection is determined by the engine manufacturer and usually lies between 15° and 30° before TDC for high-speed diesel engines.

It may be seen, then, that the fuel injection pump must force fuel into the combustion chamber via the injector(s) at a pressure high enough to enable it to be atomised, measure out the fuel charge to be distributed to each cylinder and send this fuel charge to its respective cylinder at exactly the right moment. Since these operations may occur more than a hundred times per second when the engine is operating, the fuel injection pump must be capable of accurately and reliably fulfilling its functions at a very high rate and of resisting wear under high-speed, high-pressure conditions. While being able to satisfy all the demanding requirements outlined above, the pump must remain light and compact enough to be fitted to the engine without any inconvenience.

One of the most extensively used and entirely satisfactory fuel pump systems is the 'jerk' pump, which was designed by Robert Bosch, and is now manufactured by almost every manufacturer of fuel injection equipment. Pumps of this type are made in a variety of forms— from single pumping element units operated by a special engine camshaft provided for the purpose to multi-element units with a camshaft enclosed within the pump housing. Single element pumps are usually located along one side of the engine block to which they are flange mounted, where they are readily operated by the camshaft provided by the engine manufacturer for the purpose, while the multi-element, enclosed camshaft units are generally mounted on the side of the block where their camshaft is driven (at half engine camshaft speed in four-stroke engines) from the engine crankshaft via the timing gears.

Flange-mounted injection pumps

Flange-mounted jerk-type injection pumps may be described as cam-operated, single acting, constant stroke plunger-type pumps, attached by a flange mounting plate to the engine. Usually mounted vertically, pumps of this type are generally operated by a special engine camshaft driven from the engine crankshaft by chain or gears, or by a special lobe(s) on the engine camshaft.

Fuel injection pumps of this type may be used for either single or multi-cylinder engines that operate at low, medium or high speeds.

The majority of engine manufacturers who use flange-mounted pumps have them made by such companies as American Bosch (Ambac Industries Incorporated), Bryce Berger Ltd, CAV, Robert Bosch, Nippon Denso and Diesel Kiki, which specialise in the manufacture of fuel injection equipment.

The majority of flange-mounted jerk-type fuel injection pumps are single element units, able only to serve one engine cylinder. Multi-element units are also manufactured, but for reasons of simplicity and because of their greater use by engine manufacturers, we will restrict all basic discussion to single element units.

Construction of a typical single element, flange-mounted pump

By reference to Fig 18.1, it can be seen that the typical pump consists of the following major parts:

- the pump housing
- the pumping element, consisting of two parts—a plunger and a barrel
- the plunger return mechanism, consisting of a compression spring and two spring plates

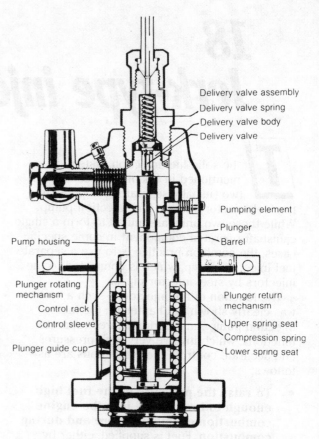

Fig 18.1 Typical single element flange-mounted fuel injection pump

- the plunger rotating mechanism, made up in two parts—a control rack and a control sleeve
- the delivery valve assembly
- the plunger guide cup.

The pump housing—The pump housing is usually manufactured from high-quality cast iron and provides the 'body' of the pump in which all components are housed, as well as a flange for attaching the pump to the engine. A fuel gallery is cast in the top portion of the housing to carry the fuel supply for the pumping element. An air bleed drilling, sealed with a threaded plug, leads upwards from the gallery to facilitate the removal of entrapped air. The lower portion of the housing is accurately machined internally to accommodate the pumping element and associated components. For ease of timing, a sight window is provided in the side of the housing through which a line on the plunger guide cup can be seen and aligned with a line inscribed on the housing.

The pumping element—The pumping element consists of two parts, a barrel and a plunger, each of which is manufactured from high-quality alloy steel, hardened, ground and lapped to the finest limits. The components are mated in manufacture and are lapped to such a degree of accuracy that they provide an efficient seal at extremely high pressures and very low speeds without the use of packing or seals of any type. Consequently, the barrel and plunger must be kept together as a mated pair during service, and never interchanged or replaced separately.

The barrel, which is a firm fit in the pump housing, is located in position by a screw and locked in place by the delivery valve holder, which screws down on top of both the delivery valve body and the flanged end of the barrel. The barrel carries two diametrically opposed holes—an inlet port and a control or spill port—through which the fuel from the fuel gallery can enter the barrel to be pressurised by the plunger. Although it is common practice to use a pump barrel that is equipped with both inlet and control ports, some pumps are fitted with a barrel with only the control port.

The pump plunger reciprocates inside the barrel, and both pressurises and meters the fuel charge that is sent to the engine cylinder. To make it possible for the pump to vary the quantity of fuel delivered per stroke, the upper part of the plunger is provided with a vertical channel extending from its top face to an annular groove, the top edge of which is milled in the form of a helix and is called the control edge. In some pumps the plunger has an additional helix on the top, and this allows the point at which injection begins to be changed.

The lower end of the plunger, which protrudes from the bottom of the barrel, is supported by the plunger guide cup and is equipped with two lugs or vanes to facilitate plunger rotation.

The plunger return mechanism—This mechanism consists of a compression spring (called the plunger return spring) and two spring plates, the lower portion of the spring and the bottom spring plate being situated inside the plunger guide cup. The lower spring plate and the end of the plunger are held against the bottom of the plunger guide cup by the action of the plunger return spring. During the delivery stroke, the plunger guide cup and plunger are raised by the action of the cam and returned to bottom dead centre by the plunger return spring.

The plunger rotating mechanism—The object of this mechanism is to rotate the pump plunger and so control the amount of fuel it delivers to the engine cylinder. It consists essentially of a control rack and a control sleeve, the top of which is provided with an integral gear ring or pinion. The control sleeve, which is fitted over the pump barrel, has two longitudinal slots in its lower end. The vanes on the bottom end of the plunger are engaged in these two slots, and the gear ring at the top of the control sleeve is in mesh with the teeth of the control rack. By this arrangement, the movement of the control rack, either manually or by the governor, will rotate the plunger even while the engine is running. It is this controlled rotation of the plunger within the barrel that gives control over the quantity of fuel delivered to the engine cylinder.

The delivery valve—The delivery valve assembly is situated in the upper part of the pump body above the pumping element, and consists of a delivery valve body, a delivery valve and a compression spring to hold the valve against its seat on the top of the valve body. Both the delivery valve body and the delivery valve itself are manufactured from highest quality alloy steel, and are hardened and ground.

The delivery valve body is a close fit in the upper section of the pump body, and sits directly on top of the pump barrel. The upper end is threaded for attaching a puller so that the body may be readily removed. The axial bore is accurately lapped, while the upper end is ground to form a conical seat for the delivery valve.

The body is held in position by the delivery valve holder, which is screwed into the top of the pump body. A sealing washer, made from one of a variety of materials including phenolic-bonded fibre, nylon or solid annealed copper, lies between the delivery valve holder and the valve body.

The most common type of valve is a mitre-faced valve with a guide extending below the valve face. The guide consists of two parts—an upper section that forms a small piston or plunger, and a lower section that carries four axial grooves on its surface to form a cross-section in the form of a cross. The guide is a very close fit in the axial bore of the valve body, and an annular groove divides the two sections. Above the mitre valve face, the top of the delivery valve is reduced in diameter to fit inside the lower end of the delivery valve spring.

Note   Some delivery valve assemblies incorporate a stop to prevent the valve lifting further from its seat than is desirable. This is also known as a volume reducer, since it reduces the total volume of fuel between the delivery valve and injector and so prevents any misfiring tendency that may result at idling speed when the addition of a minute quantity of fuel does not always raise the fuel pressure sufficiently to open the injector.

The plunger guide cup—The plunger guide cup is retained in the pump housing by a circlip, and reciprocates under the action of the cam provided for the purpose, either directly or via a tappet assembly. The plunger guide cup in turn imparts the necessary reciprocating motion to the pump plunger.

The guide cup usually has a line inscribed around its circumference, and this is used in conjunction with a line inscribed on the edge of the window in the pump housing to facilitate timing the pump to the engine.

Operation of a typical single element, flange-mounted pump

The pumping principle—Fuel from the supply system enters the pump body through the inlet connection and fills the gallery that surrounds the top of the barrel. With the plunger at the bottom of its stroke (Fig 18.2a), fuel from the gallery flows through the barrel ports, filling the space above the plunger, the vertical slot in the plunger and the cut-away area below the plunger helix (or scroll). As the plunger moves upwards the barrel ports are closed by the plunger (Fig 18.2b), and fuel is trapped above the plunger. Further movement of the plunger forces the fuel

from the barrel, through the delivery valve into the delivery line and on to the injector (Fig 18.2c).

Delivery of fuel ceases when the plunger helix passes the barrel spill port (or control port) as shown in Fig 18.2d, and the delivery valve returns to its seat. During the remainder of the stroke the fuel displaced by the plunger simply returns to the gallery via the vertical slot, cutaway area and spill port. Thus fuel ceases to be injected when the helix passes the spill port.

Metering the fuel charge—Since the plunger is driven by a cam, its stroke is constant and cannot be varied to control the quantity of fuel injected per stroke. However, the effective part of the pumping stroke can be varied to control the quantity of fuel injected per stroke simply by rotating the plunger in the barrel.

Fuel delivery begins at the instant the upper end of the plunger covers the barrel ports and continues until the helix edge uncovers the spill port, at which point fuel trapped above the plunger is allowed to return to the fuel gallery. Thus the effective pumping stroke ceases when the spill port is uncovered, and is directly controlled by the distance through which the plunger must travel before the edge of the helix passes the bottom of the spill port.

Because the edge of the helix lies at an angle to the top of the plunger, the distance through which the plunger must travel before the spill ports is uncovered is dependent on the angular position of the plunger in relation to the barrel spill port. Thus, turning the plunger in the barrel by means of the plunger rotating mechanism varies the effective plunger stroke, and hence the quantity of fuel injected may be controlled.

The position of the plunger for maximum fuel

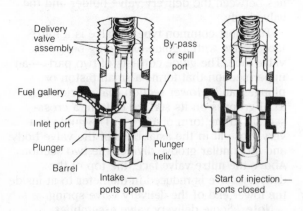
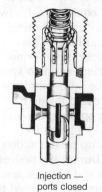
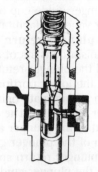

Fig 18.2 The pumping principle

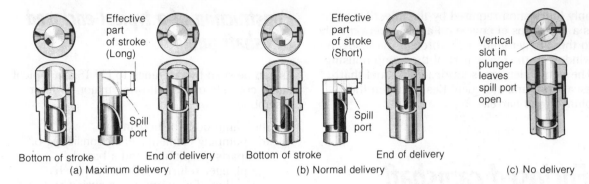

Effective part of stroke (Long)

Spill port

Bottom of stroke   End of delivery
(a) Maximum delivery

Effective part of stroke (Short)

Spill port

Bottom of stroke   End of delivery
(b) Normal delivery

Vertical slot in plunger leaves spill port open

(c) No delivery

Fig 18.3 The metering principle

delivery is shown in Fig 18.3a, while Fig 18.3b illustrates the plunger position for normal delivery.

In Fig 18.3c, the plunger is shown rotated to the point where the vertical slot aligns with the spill port. Since in this position the spill port must remain open regardless of the vertical movement of the plunger, the entire stroke is ineffective and no fuel can be delivered. This, then, is the 'engine stop' position.

Function and operation of the delivery valve—The delivery valve, often referred to as an 'anti-dribble' device, performs two essential functions:

- It acts as a non-return valve, so maintaining some pressure in the injector line.
- It reduces the pressure in the injector line to a level well below the injector opening pressure, so preventing dribble from the injector nozzle.

During the effective stroke of the plunger, the delivery valve is lifted from its seat in the body by the fuel acting against the underside of the delivery valve piston (or relief plunger) and against the end of the delivery valve. The relief plunger is lifted clear of the delivery valve body bore, and fuel from on top of the pump plunger is delivered into the injector pipe, which conveys

it to the injector and so into the combustion chamber. When the helical edge of the pump plunger uncovers the spill port in the barrel, the effective pumping stroke is terminated and the pressure of the fuel in the barrel immediately drops to feed pump pressure. The delivery valve instantly resumes its seat, due to the force exerted by the spring and the high fuel pressure above the valve in the injector pipe. Thus the delivery valve cuts communication between the pumping element and the injector until the next delivery stroke takes place, and retains some pressure in the line so that the element does not have to 'pump up' the pressure in the line to any extent before the next injection can take place.

In moving down onto its seat to act as a non-return valve, the delivery valve performs its other important function of reducing, by a predetermined amount, the pressure of the fuel in the injector line. As the delivery valve resumes its seat, the small piston on the guide sweeps down the bore of the valve with a plunger-like action, thereby increasing the volume enclosed in the injector pipe between the top of the delivery valve piston and the injector. Thus the volume enclosed in the delivery line is increased by an amount equal to the cross-sectional area of the delivery valve bore multiplied by the distance the piston section moves on its return down the bore. This **increase** in volume is directly associated with a **decrease** in pressure. The pressure is reduced to a level well below the opening pressure of the injector, thus allowing the injector needle valve to snap to its seat, instantly terminating the spray of fuel entirely without dribble.

Injection pump lubrication—The method of lubrication of the tappet and roller assembly is arranged by the engine manufacturer, since these parts are incorporated in the engine. The

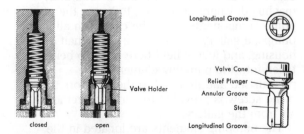

Longitudinal Groove

Valve Cone
Relief Plunger
Annular Groove
Stem
Longitudinal Groove

Valve Holder

closed   open

Fig 18.4 A typical delivery valve

only lubrication required by the injection pump is a few drops of engine oil applied periodically to the plunger guide cup, through the inspection window in the lower part of the pump housing. The pump element is lubricated by fuel that escapes, in minute quantities, between the plunger and barrel.

Enclosed camshaft injection pumps

Before the development of satisfactory rotary or distributor-type fuel injection pumps, most high-speed, multi-cylinder diesel engines were equipped with multi-element, enclosed camshaft, jerk-type fuel injection pumps of one make or another. Now, however, owing to the development of the distributor-type fuel injection pump, the multi-element pump is not used as extensively in this field as it was.

Single element, enclosed camshaft pumps are made but these are rarely seen. In general, these pumps are simply one element of a multi-element type, and may be treated in the same way. However, because balancing of fuel deliveries is not required, the control pinion is an integral part of the control sleeve and the rack and pinion have only to be correctly timed one to the other to ensure correct fuel delivery, as is the case with a single element flange-mounted pump.

Enclosed camshaft jerk pumps operate in the same way as flange-mounted jerk pumps and so may also be described as cam operated, single acting, constant stroke plunger pumps, the effective working stroke of which can be varied. Unlike flange-mounted pumps, which are operated by a camshaft running in the engine block, these pumps feature a camshaft running in bearings in the pump housing.

Like most items of injection equipment there are many makes and models of the enclosed camshaft fuel injection pump, although the basic construction and operating principles are the same. Therefore we will first consider a typical example before examining some of the various makes and models.

Construction of a typical enclosed camshaft pump

As can be seen by reference to Fig 18.5, a typical pump consists of the following major parts or assemblies:

- the pump body or housing
- the pumping elements, each consisting of two parts—a plunger and a barrel
- the plunger return mechanisms, each consisting of a compression spring and two spring plates
- the plunger rotating mechanism, made up of a control rack or control rod and a set of control sleeves
- the delivery valve assemblies
- the tappet assemblies
- the camshaft
- the control rod stop.

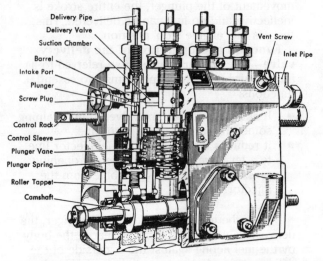

Fig 18.5 A typical multi-element enclosed camshaft pump

The pump body or housing—The body is cast from aluminium alloy and is accurately machined, where necessary, to house the other pump components. As is the case with all jerk pumps, a gallery is cast in the upper half of the housing, and fuel is held here prior to being taken into the pumping element to be pressurised and pumped to the injectors. A bleed screw is provided to allow the removal of air from the gallery.

The pumping elements are located in the upper half of the housing, where the barrels are a firm fit. They are located in their correct position by spigot-ended locating screws, which engage an elongated depression provided for the purpose in each barrel.

The lower half of the body where the camshaft runs is usually referred to as the **cambox**, and is partially filled with the lubricating oil necessary to lubricate the tappets as they reciprocate, the cam rollers as they pass over the cams, and the camshaft bearings. From underneath the housing, directly beneath each pumping element, a set of plugs screw into the housing. These are known as **closing plugs**, and must be removed to allow the tappet, plunger, spring and spring plate to be removed.

At either end of the cambox, and bolted to the housing, lie the two end plates, which are machined to accommodate the bearings that carry the camshaft.

The pumping elements—A multi-element pump must have, of course, the same number of elements as the engine it is to serve has cylinders. The elements used in pumps of this type are identical in design and operation with those used in flange-mounted pumps.

The plunger return mechanism—Each pumping element is provided with its own plunger return mechanism. This consists of a plunger return spring, an upper spring plate and a lower spring plate. Its action is straight-forward and its purpose twofold:

- to return the plunger to the bottom of its stroke when its upward stroke is completed
- to keep tension on the tappet assembly to ensure that the roller remains in contact with the cam lobe.

The plunger rotating mechanism—As is the case in flange-mounted pumps, the effective pumping stroke of an enclosed camshaft pump is controlled by rotating the plungers in their barrels. The mechanism employed to achieve this is identical with that used in multi-element flange-mounted pumps, and consists of a number of control sleeves (one for each element) and a control rack. Since individual adjustment of the fuel delivery from each element is necessary if balanced deliveries are to be obtained, the brass control pinions are clamped to the sleeves.

In almost every case movement of the control rack is controlled by the governor, which is usually mounted on one end of the pump housing and is connected to the end of the rack that protrudes from the pump housing. The

other end of the rack also protrudes from the pump housing into a 'rack stop' housing. The rack stop (see below) is an adjustable stop, the purpose of which is to limit the movement of the control rack in the maximum fuel direction and thus control the maximum amount of fuel delivered by the pump.

The delivery valve assemblies—Each element of a jerk pump has its own delivery valve assembly. These valves are identical in design and operation with those used in flange-mounted pumps.

The tappet assemblies—Each pumping element is operated by its own tappet assembly, which consists of a body, a hardened adjusting screw and lock nut and a shaft supporting a hardened roller. The shafts protrude slightly from the sides of the body and engage in two slots in the side of the bores in which the tappets reciprocate, thus preventing the tappets from turning in the housing. The top of each adjusting screw head makes direct contact with the bottom of its plunger, while the tappet rollers are held in contact with the camshaft lobes by the plunger return springs. As the camshaft rotates, the rollers follow the contour of the cams and so the necessary reciprocating motion is given to the plungers. Adjustment of the plungers' strokes is made possible by the set screws that are used to raise or lower the plungers in their barrels.

The camshaft—The camshaft is usually made of high-grade nickel alloy steel, the lobes being heat treated and precision ground. There are the same number of cam lobes on the shaft as there are pumping elements, the lobes being set to operate the pump elements in the same sequence as the firing order of the engine to which the pump is fitted. If the injection pump has a fuel lift pump mounted to it, this pump is driven from the injection pump camshaft, either by one of the cams used to operate the plungers or by a separate cam or eccentric specially provided for the purpose.

Two basic cam profiles are employed (see Fig 18.6) although there are many variations of each to give specific injection characteristics. Because of its shape, the early type of camshaft would cause injection to occur on compression stroke in either direction of the engine rotation—a very undesirable characteristic if the engine were to kick back, since it could fire and turn backwards. Consequently a design to prevent this from happening was introduced.

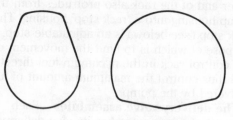

Standard cam profile   'Anti-backfire' cam profile

Fig 18.6 Basic cam profiles

The camshaft is carried in two bearings—usually single row ball races, although tapered roller bearings are used in some pumps—and generally has a thread, taper and Woodruff keyway machined on the drive end to allow the drive coupling to be fitted.

The end of the camshaft remote from the drive coupling is usually provided with either a thread, taper and Woodruff keyway or a spline to allow a governor to be driven.

Most injection pump camshafts are provided with a dot or some other means of identification on one end. In cases such as this, the camshaft can be fitted in either of two ways, and the mark provides a means of identification to ensure that it is fitted correctly.

The control rod stop—The control rod stop also known as the rack stop and the smoke stop, is fitted to the main housing of a multi-element pump to prevent the control rod (or

rack) from moving too far in the direction of maximum fuel delivery, and so limits the amount of fuel able to delivered to the correct maximum. If the control rod setting is not correct and more fuel is delivered to the engine cylinders than can be burnt completely, fuel is wasted and black exhaust smoke results. On the other hand, if the stop is incorrectly adjusted and the rack cannot move far enough, the engine will lack power due to insufficient fuel being supplied to the cylinders.

A large variety of rack stops are employed over the complete range of enclosed camshaft pumps. However, those shown in Fig 18.7 are representative of the more commonly used types.

The fixed stop shown in Fig 18.7 is adjusted with a screw secured in its correct position by means of a split pin. However, in many applications, engines require injection to continue for a longer period under cold conditions than is required under full load conditions, and a fixed stop is not suitable. For these applications, a rack stop that can be temporarily disengaged or over-ridden when starting, but will correctly limit the rack travel under normal operating conditions, is required.

The lever-actuated stop in Fig 18.7 incorporates a manual over-ride to allow extra fuel for cold starts. It is adjusted by turning the guide in its threaded socket and securing it in position by means of its lock nut. In its normal position,

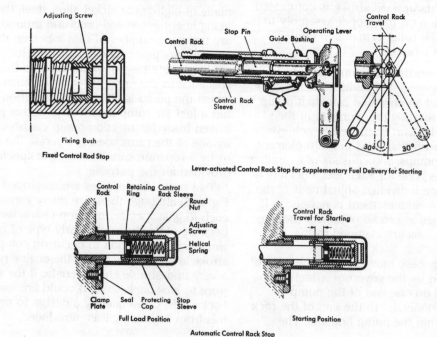

Fig 18.7 Typical control rack stops

this stop will limit the control rack travel for full load delivery. When the lever is moved forwards or backwards axially (only for the purpose of starting), the spring loaded stop will move away from the control rod so that the rod can move beyond the full load position.

The automatic control rod stop in Fig. 18.7 can be used in conjunction with idling and maximum-speed governors (see Chapter 19). The stop is adjusted by turning the guide bushing in its threaded sleeve. The guide bushing, which is secured in its adjusted position by a lock nut, sets the limit to the full load fuel delivery at pump speeds above 400 to 500 rpm. When the driver fully depresses the accelerator pedal with the engine stopped, the spring in the stop sleeve will yield to pedal pressure so that the rack will travel beyond the full load position and thus the rack will travel the greater distance required for starting. However, when the engine is running, the automatic stop will no longer yield to the pedal pressure since the governor, assisted by the spring of the stop device, will return the control rod to the driving (full load) position before the engine has reached medium idling speed.

Operation of a typical multi-element enclosed camshaft pump

Since pumps of this type use the same pumping components as flange-mounted jerk pumps, it is rather obvious that the operation of these components will be identical with that of the typical single element, flange-mounted pump.

Electric fuel pump control device

A number of automotive diesel engines fitted with in-line injection pumps are fitted with an electric fuel pump control device to facilitate the starting, running and stopping of the engine.

A typical fuel pump control device, as shown in Fig 18.8, is controlled by the engine control (stop/start) switch located in the cabin of the vehicle.

Operational modes

Engine starting (Refer to Fig 18.9a)—When the engine control switch is turned to the start position, the fuel pump control device operates and moves the governor stop lever away from the engine stop (no fuel) position, and into the maximum fuel position. This in turn allows the governor mechanism to move the rack into the maximum fuel delivery position, thereby creating optimum starting conditions.

Run (Refer to Fig 18.9b)—When the engine starts and the control switch is returned to the

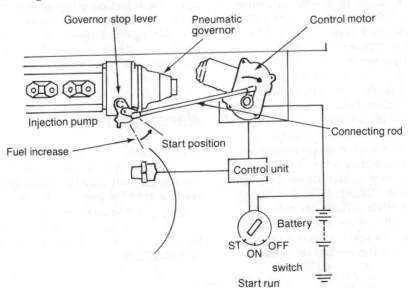

Fig 18.8 Schematic layout and circuit diagram of a typical electric fuel pump control device

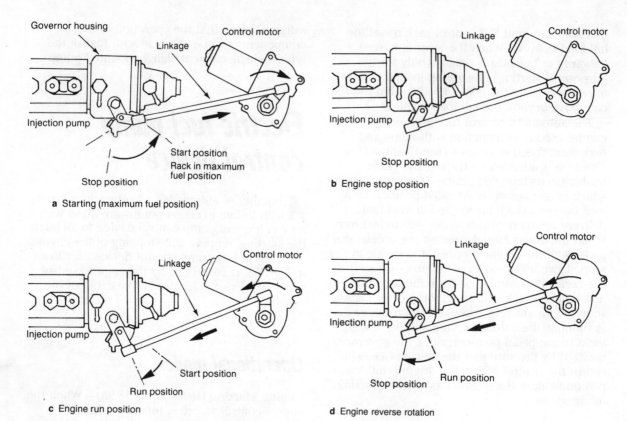

Fig 18.9 Operational positions for a typical electric fuel pump control device

run position, the control system moves, independently of the governor stop lever, and positions itself in a non-active position, about mid-way between the stop and start positions.

Stopping the engine (Fig 18.9c)—Turning the control switch to the stop position activates the control device, which moves the governor stop lever to over-ride normal governor action. The injection pump rack is moved to the 'no-fuel' position, thus stopping the engine.

Anti-reverse operation (Fig 18.9d)—Some fuel pump control devices monitor the engine oil pressure at all times. If the engine were to start in reverse rotation (which can happen in diesel engines), the oil pressure would fail to build up. Unless the control unit receives a signal indicating that engine oil pressure has reached a pre-set minimum level, it will move the governor stop lever to shut down the engine, so preventing engine damage.

Oil pressure switch—There are two oil pressure sender switches mounted on a diesel engine fitted with a fuel pump control device. One is the sender for the normal oil pressure gauge in the cabin and the other is the oil pressure signal for the fuel pump control device.

Should the engine oil pressure fall below the pre-set minimum (generally approximately 28 kPa) during operation, the control device will move the injection pump operating lever to the 'no-fuel' position within approximately 13 seconds, again stopping the engine. Under these conditions, the engine can be restarted, but will stop after another 13 seconds delay.

Manual over-ride

Should the fuel pump control device fail to function correctly in any or all of its different modes, the unit may be manually over-riden to start or stop the engine. This is carried out by removing the linkage between the control device and the injection pump, and manually moving the governor stop lever to the required operating position.

Features of various makes of multi-element pumps

As multi-element, enclosed camshaft fuel injection pumps are produced in many countries by various manufacturers who each produce many models, it would be very difficult to give detailed information regarding all of them. However, it is possible to include information regarding the design features and service procedures of some of the more common types.

American Bosch

The American Bosch Arma Corporation, now known as Ambac Industries Incorporated, manufacture a range of multi-element, constant stroke enclosed camshaft fuel injection pumps, type coded APE, that are generally very similar to the typical pump. Indeed, the metering and pumping principles involved are exactly the same.

Plunger design—Although the plungers used in the pumping elements of most American Bosch pumps are of the standard design, plungers of other design are manufactured for special applications. These special plungers include upper helix plungers, plungers with both upper and lower helices, and plungers with two identical helices, one on each side of the plunger.

Governors—There are two types of American Bosch governor commonly used—centrifugal (or mechanical) and pneumatic—although these are not the only type manufactured. In both cases, the governor housing is bolted directly to the pump housing without external linkages, to form a compact, interconnected pump and governor unit.

Mounting and driving—These pumps are usually mounted on a flat-topped bracket fastened to the crank case or cylinder block of

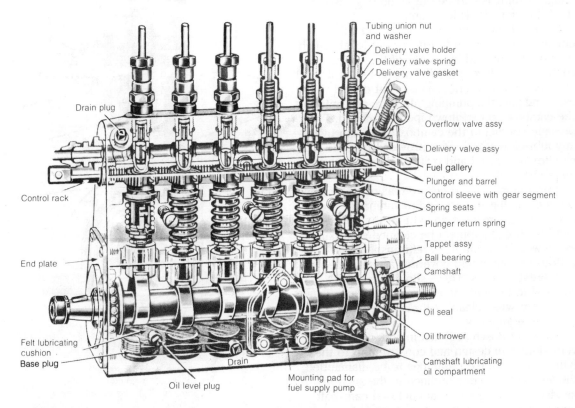

Fig 18.10 American Bosch six-element type APE pump (self-contained drive)

the engine. They may also be provided with a mounting flange for fastening to a suitable adaptor on the drive gear housing. While the pump is usually mounted in a vertical, upright position, horizontal as well as inverted mountings are permitted.

The lobes of the cams on the camshaft are symmetrical and these pumps may be driven in either a clockwise or an anticlockwise direction as required. Changing the direction of rotation of the pump camshaft, however, does reverse the firing order of the pump and so the camshaft may have to be reversed in the housing to give a suitable pump firing order. Since the cams are arranged on the camshaft in the standard firing order, reversal of the camshaft in the housing allows, irrespective of direction of rotation, the selection of a firing order that is the same as the firing order of the majority of engines, so that it is almost never necessary to cross the fuel lines from pump to injectors.

Timing—American Bosch APE pumps carry the same timing marks as the typical pump—two timing marks on the housing marked R and L, and one on the camshaft or on the coupling flange. Timing is carried out as on page 205 described for the typical pump.

Lubrication—The cambox is filled with engine lubricating oil to the correct level, which may be determined by means of the oil gauge rod or the oil level plug. To prevent the oil level from becoming too high as a result of fuel escaping past the pumping plungers and entering the cambox, an overflow pipe is provided. The presence of fuel in the cambox will not provide any lubrication problems, provided that the cambox is drained and refilled when the engine oil is changed.

Pump service—To ensure maximum efficient service, intelligent care and maintenance are essential; it is recommended that the entire fuel injection equipment be cleaned externally every week to facilitate careful inspection.

Both the pump and governor are provided with breather caps. These should be tighly covered to prevent the entrance of steam, water or solvent when cleaning the engine with pressure equipment.

American Bosch APE fuel injection pumps are very similar in design and construction to the typical pump. If the pump has to be dismantled, the general sequence outlined in the section headed 'Servicing a typical enclosed camshaft pump', further on in this chapter, may be followed.

Caterpillar

The Caterpillar Tractor Company, the American-based earthmoving equipment manufacturers, manufacture and use constant stroke, cam-operated, plunger-type injection pumps of unique concept. The pumps follow the basic jerk pump principle of operation, but require no calibration and phasing as we know it, need no test bench for setting, and can have an individual pumping element changed without removing the pump from the engine in any situation where relatively clean conditions prevail. Two of these pumps are in current production—the **compact** pump, which is gradually superseding the second type, known as the **non-compact** pump.

The compact pump

Following normal Caterpillar practice, this pump (see Fig 18.11) consists of a housing assembly made up of the housing, camshaft, cam followers, rack etc. and the pumping elements, which are

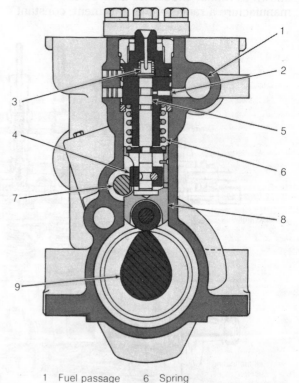

1	Fuel passage	6	Spring
2	Inlet port	7	Fuel rack
3	Check valve	8	Lifter
4	Gear segment	9	Camshaft
5	Pump plunger		

Fig 18.11 A compact fuel injection pump

retained in the housing by some means—in this case by threaded bushings that screw into the housing above the elements.

Pump element design—As shown in Figs 18.12 and 18.13, Caterpillar compact pumps feature a single port barrel (7) positively located in the pump housing by a dowel that engages in a slot in the barrel. Connected to the barrel by means of a clip (5) is the bonnet (2) which locates the check valve (4) and spring (3) and provides the connection for the injector line. The plunger (10) carries the gear segment (pinion) clamped to its lower end, and has a groove machined on its lower end to carry a washer (9). The washer provides the lower thrust face for the plunger return spring that lies between the washer and the flanged upper end of the barrel.

The housing—A one-piece steel casting forms the pump housing, and also forms a part of the governor housing. The housing carries the camshaft in three plain bearings, the two end ones being pressure lubricated, the centre one being splash lubricated. The roller-type cam followers (or lifters, as they are called by Caterpillar) reciprocate directly in the housing, while the round section control rack is supported in bearings at each end of the pump housing.

Governors—Two types of governor are used on Caterpillar engines, mechanical and hydraulic. Both governor assemblies bolt to the pump housing and connect directly to the rack.

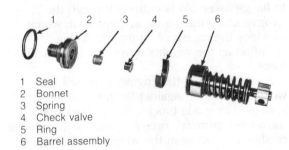

1 Seal
2 Bonnet
3 Spring
4 Check valve
5 Ring
6 Barrel assembly

Fig 18.12 Injection pump assembly

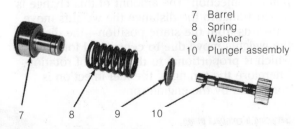

7 Barrel
8 Spring
9 Washer
10 Plunger assembly

Fig 18.13 Barrel assembly

Mounting and driving—Caterpillar compact pumps are held in a cradle by studs passing down through appropriate holes in the pump housing.

On most of the later (and higher speed) engines, the pump is driven through a variable timing unit—an hydraulically or mechanically actuated timing control unit driven by the camshaft gears. This unit automatically advances the injection timing as the speed increases and retards it again as the speed decreases.

In the hydraulic type (Figs 18.14 and 18.15), this is achieved by the use of helical splines in the drive components. Any longitudinal

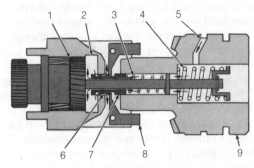

1	Power piston	5	Oil inlet passage
2	Power piston cavity	6	Drain port
3	Control valve spring	7	Control valve
4	Power piston return spring	8	Flyweights
		9	Shaft assembly

Fig 18.14 The hydraulic variable timing unit, low rpm position

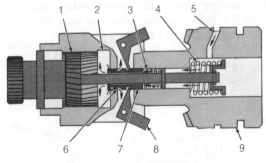

1	Power piston	5	Oil inlet passage
2	Power piston cavity	6	Drain port
3	Control valve spring	7	Control valve
4	Power piston return spring	8	Flyweights
		9	Shaft assembly

Fig 18.15 The hydraulic variable timing unit, high rpm position

movement between the two splined components will cause partial rotation of one in relation to the other.

During low engine rpm operations, the centrifugal force acting on the flyweights is not sufficient to overcome the force of the control valve spring (3) and move the control valve (7) to the closed position. Oil flows through the power piston cavity (2), but does not create enough pressure to compress the spring (4) and move the power piston (1).

As the engine rpm increases, the flyweights (8) move outwards under centrifugal force, overcome the force of the control valve spring (3), and move the control valve (7) to the closed position, blocking the oil drain port (6). Pressurised oil, trapped in the power piston cavity (2), overcomes the force of the spring (4) and moves the power piston (1) outwards.

Any outward movement of the power piston increases the force on the control valve spring, which tends to reopen the control valve, letting oil escape from the power piston cavity. As oil begins flowing from the cavity again, the return spring (4) moves the power piston inwards.

At any given rpm, a balance is reached between the centrifugal force acting on the flyweights and the control valve spring force. The resultant position of the control valve (7) will be such that it will maintain proper pressure behind the power piston. The greater the rpm, the smaller will be the drain port opening and the further outwards the power piston will be forced.

As the power piston is moved outwards, the angular relationship of the ends of the drive unit changes, due to the power piston moving along the helical spline, which in turn advances the injection pump timing.

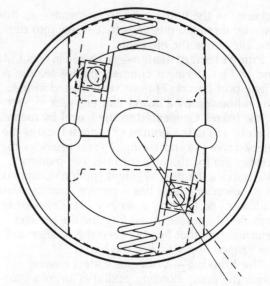

Fig 18.17 *End view of the mechanical variable timing unit showing an exaggerated angle of advance*

The mechanically actuated, variable timing unit consists of two spring loaded weights turning with the pump drive gear, which drive a flange assembly through angled grooves in their surface, which engage with square guide blocks on the flange. The weights and springs are part of the carrier assembly, which is connected directly to the pump drive gear. The flange assembly drives the pump through a short splined shaft. The drive from carrier assembly to flange assembly is entirely through the 'groove and guide block' arrangement, which provides the means of automatic timing variation as the weights move due to centrifugal force.

As the speed of the engine increases, the weights move out against the force of the springs. The guide blocks, which must ride in the angled grooves, force a change in the angular relationship between the weights and flange assembly. This change in position alters the relationship of the injection pump camshaft to the drive gear, which results in a change in the point of injection. The amount of this change is proportional to the distance the weights move outwards from the static position—the distance the weights move due to centrifugal force, which is proportional to the speed of rotation. Therefore the timing of the fuel injection is controlled by the engine rpm.

Servicing a compact pump

Because of their unique design, Caterpillar

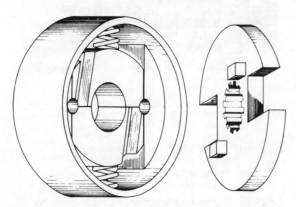

Fig 18.16 *The mechanical variable timing unit*

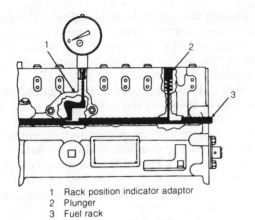

1   Rack position indicator adaptor
2   Plunger
3   Fuel rack

Fig 18.18 Centring the fuel rack

pumps are not serviced in the same manner as any others.

Centring the rack—One of the most important operations in Caterpillar pump service is what is known as 'centring the rack' (refer to Fig 18.18).

The fuel rack is positioned at the centre or zero position by depressing the plunger until it is flush with the top of the fuel injection pump housing, and moving the fuel rack forward until it contacts the plunger.

Removing a pump unit from the housing— To remove a pump from the housing, proceed as follows:

1   Using the correct tool, remove the pump unit retaining bush.
2   Remove the felt bush from the pump bonnet.
3   Centre the fuel rack.
     Note   The rack must be held in the centre position when removing or installing the pump units.
4   Using an extractor, remove the pump unit.

Dismantling and reassembling a pump unit—To dismantle and reassemble a pump unit, proceed as follows (refer to Figs 18.12 and 18.13):

1   Remove the seal (1)
     Note   All disassembly and assembly should be carried out with clean hands on a clean, lint-free cloth.
2   Remove the retaining ring (5) and separate the bonnet (2) and barrel assembly (6).
     Note   Considerable care should be taken not to drop and lose the check valve (4) and spring (3) retained in the bonnet.
3   Separate the spring (8), washer (9) and plunger (10) from the barrel (7).

4   Remove the washer and spring from the plunger.
     Note   Considerable care should be exercised when removing the spring and washer to prevent damage to the plunger surfaces.
5   Clean all parts with clean fuel and assemble in the reverse order, replacing any worn parts.
     Note   Like other pumps, the barrel and plunger are a mated pair, and must not be mixed with components of other elements. Use extreme care in inserting the plunger into the bore of the barrel.

Fuel pump timing dimension setting— Caterpillar pumps are not phased (refer to 'Final testing and setting' on page 199) as we know the term, but the firing point of each element is checked and adjusted if necessary. The length of the pump plungers is maintained within very close limits in manufacture and should be checked for correct length when the pump is dismantled during service. This is done by measuring with a micrometer, and comparing the length with the correct dimension given in the specifications. A wear plate in the cam follower (or lifter) should also be examined for wear and renewed if necessary.

Provided the plunger length is correct and the wear plate is not excessively worn, the timing dimension may be checked and the pump assembled. Checking the timing dimension involves measuring the height of the cam follower from the top of the housing at a particular point in camshaft rotation. Since the height of the cam follower governs the point of commencement of injection, if this height is correct then the timing of that element must be correct. If all elements are checked, and the heights are equal and correct when the camshaft has been turned through the correct number of degrees, then the pump is correctly phased.

The timing dimension is checked by fitting a special timing plate (or protractor) to the pump drive and a pointer assembly to the correct point on the housing. The pump camshaft can then be turned by the timing plate until the specified degree line for the particular follower being checked aligns with the pointer. At this point, a gauge spacer is fitted to the top of the housing, and the timing dimension measured from the top of the gauge to the follower wear plate by means of a micrometer depth gauge. Adjustment is made by changing the spacer under the barrel shoulder for one of a different thickness.

Refitting a pump unit to the housing—To refit a pump unit, proceed as follows:

1 Centre the fuel rack.
2 Install the correct adaptor (as used to remove the pump unit) on the pump unit.
3 Align the notches on the bonnet and barrel with the fourth tooth of the segment (pinion) gear.
4 Insert the injection pump unit into the housing. When inserting the pump, the slot (Fig 18.19) must be aligned with the locating pin in the lifter, and the notches on the bonnet and barrel must align with the locating dowels in the pump housing. With the slot and pin, the notches and dowels aligned, the fourth tooth of the segment gear will be aligned with the fourth groove of the rack. Do not try to force the pump into the injection pump housing as the locating pin can be damaged. When properly aligned, the pump can be installed without force.
5 After the plunger and barrel assembly is installed, place a new rubber sleeve over the bonnet, install the retainer bushing and tighten the bushing finger tight. If the bushing is not approximately flush with the top of the pump housing when tightened finger tight, the notch in the bonnet is not aligned with the dowel in the housing.
6 Using the correct tool and a torque wrench, tighten the retaining bushing to the torque specified in the manual specifications.
7 Install a new felt washer.

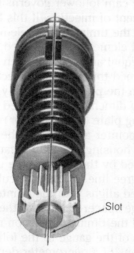

Fig 18.19 Aligning the pump

Slot

Timing the pump to the engine—The point of injection commencement of no. 1 pumping unit may be established for timing purposes, in one of two ways—by using the timing pin or by the timing dimension.

Most Caterpillar pumps feature a timing pin (or timing dowel) to aid in timing the pump to the engine. The timing pin is carried in the pump housing and, when required, is removed and fitted to a timing hole that is normally concealed with a cover. The lower end of the pin engages with a slot in the fuel injection pump camshaft at the injection commencement of no. 1 pump element.

It is suggested that the engine be turned in the direction of rotation through at least 180° to take up any backlash in the gear train, and then turned on to the injection point. Correct timing is achieved when, after turning the engine to the timing point, the timing pin will slip freely and fully into the pump camshaft groove. (The timing pin is fully engaged with the camshaft slot when the lower face of the hexagon head is flush with the top of the injection pump housing.)

If the timing has had to be adjusted, the timing should be rechecked after turning the engine in the direction of rotation through two revolutions.

Note   Be sure to remove the timing pin before turning the engine over for rechecking.

Some Caterpillar pumps do not feature a timing pin and the injection point must be found from the 'timing dimension'. As was stated above, the beginning of injection must occur when the cam follower is a specified distance from the top of the pump housing. This knowledge can be used to find the timing point for any pumping element, no. 1 in this case.

Once again the engine should be turned through 180° and then on to the engine timing point. No. 1 element should then be removed and the timing dimension checked against specifications. If it is not correct, the engine-to-pump drive should be adjusted, the engine turned through two revolutions and the timing dimension rechecked.

CAV

CAV produce a variety of multi-element, enclosed camshaft fuel injection pumps to suit a wide range of engines. Like the typical pump, all CAV pumps are constant stroke, cam-operated units using one pumping element per engine cylinder.

These pumps are made in a variety of designs, including those type coded BPE, N, NN and AA. The pumping and metering principles employed in all CAV pumps follow the standard pattern.

Plunger and barrel design—While most CAV pumps feature standard plungers and barrels, they are available, like American Bosch, with many different designs for special applications. These special designs include single port barrels and various single- and double-helix plungers.

Governors—CAV governors are specially designed for use in conjunction with CAV fuel injection pumps, and are made in a number of types. These include mechanical, mechanical–hydraulic, pneumatic and mechanical–pneumatic. Each governor comprises a self-contained unit enclosing the mechanism and linkage to protect it from dirt and interference and is constructed in such a way that it can be bolted directly onto the end of the injection pump housing.

Mounting and driving—As with most makes, CAV pumps may be mounted in one of a number of ways:

- Single- and twin-element pumps are designed for mounting on a flat engine bracket by means of bolts through the flange provided on the pump base.
- Three- four- and six-element models may be mounted in one of two ways:
  —on a flat engine bracket with bolts or studs screwed into the tapped holes in the base, or
  —on a radial bracket (of 56 mm radius) with bolts through the lugs provided on the pump sides or by means of straps and screws.

All CAV enclosed camshaft pumps should be mounted so that their camshaft will be as near horizontal as possible when the engine is running normally, to ensure that the cam rollers will be effectively lubricated. When the pump is mounted with its axis 4°–6° from the horizontal, special consideration must be given, and in this event, it is recommended that mention be made when ordering.

With AA-type pumps, a flat bracket is required on the engine, the pump being held by four bolts, which may be screwed into the mounting plate, or pass through and be fixed by nuts.

Many CAV pumps are fitted with camshafts that extend from both ends of the housing so that they may be connected to the engine drive at either end. The camshafts have 20 mm taper cone ends to connect to the drive coupling.

A positive engine drive, such as well-designed helical gearing, is advised and it is recommended that the drive be transmitted to the pump through the medium of a CAV closed slot cross coupling. This type of coupling has been specially developed for use in connection with CAV fuel injection pumps after a considerable amount of research, and is available to suit two engine shaft diameters of 20 and 25 mm. While allowing for any small misalignment between the engine drive and the injection pump, it also provides a ready method of final accurate adjustment of the timing by means of the graduations marked on the coupling flanges.

The modern closed slot coupling is particularly robust in design, and gives durability and length of service very much superior to that of the old open slot type.

Note 1   So-called 'elastic' couplings with rubber compositions or spiral band cores are unsuitable for driving these pumps, and will only produce erratic running if fitted.

Note 2   The centre of the injection pump camshaft in the BPE range is 45 mm from its flat base, or 56 mm from the radial surface of the base. With A and AA pumps the dimension from the centre line of pump camshaft to the flat base is 38 mm.

The injection pumps are supplied with their cams in the same firing order as the engine, and since the cams are symmetrical in form they can be driven in either direction. The difference between the two possible installations lies in the setting of the coupling (timing marks are provided to suit either left or right rotation) and the connections from the pump delivery unions to the injectors. In applications where the firing order of the engine is different from the pump firing order, the lines may be crossed.

Timing—CAV enclosed camshaft fuel injection pumps feature the same timing marks as the typical pump. Timing should therefore be carried out as for the typical pump described on page 205.

CAV pumps, type BPE and AA

Both the BPE and AA pumps follow the typical pattern, as can readily be seen in the case of the BPE pump, which is shown in Fig 18.20. The AA pump closely resembles this.

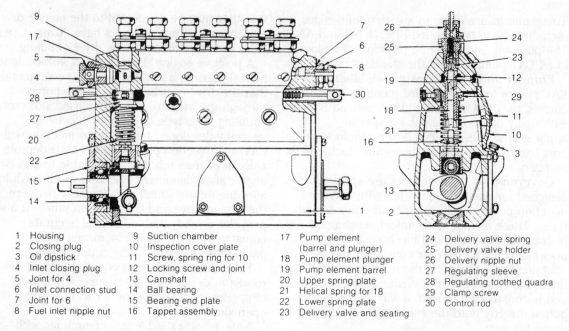

1	Housing	9	Suction chamber	17	Pump element
2	Closing plug	10	Inspection cover plate		(barrel and plunger)
3	Oil dipstick	11	Screw, spring ring for 10	18	Pump element plunger
4	Inlet closing plug	12	Locking screw and joint	19	Pump element barrel
5	Joint for 4	13	Camshaft	20	Upper spring plate
6	Inlet connection stud	14	Ball bearing	21	Helical spring for 18
7	Joint for 6	15	Bearing end plate	22	Lower spring plate
8	Fuel inlet nipple nut	16	Tappet assembly	23	Delivery valve and seating

24	Delivery valve spring
25	Delivery valve holder
26	Delivery nipple nut
27	Regulating sleeve
28	Regulating toothed quadra
29	Clamp screw
30	Control rod

Fig 18.20 CAV fuel injection pump type BPE, sectioned to show the internal construction (see Fig 18.52)

CAV N-type pumps

While pumps of this type are somewhat similar to BPE and AA pumps in general design, they possess some design features not seen on the others. Early N-type pumps, for example, had a final fuel filter incorporated in the pump body, although this has been dispensed with in the later models.

As can be seen in Fig 18.21, ND-type pumps utilise a rather unconventional pumping assembly. A standard-type volume unloading delivery valve is used in all these pumps, but the delivery valve holder is fitted with a stop that reduces the volume of fuel held above the delivery valve and limits the maximum delivery valve lift, so helping to control fuel injection. The delivery valve is retained by the delivery valve holder, a sealing washer being interposed between these components to prevent the leakage of fuel. In addition to this conventional sealing washer, a rubber sealing ring on top of a steel washer is fitted between the shoulder of the delivery valve holder and the body of the pump to prevent fuel leakage.

An important feature of the ND pump element is its calibration adjustment arrangement. The control rack and pinion have helical teeth, the pinion is an integral part of the control sleeve and an adjusting screw engages in a groove in the upper end of the control sleeve (see

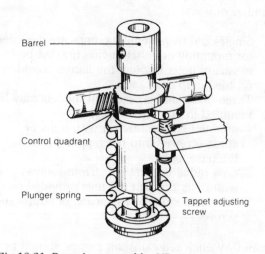

Fig 18.21 Pumping assembly, ND-type pump

Fig 18.21). Because of the helical teeth on both the control rack and the pinion, the control sleeve (and consequently, the plunger) will rotate slightly when the control pinion is either raised or lowered, so altering the position of the control edge of the plunger helix in relation to the barrel spill ports. This changes the effective length of the plunger pumping stroke, and it should be evident, therefore, that calibration adjustments are made to the elements of ND pumps by either raising or lowering the tappet adjusting screws.

On the other hand, NE pumps use elements of a standard type.

Instead of using conventional adjustable tappets as the phasing adjustment (refer to 'Final testing and setting' on page 199), N-type pumps feature a shim adjustment. Shims of various thicknesses are available and are fitted between the top of the special cam followers and the plunger foot. The thickness of the shim inserted determines the position of the plunger stroke in its barrel; hence phasing adjustments are made by increasing or decreasing the thickness of the shim used, as necessary.

CAV–Simms

While the range of 'Simms' design, multi-element, enclosed camshaft pumps, now manufactured by CAV, is considerable, those most commonly encountered belong to the following series:

- SPE-A
- SPE-B.
- SPE-M.

CAV SPE-A series fuel injection pumps

CAV SPE-A multi-element, enclosed camshaft fuel injection pumps are cam operated, constant stroke, plunger pumps suitable for engines with a maximum capacity of approximately 1.25 litres per cylinder.

Plunger and barrel design—CAV SPE-A pumps may be fitted with plungers of 6.0, 6.5, 7.0, 7.5 and 8.0 mm. As may be seen in Fig 18.23, the top of the plunger is provided with an inclined slot in its periphery, which connects via a radial drilling to an axial drilling centrally located in the plunger. The base of the plunger has an arm pressed onto it and is not provided with vanes or lugs. This arm engages with a fork, which is clamped to a square control rod, and so movement of the control rod causes the plunger to rotate in the barrel, thus varying the effective plunger stroke, without the use of a control sleeve. Adjustment of the individual fuel deliveries—calibration—is effected by slackening the clamping screw in the control fork, and sliding the latter along the control rod.

Note This type of pumping element is used in most SIMMS design pumps, but not all. SPE-B pumps use a conventional element where plunger rotation is achieved in the normal way, by means of a control rack, control pinions and control sleeves.

Governors—These pumps are normally supplied with pneumatic governors, but may be obtained with centrifugal governors.

Delivery valve design—As may be seen by reference to Fig 18.23, CAV SPE-A pumps feature a standard mitre-type volume unloading delivery valve.

Pump body design—The pump body is made up in two sections, with a horizontal joint

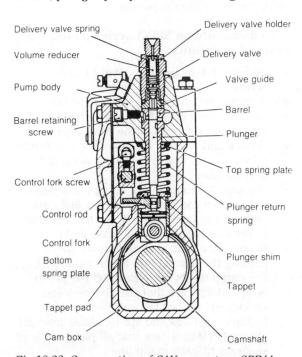

Delivery valve spring
Volume reducer
Pump body
Barrel retaining screw
Control fork screw
Control rod
Control fork
Bottom spring plate
Tappet pad
Cam box

Delivery valve holder
Delivery valve
Valve guide
Barrel
Plunger
Top spring plate
Plunger return spring
Plunger shim
Tappet
Camshaft

Fig 18.22 Cross-section of CAV pump type SPE4A

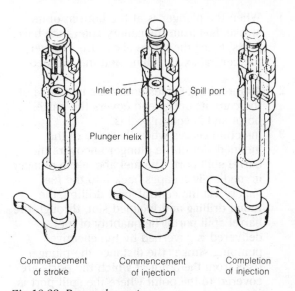

Inlet port
Spill port
Plunger helix

Commencement of stroke

Commencement of injection

Completion of injection

Fig 18.23 Pump element

between them. The upper section contains the fuel gallery and carries the elements, delivery valves, and delivery unions, while the lower section or cambox carries the camshaft and tappets. During service, the upper half can be removed for inspection of the elements without disturbing the camshaft and tappets. This two-piece body design is a feature of all CAV SPE pumps.

When a pneumatic governor is fitted to this pump, a maximum fuel stop is provided at the free end of the control rod. An excess fuel device to prolong injection for cold starting can also be fitted in this position when required.

Provision is made for fitting fuel lift pumps to the injector pump housing where they are operated by the camshaft.

The camshaft and tappets—CAV SPE-A pumps feature camshafts of large diameter to give maximum rigidity. For pumps fitted with elements of 6.0 and 6.5 mm diameter, the camshaft tapered driving end extension has a maximum diameter of 17 mm. Pumps fitted with the elements of 7.0 mm diameter and more have a driving extension of 20 mm and heavier camshaft bearings identical with those in SPE-B series pumps. The adjustment of tappets to equalise the phase angle between injections is carried out by exchanging spacing pieces of graded thickness secured in the tops of the tappets by circlips (refer to 'Final testing and and setting' on page 199).

Method of operation—By reference to Fig 18.23 it may be seen that the method of operation is as follows:

1   When the plunger is at the bottom of its stroke, fuel from the gallery enters the barrel ports, filling the space above the plunger, the central axial drilling and the inclined slot.

2   As the plunger rises under the influence of the cam, its upper end covers the barrel ports and injection begins.

3   Injection ceases when the edge of the inclined slot in the plunger uncovers the barrel spill port. The fuel above the plunger immediately escapes back into the fuel gallery via the central axial drilling, the radial drilling and inclined slot, through the barrel spill port. The quantity of fuel delivered is governed by the effective pumping stroke (the distance the plunger rises from the point at which the ports are covered to the point where the spill port is uncovered by the inclined slot), and the effective pumping stroke is varied by rotation of the plunger.

4   When the rack is moved to the engine stop position, the plunger is rotated in the 'short effective stroke' direction. In fact, the inclined slot exposes the spill port before the top of the plunger covers both ports. Thus the 'effective stroke' is non-existent, and fuel escapes from the barrel as soon as the plunger starts to rise.

Pump lubrication—On CAV SPE-A pumps provided with an oil filter, engine lubricating oil should be added to the level of the leak-off point when a new or reconditioned pump is fitted. The level should be checked weekly, and topped up if necessary. The oil should be drained and refilled every time the engine oil is changed.

On CAV SPE-A pumps not provided with an oil filler, the inspection cover should be removed and engine lubrication oil poured into the interior of the pump until it begins to run from the leak-off connection.

Pump service—Due to their unique two-piece pump housing and other design features, SPE pumps are serviced in a different sequence and manner from the more conventional pumps. The following service procedure, though specifically for SPE-A pumps, provides a general guide for many Simms pumps.

To dismantle an SPE-A series pump, proceed as follows:

• Remove the inspection cover (4 or 8 screws).
• Remove the governor complete (4 bolts).
• Remove the excess fuel case outer half (4 screws, or 2 screws in the early type).
• Unlock and remove the governor link; (either a lock nut and taper screw or nyloc nut may be used).
• Slacken the control fork screws and remove the control rod from the excess fuel case end; (in the later types, it will be necessary to remove the excess fuel case first).
• Remove the four nuts and spring washers holding the pump body to the cambox. Lay the pump on its back. Ease off the pump body, taking care that the plungers and springs etc do not fall out. Remove the plungers, springs and spring plates from all elements (taking care that all parts are kept separate to each pump line).

- Remove the delivery valve holders, volume reducers, delivery valve springs and delivery valves (again keeping separate to each pump line).
- Remove the barrel locating screws and with a soft mallet tap out the element barrels, complete with the delivery valve bodies and sealing washers. It is essential that all parts are kept separate to each pump line.
- Remove the tappets, taking care to keep them in correct order.
- Remove the roller pins, the roller bushes and the tappet rollers.
- Remove the coupling fixing nut and the coupling.
- Remove the four screws from the bearing housing remote from the coupling. Tap the coupling end of the camshaft with a soft mallet to break the liquid jointing seal and remove the bearing housing.
- Remove the four screws from the coupling-end bearing housing. Break the seal by tapping the end of the camshaft and remove the bearing housing complete with the bearing and camshaft from the cambox.

SPE-A pumps should be reassembled as follows:

- Immediately before assembly, thoroughly wash all components in petrol or kerosene and then immerse in clean fuel or test oil.
- Assemble the camshaft in the cambox so

that the end float of the camshaft lies between 0.05 and 0.15 mm. (Adjust the end float by inserting or removing shims between the bearing inner race and the shoulder of the camshaft. Pack the bearings with high melting-point grease (eg Mobilgrease MP) before final assembly, and check that the camshaft is installed to give the correct firing order.)

- Fit all the tappets in their original bores.
- Rinse the barrels in clean fuel or test oil, and fit them in the same bores in the upper body section as they were removed from.
- Fit the barrel locating screws with their sealing washers under their heads so that they engage with the slots in the barrels. When tightened, the locating scews must not pinch the barrels.
- Rinse the delivery valve assemblies, and place them in position with their sealing washers in place. The joint faces of the barrels and delivery valve bodies must be perfectly clean. Screw in each delivery valve holder after fitting the delivery valve spring and the volume reducer. Tighten the delivery valve holders with a tension wrench to a torque of 39.5 N.m.

CAV SPE-B and SPE-BN fuel injection pumps

Both the SPE-B and SPE-BN series fuel injection pumps are cam-operated, constant stroke plunger pumps of from one to six elements.

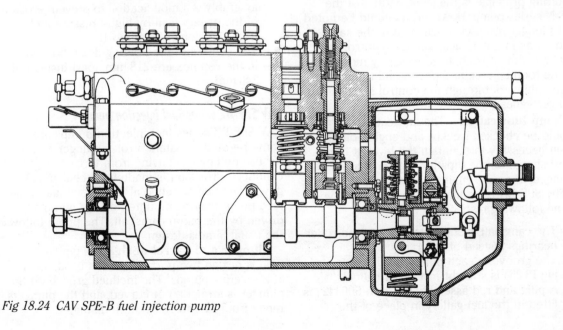

Fig 18.24 CAV SPE-B fuel injection pump

Provision is made for a governor to be fitted to either end of the pumps and, if required, a camshaft-driven fuel feed pump may be attached to the cambox.

Fig 18.24 illustrates the construction of an SPE-B series pump, and subsequent construction details will deal with this pump. The main points of difference to be found in SPE-BN pumps will be outlined further on in this section.

Plunger and barrel design—The barrel is 'typical' two-port design, while the plunger carries vanes at its lower end, which engage in slots in a conventional control sleeve to facilitate plunger rotation. The upper end of the plunger is the same as that used in SPE-A series pumps, except that the inclined groove is slightly curved.

The plunger stroke is comparatively short (7.5 mm standard) and this reduces spill disturbances and stress in the plunger spring. Plungers of 6.0, 6.5, 7.0, 7.5, 8.0, 9.0, 10.0 or 11.0 mm diameter may be fitted.

Delivery valve design—As for SPE-A series pumps, standard-type volume unloading delivery valves are used.

Pump body design—The pump body is manufactured in two parts that are bolted together, as is the case in all Simms pumps.

Camshaft and tappets—The camshaft is of large diameter (for reasons of rigidity) and the tappets are the conventional screw adjustment type—shims are not used for phase angle adjustment.

Method of operation—The pumping and metering principle is the same as that of the SPE-N series pump. However, it should be noted that plunger rotation is achieved in the same manner as in the 'typical' pump—control rack axial movement is changed to rotary movement by the toothed rack and pinion and transferred to the plungers through the control sleeves and plunger vanes.

Pump lubrication—The lubricating oil should be checked weekly, and engine oil added when necessary to maintain the level at the point shown on the dipstick. The oil should be changed when the engine oil is changed.

The SPE-BN type differs from the SPE-B type in the following respects:

- The camshaft is supported by tapered roller bearings instead of ball races.
- The groove in each pump plunger (see Fig 18.25) is straight instead of helical.
- A plug and rod assembly (part no 500342) is fitted in the fuel gallery in place of the

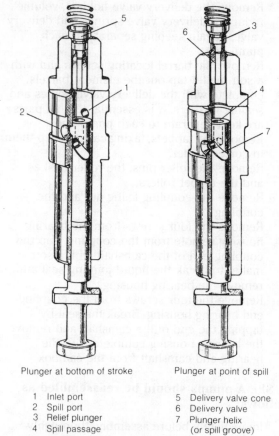

Plunger at bottom of stroke   Plunger at point of spill

1   Inlet port
2   Spill port
3   Relief plunger
4   Spill passage
5   Delivery valve cone
6   Delivery valve
7   Plunger helix
     (or spill groove)

Fig 18.25 SPE-B pump element

gallery plug. The fitting of the plug and rod assembly is a modification to prevent erosion of the gallery wall by high-pressure fuel from the element spill port.

- The centres of the holding-down bolt holes in the cambox are 218 mm apart instead of 240 mm.

CAV SPE-BZ series fuel injection pumps

CAV SPE-BZ series fuel injection pumps are cam-operated, constant stroke, plunger pumps for use on engines having from three to eight cylinders. Provision is made for a centrifugal governor to be attached at either end and a fuel feed pump may be mounted on the pump and driven by the pump camshaft. The pump follows the usual Simms design.

Plunger and barrel design—A typical Simms design two-port element is used, as in SPE-A series pumps. The inclined groove on the plunger is straight, not curved, and the plungers range from 11.0 to 13.0 mm in diameter.

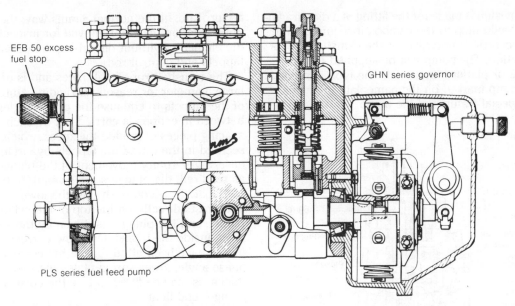

EFB 50 excess
fuel stop

GHN series governor

PLS series fuel feed pump

Fig 18.26 CAV SPE-BN fuel injection pump

Delivery valve design—Standard volume unloading delivery valves are employed.

Pump body design—The pump body follows the usual Simms design; it is made in two parts, one of which carries the camshaft and tappets, the other the elements, delivery valves and fuel unions.

Camshafts and tappets—The camshaft runs in tapered roller bearings, and is of robust design for maximum rigidity. The tappets follow the same design as those in SPE-A pumps; phase angle adjustmets are made by changing spacer shims between the tappet and plunger.

Method of operation—Pumping and metering is achieved in the same manner as was described for other SPE pumps. Like the SPE-A series pump, the SPE-BZ series does not use a control sleeve to rotate the plunger; instead, a fork on the square control rod engages the 'arm' pressed onto the end of the plunger.

Pump lubrication—The lubrication requirements of SPE-BZ pumps are the same as those already discussed for other CAV-Simms pumps.

CAV SPE-M series fuel injection pumps

The CAV SPE-M series of fuel injection pumps are all cam-operated, constant stroke plunger pumps. This series of pumps (known as the 'mini' series) is made up of three separate pumps:

- the Minipump
- the Minivac
- the Minimec.

As the names imply, these pumps are smaller and more compact than other Simms pumps, and therefore have their own particular design features. The most noticeable of these special features are as follows:

- A steel upper pump housing is used to hold the elements.
- The delivery valve holders have serrations on the outside diameter, instead of the usual hexagons.
- In some cases special plungers, which prevent fuel leakage into the cambox, are used.
- In the case of the Minivac and the Minimec, the governor is built into the main pump housing (except the 8-element Minimec).

The Minipump

The SPE-M series Minipump is suitable for use on compression ignition engines having a maximum swept volume of up to approximately 1.5 litres per cylinder. The Minipump does not have a governor attached, although provisions are made to enable either a GP pneumatic or a GMV centrifugal governor to be fitted to suit different requirements.

Provision is made for the fitting of a diaphragm-type feed pump to the cambox if required, and an eccentric is provided on the camshaft for its operation. The pump can be supplied either flange or platform-mounted. The type number of the pump marked on the nameplate indicates the special features of the pump.

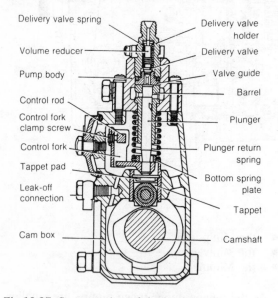

Fig 18.27 Cross-section of the Minipump

The plungers used in the Minipump are of the typical CAV–Simms design, featuring an arm pressed onto the lower end to facilitate plunger rotation. Two-port barrels are used, and these carry serrations on their outer surface below the shoulder, which match the corresponding serrations in the pump body, thus providing a means of locating the barrels.

Standard mitre delivery valves are used, but the delivery valve holders are different from those in other SPE pumps, having serrations on their outside diameters instead of the usual hexagons. This is necessary to allow the elements to lie close together, and a special spanner is necessary to remove and refit them.

In the usual manner, the pump body is made in two sections, the upper section carrying the elements etc, the lower carrying the camshaft and tappets. Unlike other SPE pumps, however, the upper section is made of steel to provide sufficient strength in a housing where the elements lie very close together.

The lower section or cambox is of light aluminium alloy, but has steel T-shaped inserts between the bores in which the tappets reciprocate. In the normal Simms way, the upper section may be removed for inspection of the elements, without disturbing the camshaft, tappets and control rod.

The camshaft runs in all races and is of large diameter for reasons of rigidity. Adjustment of the tappets to equalise the phase angles between injections is carried out by exchanging spacing pieces of graded thickness, which are secured in the top of the tappets by circlips. To ensure free rotary movement of the plungers as required, the plungers are given 0.05–0.2 mm end float. To accomplish this, the lower spring plate rests directly on the top of the cylindrical tappet, and the plunger is driven by the spacer piece inside the tappet. Once the correct spacer pieces have been fitted to give the correct phase angle, lower spring plates of various thickness may be fitted to give the correct plunger end float.

The method of operation in regard to pumping and metering follows the standard Simms pattern.

Weekly checks and regular changes of the lubricating oil are recommended as for the other SPE pumps. During overhaul, it is recommended that the camshaft bearings be packed with Mobilgrease MP.

The Minivac

The Minivac fuel injection pump is very similar to the Minipump but the lower section of the housing is made to incorporate a pneumatic governor. By incorporating the governor in the main housing, a very compact unit is produced.

The Minimec

There are two separate types of Minimec fuel injection pump—the standard type of Minimec that follows the Minipump and Minivac in design and is available in three-, four- and six-element models, and the more recently introduced eight-element type. The standard Minimec follows the typical 'mini' design, and features a mechanical governor in the rear section of the pump housing. The eight-element model, however, while still a compact pump using a mechanical governor, has many exclusive features, which must be considered.

The eight-element Minimec pump is provided with a mounting flange for mounting the pump to the engine timing case. A bracket must be attached to the governor end to help support the pump. The main points of difference between

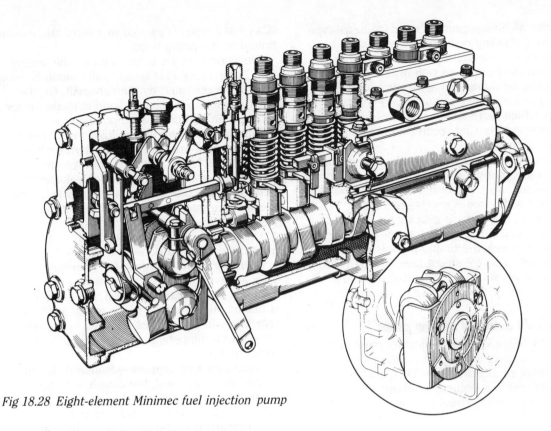

Fig 18.28 Eight-element Minimec fuel injection pump

the eight-element Minimec and the standard type of Minimec may be outlined as follows:

- The pump elements feature 'T'-flatted plungers to reduce plunger leakage into the cambox to a minimum (see Fig 18.29).
- The governor case is not an integral part of the cambox.
- The governor incorporates a cushion drive to dampen the effect of sudden accelerations in speed.
- The pump is lubricated by the engine lubrication system.

The plungers are similar to standard Simms plungers and have a straight spill groove at an angle of 45°. However, they also feature a 'T'-flat and a circular groove machined on the plunger stem below the spill groove. Any fuel leaking past the top of the plunger will collect in the 'T'-flat and groove and, as soon as it builds up sufficient pressure to do so, will escape back into the gallery via the inlet port when the plunger is at the top of its stroke. The barrel features the serrations on its outer diameter as do all the 'mini' pumps. However, the port layout differs from the standard style, with the inlet port substantially above the spill port.

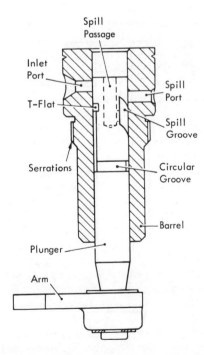

Fig 18.29 The Minimec 'T'-flat pump element

As for all Simms pumps, standard mitre-type volume unloading delivery valves are used.

The pump housing is made of a number of parts—a steel upper section, which carries the elements, an aluminium alloy cambox, a cast iron mounting flange at the drive end of the pump, which carries the camshaft bearing, and the governor housing secured to the cambox by five set screws.

The camshaft and tappets follow the standard 'mini' series design. A gear adaptor is keyed to the tapered drive end of the camshaft, and is provided with three screw holes for the attachment of the drive gear by set screws. A 'V'-slot is machined in the adaptor periphery for timing purposes; the timing mark is engraved on the mounting flange spigot face. The TDC of no. 1 engine cylinder is indicated when the 'V'-slot is in line with the timing mark.

CAV SPE-ZN series fuel injection pumps

The CAV SPE-ZN series (Majormec) fuel injection pumps are designed for use on the larger types of high-speed diesel engines (4, 6 and 8 cylinders) with a capacity of about 1.5–3.5 litres per cylinder. These pumps have been developed and produced on the same lines as the Minimec pumps. A mechanical governor

(CAV GMV type) is carried in a separate housing bolted to the pump housing.

Provision is made to mount a double-acting feed pump (CAV PLD type) to the pump housing where it is operated by the camshaft. On the eight-element pumps, provision is made for the fitting of an additional feed pump.

Plunger and barrel design—Two types of pump element are used—a standard type of element (Fig 18.30) and a 'T'-flat element (Fig 18.31), which minimises fuel leakage and dilution of the lubricating oil. Pumps featuring the second type are normally lubricated by the engine lubricating system.

The barrels are located positively in the housing by serrations.

Delivery valve design—As is normal practice, standard-type delivery valves are used.

Pump body design—Again, the pump body is manufactured in two parts, which are bolted together, while 'T'-pieces are used in the housing between the tappets as is the case in the SPE-M series pumps.

Camshaft and tappets—Standard-design components are used, the camshaft being supported in tapered roller bearings.

Operation—As for SPE-M series pumps.

Lubrication—Two methods of lubrication are employed—either the conventional system with the lubricating oil carried in the cambox, or lubrication from the engine lubrication system.

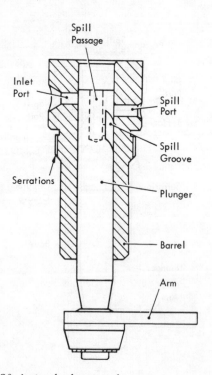

Fig 18.30 A standard pump element

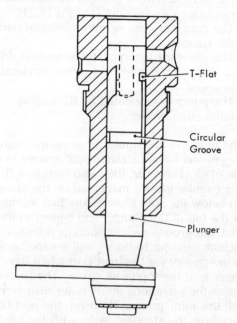

Fig 18.31 A pump element with a 'T'-flat plunger

Robert Bosch

Robert Bosch manufacture a considerable range of multi-element, enclosed camshaft fuel injection pumps.

These, which include some models with up to eight pumping elements, are single acting, constant stroke plunger pumps with outputs controlled by varying the effective working stroke.

As in the typical pump, mitre-type volume unloading delivery valves are used.

The Bosch PE..A fuel injection pump (Fig 18.32) is of conventional design, but is not the only design to which Bosch pumps are manufactured.

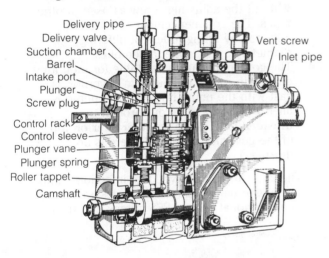

Fig 18.32 The Bosch PE..A-type injection pump

Plunger and barrel design—By reference to Fig 18.33, it can be seen that the type PE..A pump on the right is fitted with a single port element, while the type PE..B shown on the left features a conventional two port element.

The single port elements are the same as those used in Bosch flange-mounted fuel injection pumps. The barrel is provided with only one radial port and the top of the plunger has a vertical bore joined to an inclined slot or helix by a radial drilling.

Some Bosch pumps use two port barrels fitted with plungers that have an additional helix on the top for varying the point of commencement of injection.

Governors—Bosch enclosed camshaft fuel injection pumps are usually equipped with a governor, the housing bolting to the pump housing to make one combined unit. Both pneumatic and mechanical governors are used.

Mounting and driving—Bosch pumps are mounted either by bolts or by clamping straps on a flat bracket or in a semicircular cradle. The radius of the cradle is 45 mm for the PE..A, 56 mm for the PE..B and 62.5 mm for the PE..Z type pumps. The PES-type pumps are flange mounted. This type of mounting is now more and more in favour with engine manufacturers.

When mounting the pump, the camshaft should be in the horizontal position to ensure even lubrication of all cams. The pump proper (the pump element) is usually mounted in the vertical position. Should the pump be mounted

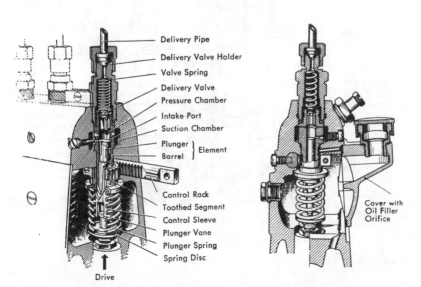

Fig 18.33 Fuel injection pumps in section; PE..B type, (left), PE..A type (right)

in other than a vertical plane, special types have to be employed.

The pumps are provided with an extended camshaft with a taper and thread at both ends so that they may be coupled to the engine at either end. To compensate for slight misalignment between the engine drive and pump, a closed slot cross-coupling is available. (This coupling will be supplied on special order only.) When mounting the pump 1.5 mm end clearance for the coupling should be secured in order to prevent axial stress on the bearings. So-called 'elastic' couplings with spiral band or rubber composition cores must not be used for driving these pumps, as correct timing would be impossible.

These pumps can be driven in either direction since the cams are symmetrical; however, by changing the direction of rotation, the injection sequence of the pump is reversed. Because of this, care should be taken to properly connect the pump elements to the engine cylinders.

Timing—Robert Bosch pumps carry the same timing marks as the 'typical' pump and American Bosch pumps—timing is carried out as for the typical pump (refer to page 205).

Bosch PES..M-type fuel injection pump

This type of pump is the smallest in the PE series. The main difference between this type and 'conventional' Bosch pumps lies in the method of transmitting the control rack movement to the pump plungers. Whereas in the conventional pumps control rack travel is transmitted to the plungers by rack and toothed segments, in the PES..M-type pump clamping pieces with a groove on the top are located on the control rack. A lever with a riveted pin protrudes from the control sleeve of each element, the pin being engaged in the groove of the clamping piece.

Further special features of the PES..M pump are:

- The bottom end of the pump plunger contacts the roller tappet assembly directly, without the adjusting screw as used in other types. The commencement of injection is adjusted by fitting rollers of differing diameter to the tappet assemblies.
- The governor housing, suitable for mounting a mechanical or a pneumatic governor, is cast with the pump housing. The pump is filled with lubricating oil through the threaded hole of the breather filter on top of the governor housing. (There is no dipstick.)

The PE..P and PES..P type fuel injection pump differ in external appearance and design from the conventional PE and PES pumps, but their

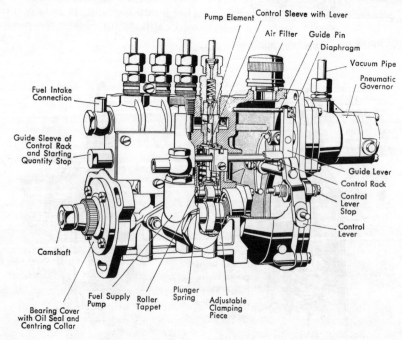

Fig 18.34 The Bosch fuel injection pump PES4M

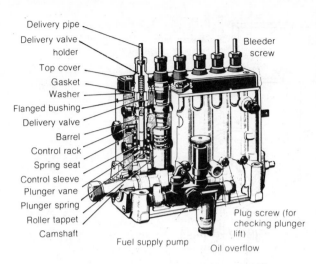

Delivery pipe
Delivery valve holder
Top cover
Gasket
Washer
Flanged bushing
Delivery valve
Barrel
Control rack
Spring seat
Control sleeve
Plunger vane
Plunger spring
Roller tappet
Camshaft
Fuel supply pump
Bleeder screw
Plug screw (for checking plunger lift)
Oil overflow

Fig 18.35 The Bosch fuel injection pump PE6P

method of operation is fundamentally the same.

They are distinguished from the other pumps in that:

- Individual plunger barrels are connected by a flanged bushing to the delivery valve and holder to form a compact, composite unit.
- Calibration adjustments are made by rotating this flanged bushing, and adjustments to the commencement of delivery are made by inserting shims under it. Thus the pump housing has no aperture for making adjustments and is totally enclosed—it has no side cover.
- A toothed control rack and toothed segments are not used to govern delivery – the control rack is of angular section and has slots that register with balls fixed to the control sleeve.
- The pump is connected to the lubricating oil circuit of the engine.

Servicing a typical single element flange-mounted pump

*T*he need for the utmost cleanliness in any service operation performed on fuel injection equipment cannot be over-emphasised. The ideal conditions as regards cleanliness can only be created in an air-conditioned, dust-proofed room, and injection pumps should only be dismantled in such a room. In addition to a suitable room, special tools, specialised equipment and pump data sheets are necessary before any pump service can be carried out, so that trained personnel with the necessary equipment and information are the only persons who should attempt pump service.

Once it has been removed from the engine, and all dirt washed from the exterior, the service procedure for a typical single element, flange-mounted fuel injection pump may be divided into six separate operations:

- preliminary testing
- disassembly
- cleaning components
- inspection of components
- reassembly
- final testing.

Preliminary testing

The main object of a preliminary test is to determine the condition of the pumping element. However, a defective delivery valve, or a faulty delivery valve body joint, any also be detected by means of this test.

The required test rig consists of a reservoir or tank to carry fuel or test oil, from which fuel flows through a flexible fuel line to the inlet connection of the fuel injection pump, mounted on a pump test bracket. This pump test bracket consists of a means of mounting the pump, and an operating lever to operate the pump plunger via the plunger guide cup and so on in the normal way. A high-pressure line carries fuel from the pump outlet (when the operating lever is operated) to a pressure gauge. A release valve is provided in the high-pressure line to faciliate bleeding air from the system.

To perform the test, the rig must be set up as described above. The fuel supply to the pump should be switched on, and the release valve opened. All air should be expelled from the system by pumping the operating lever, and the release valve is then tightened.

Before starting the actual testing, it is essential to ensure that there are no leaks in the test system, or on the pump or pump connections.

Note   The pump will not operate if the fuel control rack is in the 'engine stopped' position.

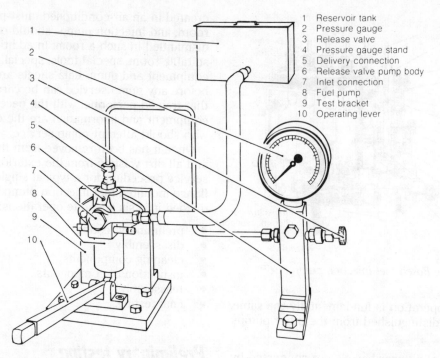

1 Reservoir tank
2 Pressure gauge
3 Release valve
4 Pressure gauge stand
5 Delivery connection
6 Release valve pump body
7 Inlet connection
8 Fuel pump
9 Test bracket
10 Operating lever

Fig 18.36 The test-rig layout

During bleeding and throughout the test, the rack should be set in the normal running position.

The test is performed in three phases:

1 Operate the pump by working the hand lever of the test rig. No actual test is made for 'suction' on this type of pump, as no inlet valve is fitted, but the intake side is checked by the feel of the pump when operating the lever. Any defect on the intake side of the pump will allow the admission of air and will be readily apparent because of the loose or elastic feeling of the lever. The reason for this can be either lack of fuel in the supply tank or a leak on the intake side of the test system.

    Any defect must be rectified before proceeding further with the test.

2 The delivery of the pump is proved and tested by the regular increase of pressure with each pumping stroke, indicated on the pressure gauge.

    During the delivery test, pumping should be stopped at any pressure reading and the indicated pressure noted. This pressure should remain steady. Any falling back of pressure indicates either a leak on the delivery side of the test system or a defect

at either the delivery valve or the delivery valve seat joint.

3 Defective parts should be corrected or removed. The final phase of this test is to check that the pump can readily provide fuel at a pressure high enough to overcome the injector opening pressure.

    Note the injector opening pressure. (This should be found in the manufacturer's manual.) Operate the pump until the indicated pressure is equal to the injector opening pressure. Further operation of the lever through one pumping stroke only **must ensure** that the pressure reading is well in excess of the injector opening pressure. This test indicates that the clearances between the pump plunger and the barrel are satisfactory, and if the pump fails to meet the requirements of the test the pumping assembly will require further checking.

Note This preliminary testing on the equipment described does not give accurate, detailed information such as is obtained with a power test bench or similar test equipment, but the results obtained are satisfactory for most purposes.

Disassembly

When the injection pump is being dismantled, the manufacturer's recommended dismantling sequence should be followed. While some manufacturers such as Ruston and the Scintilla Division of the Bendix Corporation recommend that dismantling begins at the delivery valve holder, most manufacturers of flange-mounted fuel injection pumps recommend that dismantling starts at the base of the pump. The following sequence (by courtesy of Bryce Berger Ltd) outlines a good, easy-to-follow procedure that can be readily applied to a typical single element flange-mounted fuel injection pump.

Fig 18.37 Holding the pump

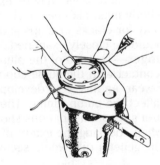

Fig 18.38 Removing the circlip

Grip the pump by the discharge union in a vice with plain jaws. Depress the plunger guide cup against its spring until it is clear of the cross-hole drilled in the pump body spigot. Insert a pin or suitable piece of wire into this hole to retain the plunger guide cup in the depressed condition.

Insert a tool into the hole breaking into the circlip groove and lift the circlip. Remove the circlip by inserting a screwdriver under it. Remove the pin holding the plunger guide cup depressed and allow it to rise gently under the

action of the plunger return spring, clear of the pump body.

Remove the plunger guide cup and the lower spring plate.

Fig 18.39 Removing the lower spring plate

Remove the plunger return spring and carefully withdraw the plunger and place it in a container of clean test oil where no possibility of scratching, burring or damaging of the lapped surfaces can occur.

Fig 18.40 Withdrawing the plunger

With the aid of pliers, pull out the control sleeve and pinion, together with the upper spring plate and retaining circlips (if fitted). Before removing the control sleeve and pinion,

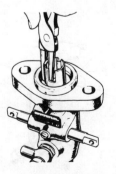

Fig 18.41 Removing the control sleeve

however, note the position of the plunger vane in the control sleeve slot and also the assembly marks on the pinion and control rod teeth.

Withdraw the control rod after removing the control rod locating screw (where fitted).

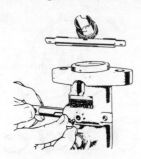

Fig 18.42 Withdrawing the control rod

Reverse the pump body in the vice and unscrew the delivery valve holder. Remove the delivery valve spring or springs, the delivery valve and the valve stop (if used), care being taken not to damage the lapped surfaces. Remove the air vent screw and washer.

Fig 18.43 Removing the delivery valve and spring

With the delivery valve extractor, remove the valve body and the discharge union washer. Remove the barrel locating screw and washer.

Fig 18.44 Withdrawing the delivery valve body

On removal of the pump from the vice, the barrel can be easily pushed out.

The plunger and barrel should immediately be mated to avoid damage and to prevent confusion when servicing a number of pumps.

Cleaning components

Wash and thoroughly clean the body, discharge union, rack, pinion, spring plates, circlip etc. Rinse all the components with lapped surfaces in clean petrol or distillate. Gently rotate and reciprocate the plunger in its barrel, rinsing frequently. After cleaning, place all parts on clean, lintless paper.

Inspection of components

Once the pump has been dismantled and cleaned, the various components must be inspected thoroughly to determine whether they should be renewed.

The body—Check the passage in the housing through which the control rod passes. If the pump is used on an engine that is subjected to varying loads, this passage may be excessively worn. The body must also be checked externally and internally for cracks or chips and all threads should be closely examined.

The plunger—The end of the plunger that comes in contact with the guide cup will eventually wear. As this wear develops, it will gradually affect the pump timing. The manufacturer's recommendations should be followed when checking the extent of this wear, although the general rule would seem to be as follows:

- Hang the plunger vertically from the lower spring plate and place a straight edge across the spring plate and plunger. With a feeler gauge, measure the distance the plunger end protrudes above the spring plate face or lies below the face.
- Check the measurement against the manufacturer's figures.
- Excessive wear indicates that a new plunger and barrel assembly is required.

Pitting or erosion of the plunger is caused by cavitation in the vicinity of the barrel ports of pumps fitted to high performing engines where the plunger speeds (and consequently, the fuel velocities) are high. This abnormal condition is permissible if the eroded area is not too close to the leading edge of the plunger helix. The most usual position for erosion on a plunger is along a line parallel to and approximately one port diameter distant from the helix. It is rare for cavitation to affect the actual helix edge.

When helix edge deterioration occurs in service, it is probably due to solid particles in the fuel abrading the edge.

If the leading edge of the plunger helix does become damaged by erosion, it will seriously affect the pump timing and fuel delivery. Any plunger showing erosion approaching that on the helix as illustrated on plunger A in Fig 18.45 should be renewed; a new plunger and barrel assembly, of course, being fitted. On any plunger, 0.8 mm is the absolute minimum distance allowable between the edge of the helix and the eroded area. The number of engine running hours between overhauls will determine whether a minimum of more than 0.8 mm should be established. The erosion on plunger B is acceptable, and the plunger is fit for further service.

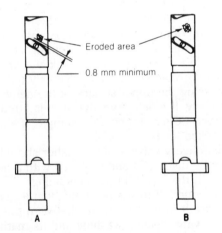

Eroded area

0.8 mm minimum

A   **B**

Fig 18.45 Eroded plungers

The plunger should also be inspected for signs of wear or scoring caused by dirt in the fuel. The wear is usually confined largely to the region that passes over the barrel ports, and consists of deep straight scratches uniformly spread over most of the length of the effective stroke. The presence of fine score marks does not indicate a faulty plunger; these fine marks appear after a short period of service and do not impair the efficiency of the plunger. It is

interesting to note that, while all abrasive particles cause wear, the most damaging particles lie in the size range from 5 to 15 microns.

Water in the fuel will also cause plunger damage in the form of corrosion, and the plunger should be examined for signs of this during the inspection.

The barrel—Like the plunger, the barrel may also be scored by particles of dirt in the fuel. Consequently, the barrel should be inspected carefully to ascertain the extent of the damage, which usually occurs in the vicinity of the barrel ports. The barrel should also be inspected for erosion, which results from cavitation—if erosion damage does occur, it usually takes the form of pits in the barrel ports.

When inspecting the barrel, particular care must be taken to check for cracks. However, it is rare to find any cracks as the barrel usually bursts instantly, without any warning, if it fails due to fracture of any type.

The upper, flat surface of the barrel is a lapped surface, and makes a fuel-tight seal with the lapped surface of the delivery valve body. This surface should be examined for scores, wear and/or corrosion. If it has sustained damage of this type, the barrel may be reclaimed by lapping on either a lapping plate or a power lapping machine, provided, of course, that the damage is not excessive.

If a barrel is to be reclaimed on a lapping plate, the following procedure is recommended:

1 Spread a small quantity of coarse lapping paste across the surface of the plate.
2 Hold the barrel so that the surface to be lapped can be pressed evenly and squarely against the surface of the lapping plate. It is probably best to use both hands. This will prevent the formation of uneven, non-parallel surfaces. Move the barrel in a smooth, figure-8 motion, occasionally rotating it between strokes, to provide a smooth flat finish.
3 When a smooth surface is obtained, wipe the plate with a clean cloth and repeat the lapping procedure using a fine lapping paste to polish the surface.
4 Thoroughly wash and flush the barrel with clean test oil, distillate or kerosene, and blow dry with clean compressed air. Coat the lapped surface of the barrel with some rust-inhibiting preparation and place it on clean lintless paper until final assembly.

5   If it is desired that the lapped surface be tested for flatness, this may be done using monochromatic light and optical flats.

The delivery valve—Both the conical face and the unloading collar of the delivery valve, together with the valve seat and bore in the delivery valve body, must be carefully inspected for signs of wear or damage, and the complete assembly renewed if the damage is too great.

If dirty fuel is being passed through the injection pump the unloading collar (or relief plunger) will become severely scored. Should this scoring be evident when the pump is dismantled, the delivery valve assembly must be renewed. This is such an important indication of dirty fuel that it is probably the first pump component at which the experienced serviceman looks when faulty filtration is suspected.

When the delivery valve unloading collar becomes worn it is possible that the amount of fuel withdrawn from the delivery line will be insufficient to adequately lower the line pressure. As a result of this, injector dribble may develop. Alternatively, if the valve seating also becomes worn, the pressure may be reduced too much, due to the fuel leaking back past the seat and collar. The fuel delivery will then be insufficient due to the line pressure having to be built up by the pump before the injector opens.

If excessive wear develops between the valve guide and the bore of the delivery valve body it is possible for the unloading collar to stick on the valve seat, thus preventing the valve from functioning correctly.

The sealing face of the delivery valve housing, which makes a metal-to-metal contact with the lapped upper face of the pump barrel, should be inspected for evidence of leakage. This will take the form of fine, irregular radial tracks on the surface.

If the tracks are not very deep, they may be removed by lapping on a lapping plate as described previously. If, however, the marks are too deep to allow reclaiming by this method, the complete delivery valve assembly should be renewed.

The upper surface of the delivery valve that is in contact with the delivery valve sealing washer should also be checked for any signs of damage. If there is severe damage the delivery valve assembly must be renewed, but minor damage at this point can be rectified in the following way:

1   Cover the precision lapped surface of the delivery valve with masking tape or some similar protective covering that can be easily removed.

2   Carefully secure the delivery valve body in a lathe fitted with a three-jaw chuck, placing small pieces of soft metal such as copper between the jaws of the chuck and the delivery valve threads. Wrap a piece of fine emery cloth (320 grit) around a piece of flat metal. As the body is rotated in the lathe, bring the emery cloth into contact with the washer mating surface, at the same angle as that to which the surface was originally ground. This angle is important, and must not be changed.

3   When a perfectly smooth, clean surface is obtained, polish it with a piece of crocus cloth using the above procedure.

4   Remove the delivery valve body from the lathe and remove the protective covering. Thoroughly wash the body in clean fuel or kerosene, blow off with clean compressed air, and place it on clean lintless paper.

The delivery valve holder should also be inspected, particular attention being paid to the threads and the lower surface of the holder that comes in contact with the delivery valve sealing washer. If this surface needs attention, it may be gripped in a three-jaw lathe chunk and dressed in a similar manner to that previously outlined for dressing the upper surface of the delivery valve body. If a delivery valve stop is fitted as in the Bryce CC size pump, for example, it should be checked for tightness of fit.

The delivery valve sealing washer should always be discarded and replaced with a new one when the injection pump is being reconditioned. If this is not done, fuel leakage could result from 'channeling' between the delivery valve sealing washing and its mating surfaces.

The delivery valve spring and stop (if fitted) should be carefully tested and examined for any possible defects. A broken or weak spring has a marked effect on the fuel delivery to the engine, and thus has a serious influence on the engine's performance.

The plunger guide cup—The base of the plunger guide cup should be checked for wear at the point where it makes contact with the engine tappet. In time a depression will develop in the guide cup surface, and as this progresses it will gradually affect the pump timing. The manufacturer's specifications should be checked to ascertain the maximum allowable wear, but a figure of 0.38 mm would seem to be generally acceptable. Once the plunger guide cup becomes worn to excess, it must be renewed.

The plunger return spring—The plunger return spring should be carefully inspected for cracks, fractures or any other damage, and tested on a spring tester to determine whether its tension is sufficient for efficient operation.

The control sleeve—If the teeth on the control sleeve pinion show signs of excessive wear, the control sleeve must be renewed. It must also be renewed if the vertical slots in its lower end are excessively worn.

The control rack—The teeth on the control rack should be checked for wear, and the rack renewed if necessary. The amount of movement between the rack and pinion has a direct bearing on fuel metering efficiency, and so must directly affect engine performance. For this reason, it is desirable to keep the amount of movement between the rack and pinion to a minimum. The body of the rack must also be inspected for wear to ensure that it is not a sloppy fit in the pump housing.

As linkage wear will have a detrimental effect on engine performance, the clevis pin holes in either end of the control rack must be examined for signs of excessive wear, the presence of which will also indicate the necessity for renewal of the control rack.

Erosion plus or pressure screws—Erosion plugs, or pressure screws, are used in some fuel injection pumps (Bryce CC and Bendix Scintilla FD and FDS) to prevent erosion damage to the pump body. Body erosion, which seems most likely to occur in large-capacity pumps working at high pressures and speeds, takes place on the plunger upstroke when the metering edge of the helix reaches the lower edge of the spill port. At this point, a high-velocity jet of fuel is released from the barrel into the fuel gallery in the pump body. The force with which this jet of fuel strikes the gallery wall is so great that in time it will erode the hardest metal. To prevent erosion damage to the pump body, a hard, erosion-resistant, replaceable metal plug— known as either an erosion plug or a pressure screw—is fitted to the pump housing at a position where it will bear the attack of the eroding fuel. After the fuel has struck the plug its force is lost, and the high-pressure fuel disperses throughout the rest of the fuel in the gallery.

If inspection of the plug reveals signs of erosion, the depth of the affected area should be measured. The maximum allowable depth of erosion varies, but 0.5 mm would seem to be a fairly acceptable figure. Continued erosion of the erosion plug beyond service limits may result in irreparable damage to the pump body.

Reassembly

Before reassembly, all pump components must be thoroughly cleaned and oiled. The plunger and barrel should be reassembled while completely coated with clean test oil. The parts of the lower section of the pump should be oiled with test oil or clean lubricating oil. The injection pump should be reassembled by working in the reverse order to the dismantling sequence, particular attention being paid to the following points:

- The barrel must be placed in the pump body so that its locating groove aligns with the locating screw hole. When the barrel locating screw is fitted it should not bottom on the barrel, which should be left free to move slightly up and down. If this slight movement cannot be felt, it indicates that the locating screw spigot is not correctly engaged in the barrel locating groove, and the barrel must be repositioned.
- When the control sleeve is fitted, take care to ensure that timing marks on the control sleeve pinion and the control rack are carefully aligned.

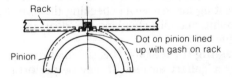

Fig 18.46 Rack–pinion timing

- Start the plunger squarely when entering it in the barrel, so that it will move down the barrel under its own weight. When the plunger is inserted in the barrel, turn it so that the marked end of the plunger vane fits into the marked control sleeve slot (see Fig 18.47). The mark on the plunger vane must coincide with the mark on the control sleeve to properly locate the plunger helix in relation to the barrel spill port.

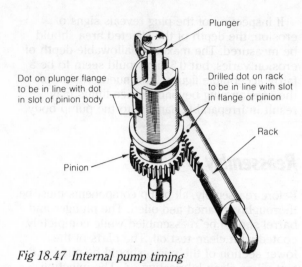

Plunger

Dot on plunger flange
to be in line with dot
in slot of pinion body

Drilled dot on rack
to be in line with slot
in flange of pinion

Rack

Pinion

Fig 18.47 Internal pump timing

- Position the plunger guide cup retaining circlip so that the ends are not less than 20 mm from the removal slot.
- When the injection pump is fully reassembled, hold it horizontally with the control rack vertical. The rack should settle to its lower extreme due to its own weight.

Final testing

When the injection pump has been reassembled, it must be subjected to several tests before it may be declared ready for use. For the sake of clarity the testing may be divided into two phases—static testing and dynamic testing.

Static testing

Static testing usually involves three operations (i) testing for leaks, (ii) checking the spill cut-off point, and (iii) checking the 'dead rack' setting.

Testing for leaks—This is sometimes referred to as a 'gallery air test'. It may be performed either with the pump completely assembled, or provided that a suitable bridge piece is bolted to the underside of the flange to support the plunger foot, without the guide cup, spring, pinion, and associated components. This test is carried out as follows:

- Ensure that the air vent is fitted to the pump body, and connect a compressed air line to the fuel inlet connection of the pump.
- Apply air at a pressure of 350 kPa and immerse the complete pump in test oil. This pressure should not be sufficient to open

the delivery valve, and no air should escape from the delivery valve holder.

- Watch closely for bubbles, which indicate that air is escaping from the pump. Bubbles issuing from the top of the pump could mean that the delivery valve washer is leaking, and bubbles issuing from the bottom of the pump could indicate a leak at the barrel-to-housing seat. It is possible that some air will escape through the clearance between the plunger and barrel, but this can be disregarded at this stage since excessive clearance will show up during subsequent tests.

Any defects revealed by this test should be corrected before proceeding further.

Checking the spill cut-off point—Before checking the spill cut-off point or checking the dead rack setting, the injection pump should be fitted to a cambox and the cambox mounted on a calibrating machine

Note A cambox is a device designed to operate a flange-mounted fuel injection pump 'on the bench'. It consists of a housing, a camshaft and one or more tappets (or cam followers). The number of cam lobes and cam followers depends on the number of pumps it is designed to operate. When the pump is mounted on the cambox and the camshaft is turned, the pump is operated as it would be on the engine. In this way the pump can be tested on the bench.

A calibrating machine or injection pump test bench is a machine on which injection pumps are fitted to ascertain their performance and to make any adjustments. Basically, a test bench consists of:

- a horizontal mounting table with suitable clamps, adaptors and fittings to carry the large variety of injection pumps
- a horizontal drive coupling shaft, some distance above the mounting table, driven through a variable speed mechanism by an electric motor
- a revolution counter connected to the drive coupling to register the pump camshaft rpm
- a test oil supply tank and fuel supply system—the test oil supply pressure is variable by means of a control, and a pressure gauge is included to indicate the pressure of the supply to the pump
- a set of special test injectors mounted to discharge into a set of graduated test tubes— an automatic cut-off device allows only a

Fig 18.48 Hartridge series 3000 fuel pump test stand

specified number (usually 100) of injections to enter the tubes, after which the spray is deflected away
● flexible hoses to connect the test oil supply to the pump under test and steel injector lines to run from the pump to the injectors.

Once the cambox is mounted on the calibrating machine, and the calibrating machine drive is connected to the cambox, proceed as follows:

1 Connect the fuel supply line from the calibrating machine to the fuel pump inlet connection.
2 Remove the delivery valve holder, delivery valve and spring. Do not remove the delivery valve body.
3 Refit the delivery valve holder, and tighten to the correct tension.
4 Fit a spill pipe to the delivery valve holder.

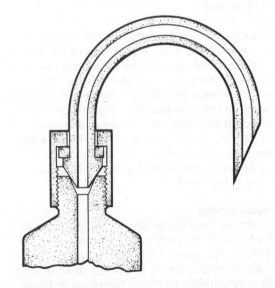

Fig 18.49 A spill pipe

5 Move the control rack to the full fuel position and ensure that the plunger is at the bottom of its stroke. Fuel will be seen to issue from the spill pipe.

6 Rotate the drive shaft of the calibrating machine slowly by hand until the flow of fuel from the spill pipe just ceases. This is spill cut-off point—the point at which the plunger just covers the barrel ports and injection begins.

On pumps provided with timing windows, the timing line on the plunger guide cup should coincide with the line inscribed on the pump window at spill cut-off point. If the pump element assembly has been renewed during reconditioning, the lines may not coincide, and the pump body will have to be re-marked.

If the pump is not provided with a sight window, the spill cut-off point should occur when the bottom of the plunger guide cup is a certain distance (specified by the manufacturer) from the base of the pump.

Checking the dead rack setting—With the pump mounted on the calibrating machine as described previously, proceed as follows

1 Move the control rack to the full fuel position.

2 Rotate the calibrating machine drive shaft slowly by hand until the fuel flow from the spill pipe ceases.

3 Slowly move the rack towards the stop position until fuel flow begins again. This must occur before the rack reaches the end of its travel.

This final test is important, since it gives an indication of the relative positions of the rack and the control sleeve. If fuel does not begin to flow as the rack is moved towards the stop position, it indicates that the plunger is not being rotated to the stop position. The engine would not stop if the pump were fitted in this condition.

If the fuel begins to flow before the rack has been moved far from the full fuel position, it may be an indication that the rack-to-pinion timing is incorrect.

Dynamic testing

The aim of dynamic testing is to determine if, under operating conditions, the pump is delivering the correct quantity of fuel at all rack positions. Since there is no means of adjustment, it is only possible to check the pump delivery at various rack positions; if the pump fails to make the correct fuel delivery, this is an indication of a fault in the pump.

To check the pump delivery a test sheet, which may be supplied by either the pump manufacturer or the engine manufacturer, is necessary. This sheet specifies the amount of fuel that must be delivered by the pump in a specified number of injections (usually 100 or 200), at specified control rack settings, for various specified cambox rpm.

To perform the test, proceed as follows:

1 Mount the pump on a suitable cambox and fit the assembly to a calibrating machine. Do not remove the delivery valve and spring.

2 Connect a high-pressure fuel delivery line from the pump discharge union to the machine's standard test injector. (A test injector is a special injector designed for testing purposes, and a standard injector should not be used unless authorised in the workshop manual.)

3 Ensure that the graduated test tube, mounted below the injector to measure the fuel delivery, is empty.

4 Ensure that the calibrating machine tank is filled with an approved test oil (such as Shell Calibrating Fluid 'B'), and the test oil temperature is as specified by the pump manufacturer.

5 Connect the fuel supply line to the pump and bleed all air from the system.

6 Run the pump at maximum rpm for not less than 10 minutes to ensure that all pump components are at operating temperature.

7 Adjust the pump control rod to the first specified setting (in mm) given on the test sheet. Although a graduated rack setting device is usually used to set and secure the control rack in the desired position, an accurate result may also be obtained by holding the rack by hand and using a depth gauge to set the rack in the specified position.

8 Start the calibrating machine and adjust its speed to that specified on the test sheet for the particular control rack setting.

9 Allow the pump to operate for the specified number of injections. To facilitate accurate counting of the injection strokes, which would be impossible at speed anyway, calibrating machines are fitted with an automatic timing device, which when engaged, allows only that fuel supplied by the required number of pumping strokes to enter the test tube and be recorded.

10 Note the quantity of fuel in the test tube.

Engine type	Pump type	Camshaft speed rpm	Rack setting at full load	Second position			Third position		
				Rack setting	Max	Min	Rack setting	Max	Min
ATX	FATC/B 20000A and all future	250	32	28	142	128	24	92	82
ATC	FATC/B	250	35	30	167	151	25	102	93

Note: All distances are in mm; all volumes are in cm^3 per 100 strokes

11 Adjust the control rack setting and the calibrating machine rpm to conform with the remaining settings specified on the test sheet, and measure and note the quantity of fuel delivered in each case.

The fuel deliveries at the various rack and rpm settings must correspond with those specified on the test sheet. Failure of the fuel deliveries at all or any setting to conform to specifications is an indication of a fault in the pump, and this must be rectified. Provided there are no leaks or external faults, the trouble will probably be found to lie in the pumping element or the delivery valve. High readings or low readings on all tests may be due to incorrect timing of the control rack to the pinion.

Once the pump has been dismantled, the fault rectified and the pump reassembled, it should be subjected to all final tests before refitting to the engine.

Here is an example of a test sheet (by courtesy of Ruston Engine Division, English Electric Diesels Ltd):

Installation of typical single element, flange-mounted pumps

Mounting and driving arrangements—
Fig 18.50 illustrates a typical mounting and driving arrangement for a single element, flange-mounted fuel injection pump.

Clearly, there are two main sections:

- cam
- tappet assembly, consisting of a roller, a tappet shell, an oil deflector and (in this

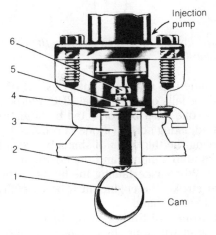

Fig 18.50 Details of a typical pump mounting

case) a tappet adjusting screw with a lock nut.

The tappet assembly is located directly above the cam, the profile of which is designed to give the required injection characteristics. The tappet adjusting screw and lock nut are provided so that the piston of the plunger stroke in the barrel may be adjusted as required. Some manufacturers, however, utilise shims placed between the mounting flange of the pump housing and the engine block instead of the adjusting screw to adjust the position of the plunger stroke.

Setting the plunger stroke—When a flange-mounted fuel injection pump is fitted to an engine, it is essential that the plunger rises to the correct height in the barrel. Not only must the plunger cover the ports before the unit can pump, but the helix must rise high enough to allow fuel to spill from the barrel ports at the completion of the stroke, full fuel delivery will not be achieved. Again, if the plunger guide cup does not rise high enough, it may touch the retaining circlip at the bottom of its stroke.

Should the plunger rise too high in the barrel, however, it will strike the bottom of the delivery valve body and serious damage may result.

For ease of checking the position of the plunger stroke, a reference line is provided around the plunger guide cup and this should remain visible in the inspection window in the pump body throughout the entire pump stroke.

Fig 18.51 Inspection window showing the three critical positions

If the guide cup reference line rises out of sight at TDC, or falls out of sight at BDC, the position of the stroke must be corrected by adjusting the tappet adjusting screw (if one is provided), or by varying the thickness of shims between the pump mounting flange and the engine mounting bracket. When ascertaining the limits of the plunger stroke, the engine must be carefully turned over by hand.

Installing and timing procedure—The following instructions (by courtesy of Ambac Industries Incorporated) for fitting and timing a flange-mounted fuel injection pump to an engine refers specifically to American Bosch single element, flange-mounted fuel injection pumps, but are given to provide a guide to the general procedure. Detailed instructions for fitting and timing the fuel injection pump to a specific engine are given in the engine workshop manual, and they should be carefully followed and strictly adhered to in all cases.

1   Turn the engine crankshaft until the cam is in its lowest position, at which the roller of the tappet assembly rides on the base circle of the cam.
2   Carefully wipe all dirt from the face of the engine mounting block and the pump mounting flange. Dirt will prevent the pump from seating squarely. Use a lint-free wiping cloth for this purpose. **Never use waste**. Then mount the injection pump on the top of the operating mechanism. Tighten the hold-down nuts or bolts evenly and securely.
3   Turn the engine crankshaft until the piston of the cylinder that is to be served by the injection pump to be timed reaches the position at which fuel injection is to begin.

This position is usually marked on the flywheel of the engine by the manufacturer and complete details of the flywheel markings are contained in the engine instruction manual. At this position, the line on the plunger guide cup and those on the inspection window should register. If they do not, then the adjusting screw on the tappet assembly must be raised or lowered accordingly, until exact registration is obtained. Be sure to tighten firmly the lock nut of the tappet adjusting scew.

4   Next, turn the engine crankshaft until the pump plunger reaches the top of its stroke, at which the roller of the tappet assembly rides on the highest part of the cam. The movement of the plunger can be followed by simply observing the reference mark on the plunger guide cup visible through the pump housing window.

Important note   With the plunger in either extreme position, top or bottom, it is of the utmost importance that the reference line mark on the plunger guide cup remains visible below the upper, and above the lower edge of the inspection window in the pump housing. Otherwise considerable damage to the injection pump may result.

Multiple pump applications

If a multi-cylinder engine is to operate smoothly and efficiently, it is essential that all engine cylinders receive exactly the same amount of fuel for each power stroke. In the case of a multi-cylinder engine that employs an individual injection pump for each cylinder, it is necessary to make sure that all injection pumps deliver the same amout of fuel at any throttle position, when the individual pump control racks are connected together. Ensuring an equal and correct fuel delivery from each pump (when separate pumps are used), or from each element of a multi-element pump, is known as calibrating the pump(s).

Calibrating the pumps

During overhaul the individual pump deliveries should have been checked, so the calibration should only entail marking the racks in such a way that they may be connected together on the engine to give equal deliveries, regardless of throttle position.

To calibrate two or more single element flange-mounted pumps, proceed as follows:

1 Mount the pumps to a suitable cambox, and mount the cambox in a calibrating machine.
2 Connect a fuel supply to each pump, and connect each pump outlet to one of the machine's test injectors.
3 Move each pump rack to the engine stop position.
4 Attach each pump rack to a common control rod (or balancing shaft) via adjustable linkages. Ensure that the common control rod scale reads zero. (A common control rod is simply a rod, free to move in the horizontal plane, one end of which is graduated in mm. The individual pump racks are connected to this rod so that they may be moved together.)
5 Operate the calibrating machine at the rpm specification shown on the test sheet and move the common control rod to the highest rack position shown on the test sheet.
6 Note the fuel delivered by each pump over the specified number of injections.
7 Adjust the control rack of any pump with an incorrect delivery by means of the adjustable linkage. Continue to test and adjust the control racks until the fuel deliveries are equal and correct.
8 Mark each control rack with a scriber or centre punch level with the face of the pump body. This will ensure that each pump rack can be returned to this exact position when required. (Ruston pumps use a pointer and a line engraved on the rack. Once the pumps are balanced, various spacers may be fitted between the housing and the pointer until the pointer and line coincide.)
9 Adjust the common control rod to any of the other positions specified on the test sheet. The deliveries from all pumps should be equal and correct.

Timing the pumps on the engine

Once the pumps have been calibrated, they may be fitted to the engine. Once fitted, it is very important that each pump is timed to the engine so that firing in each cylinder starts at the correct point. To ensure that the timing is correct, proceed as follows:

1 Turn the engine over slowly by hand in the direction of rotation, until the line on the plunger guide cup of no. 1 pump coincides with the line on the housing. This is spill cut-off point—the point at which injection begins—and the no. 1 cylinder injection marks (usually on the flywheel and housing) should coincide at this point.
2 If necessary, adjust the tappet or fit the necessary shims to ensure correct timing.
3 Turn the engine over until the next pump (in firing order) reaches spill cut-off point.
4 Check the injection marks, and make any necessary adjustment.
5 Continue the above operation until all pumps are timed.
   Note When pumps without a sight window are used, the spill cut-off point of each pump may be found by connecting the fuel supply, removing the delivery valve and spring, and fitting a spill pipe as was described previously (see 'Checking the spill cut-off point' on p 188).

Balancing the fuel deliveries

As has been mentioned, even distribution of the fuel to the engine cylinders is essential for efficient running. For this reason the fuel deliveries should be balanced after the pumps have been fitted and timed. The recommended procedure varies from one engine to the next, and the manufacturer's manual should be consulted when balancing the pumps. However, in the absence of the correct manual, the following procedure should ensure satisfactory results:

1 Move each pump rack to the position marked during calibration.
2 Connect the control rod and linkages to the pump racks, adjusting the linkages if necessary to ensure that the racks all lie at the marked position.
3 Operate the linkage system to ensure that full rack travel will be possible with governor and/or throttle movement.
4 Move the linkages back and forth a couple of times, and move one rack to the marked position.
5 Check that all racks lie at the marked position. Make any adjustment necessary.

The above procedure should ensure that all pump deliveries are very close to being equal. A final check may be made in one of the following two ways:

- Remove the exhaust manifold and start the engine (the fuel lines etc must be fitted and bled first, of course). Pass your hand through the exhaust gases issuing from each cylinder. The gases should be at the same temperature. If one cylinder is running a low temperature, it is an indication of a low fuel delivery to that cylinder and the pump rack must be adjusted accordingly. Careful, fine adjustments and frequent checking should ensure accurate balancing of the fuel deliveries.

- Remove the injectors, swing the pipes away from the engine and fit the injectors to the pipes. Bleed the fuel lines and mount a calibrating tube under each injector. Turn the engine over by hand for approximately 100 injections. The quantities of fuel in the tubes should be equal, and the individual racks should be carefully adjusted to give this result.

The individual fuel injection pumps of a large slow-speed diesel engine are checked for balanced fuel delivery by comparing the temperatures of the exhaust gas from each cylinder. As the temperature of the exhaust gas for each cylinder is proportional to the amount of fuel being burnt in that cylinder, it should be obvious that the injection pumps' deliveries are in balance when the exhaust gas temperatures are the same. Since a standard thermometer could not be used due to the high temperatures involved, an instrument known as a **pyrometer** is used to indicate the temperatures of the exhaust gases.

If the pyrometer readings are not the same when the engine is operating under load, adjustment must be made to the control rod settings of the pumps until the readings are equal. If, for example, the exhaust gas temperatures of a six-cylinder engine were 225°C, 250°C, 250°C, 250°C, 250°C, 250°C, it would be necessary to adjust no. 1 pump rack to give more fuel. By slowly increasing the fuel delivery from no. 1 pump, the temperatures could be made all equal at 250°C. This would indicate that each pump was delivering the same amount of fuel.

Note A leaking exhaust valve can cause high exhaust gas temperature, and care should be taken to avoid confusing this fault with excessive fuel delivery.

Servicing a typical enclosed camshaft pump

Again, it should be stressed that dust-free conditions, special tools and equipment, detailed information and data, and knowledge and experience are necessities before any major service is attempted on a fuel injection pump. If these prerequisites are not available, the pump in question should be forwarded to the nearest manufacturer's service depot or agent.

Once again, the complete service procedure may be divided into six separate operations:

- preliminary testing
- disassembly
- cleaning components
- inspection of components
- reassembly
- final testing and setting.

Preliminary testing

When the pump has been removed from the engine and all dirt and grease have been washed from its exterior, it should first be subjected to a preliminary test. Although this test may reveal other pump defects, its primary objective is to determine the extent of wear between the barrel and plunger of each pumping element. The following test procedure, which is recommended by CAV for testing the elements of SPE-A series fuel injection pumps, may be used as a general guide for testing the elements of all multi-element pumps in the absence of specific instructions from the manufacturer.

1 Before dismantling the pump, set it up on the calibrating machine. Make sure that the pump governor is inactive and that the control rod of the pump is locked in the maximum fuel delivery position.
2 Run the pump at 600 rpm and note the amount of fuel delivered by 200 pumping strokes.
3 Run the pump at 100 rpm and note the amount of fuel delivered by 200 pumping strokes.
4 The decrease in the amount of fuel delivered at 100 rpm **should not be more than 40 per cent**. For example, an element delivering

16 cm^3 for 200 strokes at 600 rpm should deliver at least 9.6 cm^3 at 100 rpm.

5   These readings must be taken under standard test conditions with test nozzles set at an opening pressure of 175 atmospheres.

Note 1   The fall in delivery between 600 rpm and 100 rpm is not entirely due to wear, the delivery characteristic of the pump being such that, even with new elements, there is a certain output drop between the two test speeds.

Note 2   The above test should only be carried out if it is certain that no dirt has entered the pump gallery or delivery unions.

In applying this test, it will inevitably be found that a large proportion of elements show a delivery drop very close to the permitted 40 per cent for A size pumps, and the question will arise whether such borderline cases are suitable for further use. The answer is that elements showing 40 per cent drop are still capable of satisfactory service, even for considerable engine hours, but are approaching the end of their useful lives and their future use depends on whether it is desired to rebuild the pump to 'as new' standards, or whether it is merely desired put the pump into running order and to secure the use of the remainder of the useful lives of the elements.

It must be stressed that wear on a pump element and its useful life are dependent almost entirely on the efficiency with which the fuel is filtered. With filtration of reasonable efficiency, the useful life should be approximately 250 000–325 000 km on road vehicles, or about 8000–10 000 hours running time. These figures can be exceeded with good filtration, while with poor filtration they will be very considerably reduced.

The individual elements of a multi-element pump may also be tested for wear in the same manner as the element of a single element pump. However, as this procedure takes a considerable amount of time, it is rarely used in modern practice when a power-driven test bench is available.

During preliminary testing, particular attention should be given to the control rod to ascertain whether it moves freely back and forth. If it does not move freely, the reason for its stiffness (a tight element plunger or excessive dust in the control rod bushes) should be determined. Also other defects such as a faulty delivery valve (a sticking delivery valve will give erratic calibration

deliveries) or a leaking delivery valve seat joint should be watched for. Any defects revealed during preliminary testing should be noted so that either reconditioning or replacement of parts may be carried out at the appropriate time in the servicing sequence.

Disassembly

When a multi-element fuel injection pump has to be dismantled, the procedure recommended by the manufacturer of the pump concerned should be strictly adhered to. The following disassembly sequence (by courtesy of CAV) is specified for CAV type BPE injection pumps, but may be used (with slight variations) as a general guide for most makes and models of multi-element pump (with the exception of CAV SPE Series pumps) if specific information from the manufacturer is not available.

To dismantle a CAV BPE fuel injection pump, proceed as follows (refer to Fig 18.53):

1   Attach the pump to the mounting plate of a universal pump vice.
2   Disconnect and remove the governor mechanism.

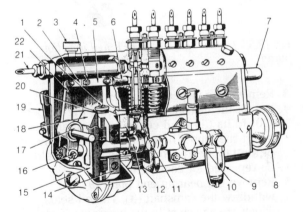

1	Adjusting nut	12	Closing plug
2	Outer link fork	13	Camshaft
3	Oil lubricator	14	Flyweights
4	Screw for link forks	15	Bell crank pin
5	Inner link fork		retaining cage
6	Control rod	16	Coupling cross-head pin
7	Control rod stop	17	Eccentric
8	Drive coupling	18	Bell crank lever
9	Preliminary filter	19	Control lever
10	Plunger-type feed pump	20	Governor spring
		21	Fuel inlet connection
11	Tappet screw	22	Floating lever

Fig 18.52 CAV fuel injection pump, type BPE

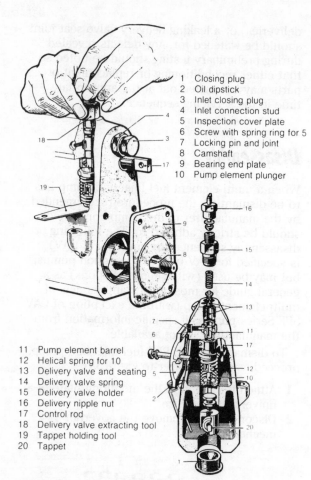

1   Closing plug
2   Oil dipstick
3   Inlet closing plug
4   Inlet connection stud
5   Inspection cover plate
6   Screw with spring ring for 5
7   Locking pin and joint
8   Camshaft
9   Bearing end plate
10 Pump element plunger

11 Pump element barrel
12 Helical spring for 10
13 Delivery valve and seating
14 Delivery valve spring
15 Delivery valve holder
16 Delivery nipple nut
17 Control rod
18 Delivery valve extracting tool
19 Tappet holding tool
20 Tappet

Fig 18.53 Dismantling a CAV BPE fuel injection pump

3 Remove the inspection cover plate (5).
4 Rotate the camshaft (8) to bring the tappet (20) to its top dead centre position and insert the tappet holder (19) under the head of the tappet adjusting screw. This should be repeated for each element.
5 Remove the bearing end plate (9) and withdraw the camshaft (8). The oil seals, which are a push fit in the bearing end plates, should not be removed unless it is intended to renew them, as removal may cause damage. Should renewal be necessary, the outer race of the bearing must first be extracted, followed by the shim washer and then the oil seal. When replacing, the spring loaded lip of the seal should be facing towards the bearing.
6 The pump half coupling need only be removed from the camshaft if it is fitted at the opposite end to the bearing end plate.

Should this be done, care should be taken that the shaft position is marked so that on reassembling the pump, the firing sequence will be correct. The removal of the coupling from the taper of the camshaft should never, at any time, be done with the use of a hammer. A properly fitting extractor should be used for this purpose.

7 Unscrew the closing plugs (1) at the base of the housing and push up the tappet (20) until it is possible to withdraw the tappet holder, after which the tappet assembly (20), the lower spring plate, the plunger spring (12) and plunger (10) may be withdrawn through the holes in the base of the pump housing.
8 The next step is to remove the delivery valve assembly. This is done by unscrewing the delivery valve holder (15), withdrawing the spring peg if fitted (not fitted to BPE pumps), the delivery valve spring (14) and the delivery valve. The delivery valve housing and its joint are now removed by means of the extracting tool (18).
9 To remove the pump barrel, unscrew the locking pin (7) and push the barrel from below by means of a fibre or soft brass drift.
10 It is seldom necessary to remove the control rack. In the event of it having to be removed, this is usually done by removing a screw from the housing since the spigot end of the screw retains the rack by engaging in a groove machined in the rack.

Cleaning components

All pump components should be thoroughly cleaned with a soft brush (not rag), washed in clean kerosene, petrol or other suitable cleaning medium and blown dry with clean compressed air. Special care should be taken when cleaning to make sure that the fuel gallery in the pump body is cleaned with the appropriate brush. To prevent rusting of highly-finished surfaces, all steel components should be smeared with light oil.

Inspection of components

Once the pump has been dismantled and cleaned, the various components must be inspected thoroughly to determine whether they are fit for further service or whether they should be reconditioned or renewed.

The body—Check the passage in the body (or bushes if used) through which the control rod passes for excessive wear, and examine the body closely for any signs of cracks. An internal crack in the upper half of the pump body in the vicinity of the fuel gallery could allow fuel to leak from the gallery and cause either an excessive build-up of fuel in the cambox of the pump or crankcase dilution, if the pump cambox components are lubricated by the engine lubricating system. Also check all threads in the pump body for damage or excessive wear, and ascertain the fitness of the barrel seats situated just below the fuel gallery.

The elements—Although the elements have been subjected to a delivery drop test during preliminary testing, they should also be given a visual examination after the pump is dismantled. Elements that have passed the delivery drop test should not be rejected for visual defects unless these are serious. Too much importance should not be attached to fine scratch marks, which are usually visible on the lapped surface of the plunger adjacent to the control edges. These are caused by very fine particles of abrasive matter in the fuel and will appear after a comparatively short period of service, but they have little effect on the pumping efficiency of the element unless they are deep and extensive.

It should be noted that wear is mainly concentrated at the upper end of the element and that the lower or guide portion of the barrel and plunger wears very little, so that after long periods of service there is no serious increase in the leakage of fuel oil from the elements. Normal element wear does not seriously diminish power output of the engine over its normal speed range, since the loss of fuel delivery can be corrected by adjustment of the maximum fuel delivery stops. It may, however, cause difficult starting owing to reduced delivery at cranking speeds and erratic idling arising from the same cause.

Note A difference in wear between the various elements will make calibration difficult. As each element consists of a barrel and plunger, these components must be examined closely for individual faults.

The plungers of a multi-element pump are inspected in a similar manner to those of single element pumps, the depth of score marks and the extent of erosion damage, if any, being carefully noted. The foot of the plunger should be examined for chips, cracks or wear.

The barrels should be inspected for deep scoring, erosion damage and cracking. Particular attention should be paid to the upper flat surface of the barrel for signs of **fretting**, a form of corrosion caused by 'breathing', which takes place between the upper flat surface of the barrel and the delivery valve body. Fretting marks may be removed by lapping.

Delivery valve assemblies—Mitre-type, volume unloading delivery valves are used in most makes of multi-element fuel injection pumps. The inspection and servicing procedure is the same as for the delivery valve used in a single element pump.

Tappet assemblies—Each tappet assembly should be carefully inspected for damage and wear, careful attention being paid to ensure that the roller is free to rotate on its shaft. If the roller is not free, it tends to skid over the cam lobe and so causes wear. The hardened roller should be examined for chips or cracks and the hardened top of the tappet adjusting screw should be inspected for damage and dressed if required.

When stripping the tappet assembly, the manufacturer's recommendations should be followed if in doubt. The following procedure, which is recommended for CAV SPE-B series pumps, but is not correct for all pumps, is given as an example:

1   First remove the tappet screw, place the tappet upside down on a block of wood and drive out the vertical dowel securing the roller pin. Press out the latter and remove the inner roller on which the cam roller revolves.
2   When the tappet assembly is reassembled, care must be taken that the tappet adjusting screw does not protrude too far from the

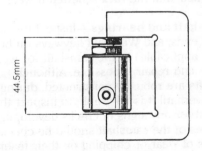

Fig 18.54 Method of setting tappet adjustment

body. To prevent this from happening, a tappet adjusting gauge (Fig 18.54) is generally used. The use of this gauge ensures that the tappet setting of the element assemblies will be the same, with the result that phasing is made easier and there is no chance of any plunger being driven far enough up its barrel to strike the base of the delivery valve body when the pump is operated.

Plunger return springs—The plunger return springs should be inspected for signs of cracks, fractures, distortion or any other damage, as a plunger spring that breaks in service will cause the engine to misfire in the cylinder concerned. They should be tested on a spring tester or have their free length checked to determine whether they are fit for further use.

Control sleeve and quadrants—If the teeth of a control quadrant (or control pinion) show signs of excessive wear or are damaged, the quadrant should be renewed. In addition, if the plunger vane slots in the lower half of one of the control sleeves are damaged, then that control sleeve should be renewed. If a new control quadrant has to be fitted to either the original or a new control sleeve, care should be taken to ensure that the clamping screw of the new quadrant bears the same relative position to the slots in the control sleeve as did the clamping screw in the previous quadrant. If this is done, time will be saved when the pump is being calibrated.

Control rack or control rod—Examine the teeth of the control rack itself for signs of excessive wear or damage. Make sure that the control rack is not bent and that it moves freely, but has no slop, in the pump housing. Excessive backlash between the teeth of the control rack and the control quadrant, or a sticking control rack, will not allow efficient metering and delivery of fuel and the engine will run erratically. The governor link pin hole in the rack should be checked, and the rack renewed if wear is evident.

Camshaft and bearings—Inspect the threads, tapered ends, and Woodruff keyways on both ends (if applicable) of the camshaft for wear or damage, and repair if possible. Although camshafts are robustly constructed, during pump overhaul it is advisable to inspect the camshaft for any signs of cracks developing. The lobes of the camshaft should be checked for signs of wear or chipping on their leading edges and the camshaft either reground or

renewed if wear is excessive. The camshaft bearings (either ball or roller) should be checked for signs of excessive wear. The balls or rollers and the races should be inspected for pitting, flaking, rusting or surface cracks, and the bearing cages inspected for damage. If in doubt, replace the bearings.

Reassembly

When the pump is being reassembled, great care should be taken to ensure that all joint faces and other parts are entirely clean. They should be rinsed in clean petrol, kerosene or fuel, allowed to drip, smeared with light lubricating oil, and finally brought together entirely without the use of cotton waste or cloth wipers of any kind. In the absence of the correct manual, the reassembly sequence for a CAV type BPE fuel injection pump, which is given below, will serve as a general guide for most multi-element, enclosed camshaft fuel injection pumps, with the exception of CAV SPE Series pumps.

1 Refit the first element barrel, making sure that the locating slot in its upper end aligns with the hole for the barrel locating screw. Tighten the barrel locating screw after ensuring that its sealing washer is in place. When the screw is tight, the barrel should be free to move slightly up and down in the pump body. Repeat for the other barrels.

2 Place the first delivery valve body in position, after ascertaining that the top face of the barrel and the lower face of the delivery valve body are perfectly clean. Fit the sealing washer to the delivery valve body, drop the valve and its spring into position, and tighten down the valve holder after ensuring that the valve spring seats correctly on the valve. Tighten with a tension wrench to the manufacturer's specification. Repeat for the other delivery valve assemblies.

3 Invert the pump and insert the control rack and its retaining screw.

4 Set the control rack in middle position so that the centre punch marks on each end of the rod are the same distance from each side of the pump housing.

5 Refit the control sleeves (with the control pinions fitted) and the upper spring plates. Make sure that when the teeth of each

control pinion are in mesh with the teeth of the control rack, the gap in the control pinion is in line with the mark on the plunger vane slot at the lower end of the control sleeve—in other words, the gap in the control pinion is at right angles to the control rack. Check to ensure that the tommy-bar holes in the control sleeve are on the same side as the control pinion clamping screw.

6 Insert, in turn, the plungers, with the springs and lower spring plates in place, into the barrels. Take care to ensure that the marks on the plunger vanes lie opposite the marks on the plunger vane slots in the control sleeves.

7 Insert the first tappet assembly, and press against the plunger return spring until a tappet holder (or tappet bridge) can be located between the head of the tappet adjusting screw and the pump housing. Repeat for the other tappet assemblies.

8 Refit the correct bearing end plate to the no. 1 end of the pump body. (The no. 1 end of the pump body is the left-hand end when looking directly at the inspection window of the pump.) Install the camshaft and refit the bearing end plate to the no. 2 end of the pump body, taking care not to damage the oil seals.

   Note If the camshaft ball races have been renewed, check that these are pressed hard against their abutments on the shaft and in the bearing housings before installing the camshaft. Also a trial assembly of the camshaft should be made to check that the amount of preload is to the manufacturer's specifications. If preload is incorrect, adjust by inserting or removing shims behind the bearing inner races, equalising the shims at each end.

   Care must be taken when refitting the camshaft to make sure that the correct firing order is maintained. If, for example, a four-lobe camshaft, correctly fitted, has a firing order of 1-3-4-2, it will have a firing order of 1-2-4-3 if reversed. In order to eliminate any confusion as to how the camshaft should be fitted, a notch or cut is usually made on one of the threaded ends of the camshaft of four-, six-, and eight-element pumps. The formula given on the pump body indicates the correct way to fit the camshaft, and it should be fitted with the notch at the specified end. However, if the firing order of

the pump is ascertained before dismantling, there should be no chance of the camshaft being incorrectly fitted.

9 Rotate the camshaft and remove the tappet holders from beneath the heads of the adjusting screws as each tappet is at the TDC position.

10 Refit the inlet connection union nut, inlet closing plug and oil dipstick (if fitted).

11 Smear the mitre joint face of the closing plugs with white lead or other sealing compound and tighten up hard. On some types of pump, such as CAV type AA, a new set of closing plugs should be fitted.

12 Fill the cambox to the prescribed level with first quality engine lubrication oil.

13 Replace the inspection cover plate.

14 Refit the governor mechanism.

Final testing and setting

After the pump has been reassembled, it must:

- be tested for leaks or other defects
- be phased and calibrated
- have the various stops adjusted.

Testing for leaks—The multi-element, enclosed camshaft fuel injection pump should be subjected to a gallery air test in a similar manner to the single element pump. As the main purpose of this test is to reveal leakage from the delivery valve assembly and the barrel to body seating, it is best carried out during reassembly, or after the barrels, delivery valve assemblies and plungers have been fitted. The advantage gained by conducting the gallery air test at this stage of reassembly is that a confirmed leak between the barrel and body can be eliminated with less work than would be the case if the pump were fully assembled. However, if the gallery air test is to be carried out before the pump is completely assembled, a device must be made to support the plungers in the barrels so that they will not be forced out by the air pressure.

The control rack should be checked for freedom of movement and if it is found to be sticking or does not move with a free, smooth action, the cause of the trouble must be found and corrective action taken.

Phasing and calibrating—There are, of course, two separate operations included under this heading—phasing, which is setting the pump so that each element will be correctly timed to its engine cylinder, and calibrating, which is setting the fuel deliveries from the elements to give equal and correct injection quantity.

Phasing involves checking, and adjusting if necessary, the number of degrees of pump camshaft rotation between successive injections. On a pump fitted to a four-cylinder engine with a firing order of 1-3-4-2, fuel delivery from no. 3 element must start exactly 90° of pump camshaft rotation after the beginning of fuel delivery from no. 1. If no. 3 is corrected to give the exact phase angle, no. 4 may be adjusted to start injection 90° after no. 3, but greater accuracy is obtained by using no. 1 as the reference point and timing no. 4 at 180° after no. 1.

Phasing may be carried out by using either a hand- or motor-driven phasing and calibrating machine, commonly referred to as a test bench or test stand. Because of their many advantages, motor-driven test benches, which are made by various companies such as Hartridge and Merlin, are now used in all modern pump rooms. In keeping with modern practice, therefore, the phasing procedure given applies to power-driven test benches.

The actual method employed to phase a pump will depend on the type of test bench used and the equipment available, but should be one of the following three:

- low-pressure phasing
- high-pressure phasing
- electronic (stroboscopic) phasing.

Low-pressure phasing—This method (phasing by spill cut-off) may be carried out as follows:

1 Mount the fuel injection pump on the table of a test bench fitted with a degree plate for measuring the angular rotation of the pump camshaft. Connect the pump drive coupling to the test bench drive, making sure that there is a small end clearance. Attach the rack setting gauge (if required) to the end of the control rack, set the rack in its middle position and lock it.

    Note   A rack setting gauge is an accessory to the test bench, and is fitted to the end of the rack where it indicates the distance the rack is moved from the stop (no fuel) position.

2 Fit the fuel supply line from the test bench to the inlet connection of the pump, turn on

the fuel supply and bleed all air from the pump. Turn off the fuel supply.

3 Disengage the power drive mechanism so that the pump camshaft may be turned by hand. Rotate the pump camshaft until no. 1 plunger is at TDC. This position may be verified by checking the drive end of the pump—at TDC, a mark on the camshaft or the pump half coupling lines up with a datum line scribed on the end of the pump body. Where there are three lines, the two on the outside are timing marks and are usually marked R and L to indicate the direction of rotation; the middle line is the TDC mark.

4 Check the clearance between the top of the plunger and the base of the delivery valve body—the recommended clearance will be specified by the pump manufacturer, and must be strictly adhered to. In this discussion, we will assume that the correct clearance is 0.5 mm since this is not an uncommon figure, and each of the following three methods of checking this clearance will relate to it:

- With the plunger at TDC, carefully screw up the tappet adjusting screw until the top of the plunger contacts the base of the delivery valve body. Back off the tappet adjusting screw half a turn. In cases where this method is recommended, the pitch of the tappet adjusting screw is usually twice the recommended clearance (or almost exactly so), and the required clearance can be obtained by backing off the tappet adjusting screw half a turn.

- Adjust the tappet adjusting screw so that a 0.5 mm feeler gauge will just fit between the bottom of the plunger and the top of the tappet screw when the top of the plunger is in contact with the base of the delivery valve housing.

- Remove the delivery valve assembly and use a depth gauge micrometer to measure the distance between the top of the plunger and the top face of the pump body. Then measure the distance between the top of the barrel and the top face of the pump body—since the delivery valve body seats on the barrel, this measurement should be the same as that from the base of the delivery valve body to the top face of the pump body. Therefore, by subtracting the second distance from the first, the clearance measurement can be obtained.

5 Unscrew the delivery valve holder and remove the delivery valve and its spring from no. 1 element. Refit the delivery valve holder and attach a gooseneck spill pipe to it.

6 Turn the test bench drive shaft by hand using the tommy bar supplied until no. 1 plunger is at the bottom of its stroke.

7 Turn on the fuel supply. Fuel will flow from the fuel gallery, through the barrel ports and out of the spill pipe.

8 Begin rotating the pump camshaft slowly by hand, in the direction of rotation, until spill cut-off point is found.

   Note   There are two spill cut-off points per revolution of the pump camshaft—one as the plunger is moving up the barrel, and one as the plunger is moving down. It is important, therefore, to watch the tappet to make sure that the spill cut-off point is found when the plunger is on the up-stroke.

   Check the coupling end of the pump to make sure that the appropriate timing line coincides with the datum line on the drive end of the pump body. If the pump rotates in a clockwise direction when driven by the engine, the timing line R should coincide with the datum line on the pump body. If the pump rotates in an anticlockwise direction, timing line L and the datum line should coincide.

9 The point at which spill cut-off occurs should be checked several times and care taken to ensure that the appropriate timing line coincides with the datum line on the pump body. If the specified timing mark does not coincide exactly with the datum line of the pump body when the spill cut-off point has been accurately found, a new timing mark should be scribed on either the camshaft or the drive coupling so that the pump may be accurately timed when fitted to the engine.

   Note   The position and number of reference marks on either the periphery of the pump camshaft or drive coupling and the pump body varies for different makes and models. For example, some pumps have one line only instead of three. Usually, with markings of this nature, the spill cut-off point is reached when the scribed line on either the camshaft or the drive coupling coincides with the datum line on the pump body.

10 Hold the camshaft securely and rotate the graduated phasing dial until the zero mark on the dial coincides with the datum line on the front panel of the test bench. To facilitate this operation, the phasing dial may be either frictionally secured to the test bench drive shaft or held positively by a locking screw. After setting the phasing dial, recheck to make sure that spill cut-off occurs when the zero mark on the phasing dial coincides exactly with the datum line on the front panel of the test bench.

11 Turn off the fuel supply to the pump and remove the gooseneck spill pipe and the delivery valve holder from no. 1 element. Wash and refit the delivery valve, spring and delivery valve holder to no. 1 element.

12 Remove the delivery valve holder, delivery valve and spring from the next element in firing order of the pump, ie no. 5 element from a six-element pump, firing 1-5-3-6-2-4. Replace the delivery valve holder and attach the gooseneck spill pipe.

13 Turn on the fuel supply to the pump and rotate the test bench drive shaft in the correct direction by hand until spill cut-off occurs. The appropriate degree marking on the phasing dial—60° for a six-element pump or 90° for a four-element pump—should now coincide with the datum line on the front of the test bench panel. If this is not the case, the element tappet adjusting screw must be adjusted so that spill cut-off does occur when the appropriate lines coincide. If spill cut-off occurs too soon, the tappet adjusting screw must be screwed down into the tappet; if spill cut-off occurs too late, it must be screwed upwards.

   Note   If tappet adjusting screws are used—in some particular pumps they are not—the lock nuts must be securely tightened after each adjustment has been made.

14 Turn off the fuel supply, remove the gooseneck spill pipe and delivery valve holder, wash and refit the delivery valve components, and check the angles between spill cut-off points of the remaining elements (in correct firing order) as previously described, making any necessary adjustments.

   Note   The normal tolerance allowed between successive injections is ½°.

15 After all adjustments have been made, a final

check should be carried out, starting at no. 1 element and proceeding in the correct firing order, to make sure that the spill cut-off point of each element occurs at exactly the specified number of degrees of camshaft rotation.

High-pressure phasing—This is similar to low-pressure phasing in that the spill cut-off point of each element is found when the flow of fuel from the top of the delivery valve holder ceases. It differs from the low-pressure phasing, however, in that the pressure of the test oil used is much higher and that it is not necessary to remove the delivery valve and spring in order to find the spill cut-off point.

To enable high-pressure phasing to be performed, the test bench must be fitted with a fuel supply pump capable of delivering fuel to the injection pump at a pressure approaching 2000 kPa. When the high-pressure fuel enters the fuel gallery of the injection pump and the top of the element plunger is below the barrel ports, it flows from the gallery into the barrel via the ports and acts on the underside of the delivery valve. The pressure of the fuel exerts a force on the delivery valve that overcomes the force exerted by the valve spring and lifts the valve off its seat, allowing fuel to flow from the top of the valve holder. When the plunger moves up the barrel and closes the inlet and spill ports, high-pressure fuel can no longer flow from the pump gallery and so the delivery valve returns to its seat, the flow of fuel from the valve holder ceases and so the spill cut-off point is determined.

High-pressure phasing techniques vary according to the make and model of the test bench used. The following procedure, however, although specified for the Merlin M6 Calimaster test bench, should serve as a useful guide for most high-pressure phasing operations:

1 Make sure that the storage tank of the test bench is filled with clean test oil and prime the high-pressure lift pump (if necessary).
2 Mount the pump on the test bench and connect the flexible pipe from the pressure connection of the machine to the fuel inlet connection of the injection pump, using an adaptor if necessary.
3 Attach the gooseneck phasing pipe to no. 1 pump element and attach a small drain funnel with an affixed length of suitable tubing to the straight, lower section of the gooseneck phasing pipe. Insert the free end of the tubing into the return connection of the test bench so that test oil flowing from no. 1 element will be returned via the tubing to the storage tank of the test bench.
4 Connect the injector pipes to the remaining elements of the pump.
5 Slacken the phasing dial locking screw so that the phasing dial may be rotated freely on the spindle and set the spindle clutch of the test bench in the off or neutral position for phasing.
6 Use the tommy bar provided to rotate the spindle by hand until no. 1 plunger is at TDC.
7 Check the clearance between the top of the plunger and the base of the delivery valve housing as described previously and adjust this clearance if necessary. Rotate the drive spindle of the test bench by hand in the correct direction of rotation until no. 1 plunger is at BDC.
8 Set the test bench controls for pressure feed according to the instruction book, start up the motor and adjust the pressure regulating valve to give a test oil pressure of approximately 1700 kPa. Fuel will flow from the gooseneck phasing pipe.
9 Rotate the spindle slowly by hand, in the correct direction of rotation, until spill cut-off occurs. Note the position of the mark on either the pump half coupling or camshaft periphery in relation to the datum mark on the end of the pump body, and re-mark if necessary.
10 Set the appropriate mark on the phasing dial to the datum line on the front panel of the test bench, and tighten the phasing dial locking screw. Recheck that the timing mark and the datum line coincide at spill cut-off point.
11 Stop the test bench motor, remove the gooseneck phasing pipe from no. 1 element and fit the injector pipe. Remove the injector pipe from the next element in firing order, and attach the gooseneck phasing pipe. Start the test bench motor and adjust the test oil pressure to 1700 kPa. Find the spill cut-off point for this element as for no. 1 element, and make any necessary adjustments to give the correct phase angle.
12 Check the spill cut-off points of the remaining elements in correct firing order, and make any necessary adjustments to the phase angles. When all adjustments have

been made, a final check should be carried out, starting at no. 1 element and proceeding in firing order, to make sure that all phase angles are correct.

Electronic (or stroboscopic) phasing—
This refers to the modern trend in phasing systems. With this system, the pump is run on the test bench and the spray from the injectors triggers an electronic circuit. When the circuit is triggered, a neon light is flashed and this light falls on the phasing dial. Due to the very short duration of the light, the phasing dial appears to be stationary every time the light flashes and the angular rotation between successive firings can be instantly and conveniently seen. Thus the phasing operation is carried out with the pump operating, all phase angles can be very quickly seen, and rapid adjustments can be made and checked.

Calibrating—As explained previously, it is imperative that each cylinder of a multi-cylinder engine be supplied with exactly the same amount of fuel if the engine is to operate smoothly, and that each cylinder receive the correct amount of fuel if the engine is to operate efficiently. In order to achieve this objective, the pumping elements of a multi-element enclosed camshaft pump must be adjusted so that the amount of fuel delivered by the elements is equal and correct for any control rack setting. This operation is called **calibrating** and before it is attempted, a test sheet for the pump concerned and a power-driven test bench must be available.

To calibrate a multi-element enclosed camshaft pump proceed as follows:

1   Mount the pump on the table of the test bench, ensuring that there is a minimum clearance of 0.5 mm between the pump coupling and the drive coupling of the machine. Connect the fuel supply line to the pump inlet connection and the delivery lines from the various elements to the pump's test injectors. Fill the cambox with lubricating oil and attach a rack travel indicator to the end of the control rod. Disconnect the governor linkage if a governor is fitted.

2   Turn on the fuel to the pump, move the control rod to its mid-travel position and run the pump at approximately 400 rpm. Either open the bleed screws on the pump body or loosen the delivery lines and bleed

all air from the pump. Continue to run for about 10 minutes to ensure that the pump is at operating temperature.

3   Stop the test bench. Move the control rod to the no-fuel position first, and then advance it towards the maximum fuel position for the distance first specified (in mm) on the test sheet. Start the test bench and adjust its speed to the lower figure stated on the test sheet. Operate the test bench controls to direct the fuel delivered by the test injectors of the machine into the graduated measuring test tubes and to automatically cut off at the specified number of 'shots'. Stop the test bench and compare the quantity of test oil in the test tubes with the figure given on the test sheet for a pump using the corresponding diameter plungers.

     Note   The 'size' of a pump (A size, B size etc) refers to the stroke of the pump. Each size pump may be fitted with a variety of plungers to provide a variety of deliveries to suit engines of different sizes.

If the variation in the reading of the test tubes is not within the limits specified by the test sheet, adjustment must be made. To do this, slacken the screw clamping the pinion to the control sleeve, and, by inserting a small pin in the tommy-hole in the regulating sleeve, move it in the required direction as follows:

• For elements with a right-hand helix, rotate the sleeve to the right to increase delivery and to the left to decrease delivery.

• For elements with a left-hand helix, rotate the sleeve to the left to increase delivery and to the right to decrease delivery.

When corrected, the pump should be run again and the variation in delivery of the elements should not exceed $\pm 2\frac{1}{2}$ per cent.

4   With the control rod settings as specified by the test sheet (eg 6, 9 and 12 mm), run the test bench at the specified speeds (eg 200 and 600 rpm) for each setting and compare the amount of fuel delivered by each element for the specified number of injections with the amount specified on the test sheet. If the deliveries are within the limits stated by the test sheet, no further adjustment is necessary. If, however, the deliveries are not within the limits specified, compromise adjustments should be made. It may be necessary, for example, to set one

element's delivery near to maximum when the pump is run at high speed to ensure that the delivery at low speed does not fall below minimum.

5 Check the fuel cut-off position by moving the control rod to the no fuel position and running the test bench. When the control rod is in this position, fuel should not be delivered by any element.

A typical test sheet for an A size fuel injection pump is shown below:

Camshaft speed (rpm)	Plunger diameter (mm)	Delivery at given control rod openings					
		7 mm		9 mm		12 mm	
		Min	Max	Min	Max	Min	Max
	6.5	1.4	1.8	2.3	3.2	4.4	5.2
200	6.0	0.6	1.0	1.4	2.1	3.1	3.6
	5.0	0.3	0.9	1.1	1.7	2.1	2.7
	6.5	1.9	2.5	2.6	3.4	4.9	5.7
1000	6.0	1.0	1.4	1.5	2.3	3.3	3.9
	5.0	0.9	1.2	1.2	1.9	2.4	3.0

Delivery quantities in cm^3 per 100

Note The various stops to be adjusted are incorporated in or connected with the governor, and the adjustment of these stops will be covered in Chapter 19.

Installation of typical enclosed camshaft pumps

Mounting and driving arrangements—Many enclosed camshaft pumps are mounted by means of bolts or clamping straps that secure them to either a flat bracket or a semicircular cradle on the side of the engine. Some pumps, however, are secured to the engine by means of a mounting flange. The mounting flange carries a large spigot concentric with the pump shaft to ensure positive location of the pump drive gear when the pump is in position bolted to the engine timing gear case.

When the pump is in position on the engine, the camshaft should lie in a horizontal plane to ensure adequate lubrication of all the camshaft lobes. To this end, the pump is usually mounted with the pumping elements in a vertical plane. Should the pump be required to lie in anything but a position with the elements vertical, a specially adapted pump must be used.

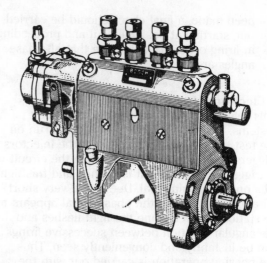

Fig 18.55 Pump with mounting flange

Many pumps are designed so that they can be driven from either end as necesary or can be used to drive some accessory mounted behind them. In pumps of this type, both camshaft ends protrude from the pump housing and are both machined to accommodate a drive coupling.

A positive drive is required for fuel injection pumps. So-called 'elastic' couplings, which drive through a rubber composition core, are best avoided as they fatigue with age and correct timing becomes impossible. However, special couplings designed for the purpose are necessary since they compensate for slight misalignment between the engine drive and the pump. When mounting the pump, a small end clearance (approximately 1.5 mm) should be given to the coupling in order to prevent axial stress on the pump bearings.

Most multi-element pumps can be driven in either direction, since the cams are symmetrical. However, changing the direction of rotation reverses the injection sequence of the pump and care should be taken to correctly connect

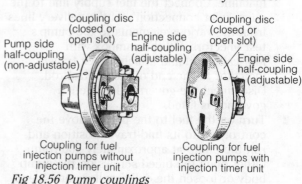

Fig 18.56 Pump couplings

the pump elements to their respective engine cylinders.

Timing the pump to the engine—The best procedure for timing a multi-element, enclosed camshaft pump to a particular engine is given in the engine manufacturer's manual, and this should be carefully followed. If, however, a workshop manual is not available, the following timing procedure will be found to be correct in most cases:

1 Turn the engine over by hand, in the direction of rotation, until no. 1 cylinder is on compression stroke. Continue turning slowly until the no. 1 injection mark is in alignment with the pointer. (The injection timing mark is usually on the flywheel, and quite often a cover plate has to be removed to expose it.)

2 Ensure that the timing mark on the boss of the pump half of the drive coupling is in line with the timing mark on the pump body. If the half coupling boss carries two timing marks—L for left-hand rotation and R for right-hand rotation—the mark that correctly applies to the pump's rotation should be made to line up with the timing mark on the pump body.

3 If zero marks are punched on the coupling components, these marks should now be lined up.

4 Couple the pump to the engine drive by tightening the coupling bolts and secure the pump firmly to the engine.

The injection pump should be correctly timed to the engine. If, however, a further check is required, or if there were no timing marks on the coupling, spill timing may be carried out. This is done by finding spill cut-off point for no. 1 pumping element, as for a flange-mounted pump. At this point, no. 1 cylinder should be at injection point—or on compression stroke, with the injection mark in alignment with the pointer. If the spill timing check indicates that it is necessary to advance or retard the injection timing slightly, adjustment can usually be made by means of the adjusting slots and set screws provided in the drive coupling. In order that accurate adjustment may be made, the coupling flange is usually graduated so that each division is equal to 1 degree of pump camshaft rotation.

19 Governors

Any device that automatically exerts control over engine speed may be termed a governor. As this broad statement indicates, governors take many forms and exert their control in many different ways.

The obsolete 'hit-and-miss' governor, which cut off the fuel supply and operated a decompressor when engine speed reached a predetermined point, and reversed the procedure when the engine speed fell to a certain point, represented one of the cruder governors. At the other end of the scale are the sensitive hydraulic governors, which exercise a delicate control over the quantity of fuel injected, increasing or decreasing it as required to maintain the engine speed at the required level, and electronic governors, which react to other variables such as ambient temperature, turbocharger boost, fuel temperature, and many other factors, to maintain precise speed control and low emission operation.

In this chapter, governor principles will be considered together with examples of various types of governor as fitted to jerk-type injection pumps. Generally, governors that are specific to or are an integral part of other fuel injection systems will be discussed in the relevant chapters. Electronic governor systems, however, are an exception, with the Detroit Diesel Allison system used in conjunction with unit injectors being featured because it embodies almost all of the features of electronic governing in the one system.

While governors are desirable on petrol engines in many applications, the inherent characteristics of the diesel fuel injection pump dictate that a governor must be used on all diesel engines. At a fixed control rack setting, the fuel delivery increases as the engine (and pump) speed increases. Should the engine load be lightened in some way, the engine speed will increase and the quantity of fuel injected will increase as the engine speed climbs. This increase in the quantity of fuel injected as the pump speed increases is known as the rising

characteristic of the injection system and unless some governing device is fitted to the engine, overspeeding must result.

On the other hand, if the rack is moved to the idle position, the engine speed will slacken and the quantity of fuel injected will fall off with engine speed. This action will cause the engine speed (and injection rate) to progressively fall until the engine stops through lack of fuel. Again, a governing device is necessary to prevent this from happening.

In certain applications, a constant speed is required from the engine. Without a governor, but with the throttle set, the load will either be heavy enough to cause the engine to lose speed, or too light and the engine will gain speed. Should the engine start to lose speed because of the load, it will continue to lose speed because of the injection pump characteristics. For the same reason, if the engine speed increases due to lack of load, it will continue to increase.

From the above it should be obvious that a governor is required on a diesel engine to:

- prevent stalling and overspeeding, and/or
- maintain engine speed relatively constant regardless of load variations.

A third governor function is often associated with the first one. This is to maintain engine speed relatively constant at any speed selected by the operator and set by means of a throttle control.

The governors generally employed are classified by the function or functions they fulfil as:

- constant speed governors
- variable or all-speed governors
- limiting or idling and maximum speed governors.

Apart from their governing characteristics, governors may further be classified by the principle on which they operate as:

- mechanical (or centrifugal)

- pneumatic
- hydraulic
- electronic.

In addition, some manufacturers make governors that operate on a combination of two of the above principles, mechanical–hydraulic and pneumatic–mechanical governors, for example.

Governor terminology

Before examining the operation of the various types of governor, there are a number of terms, peculiar to governors and their operation, that should be defined.

Speed droop—A very basic fact concerning governors is that they operate when a change in the engine speed occurs. The change in speed necessary to bring about the governor's automatic action may be temporary, lasting only until the governor makes the necessary corrections and restores the engine speed, or it may be permanent, the governor action allowing a small speed change only.

A change in engine speed will occur when a load is applied to an unladen engine running with a fixed throttle or when the load is removed from a loaded engine that also has its throttle set. The change in speed will depend on the governor—some governors react to a small speed change, some require a considerable change. The change in engine speed necessary to cause the governor to operate is termed **speed droop**, and is usually related to full throttle operation.

The speed droop of a particular governor is usually specified as a percentage calculated from the following formula:

$$\text{speed droop (\%)} = \frac{(\text{no load speed} - \text{full load speed})}{\text{full load speed}} \times 100$$

Speed droop is a measure of the efficiency of operation of a governor—the smaller the speed variation, the more exacting the control exerted by the governor. The speed variation cannot be eliminated completely in any governor, however, since it is the speed variation that brings about governor action in the first place, but it can be only a temporary speed change, which is quickly corrected.

For general applications, the governors employed have a characteristic permanent speed droop. For tractors, this may lie between 8 and 13 per cent; for general industrial and automotive applications 5–10 per cent is satisfactory, but for engines driving AC generators a temporary speed droop of $\pm 0.5 - \pm 1.5$ per cent is generally specified.

Hunting—When a governor exhibits an overcorrective tendency and causes the engine speed to continually increase and decrease about the mean governed speed, the engine is said to 'hunt'. Sensitive governors are more inclined to promote hunting than the less sensitive ones.

Sensitivity—This term refers to the smallest change in speed necessary to cause the governor to produce corrective movement of the fuel control mechanism, and is expressed as a percentage of the governed speed.

Deadband—This is the narrow speed variation during which no measurable correction is made by the governor.

Isochronous—An isochronous governor is one that is capable of maintaining a constant engine speed for any load between no load and full load without any alteration being necessary to the control setting. With a governor of this type, the speed droop is only temporary.

Stability—The stability of a governor is its ability to maintain a definite engine speed under constant or varying load conditions, without hunting. After a sudden load change, a governor showing good stability will attain a steady speed very quickly.

Torque control—In certain applications, the injection rate, as determined by the governor and maximum fuel stop, is not sufficient over the operating range to allow for sudden load increases. To provide sufficient engine torque to cater for such overload conditions the maximum fuel delivery is not fixed, but is varied as a function of engine speed to give maximum injection rate at less than full speed. It is usual for such governors to produce maximum engine torque at 55–65 per cent of rated engine speed, thereby giving a very useful reserve of torque to overcome the temporary overload conditions that cause the engine speed to continue to drop, irrespective of corrective governor action.

If the rack stop were set to the maximum amount of fuel that could be burned efficiently in a naturally aspirated engine, the engine would exhaust excessive smoke and give high fuel consumption at high speeds. This is brought about by two factors—as the engine speed increases so the amount of fuel that can be burned falls slightly due to the fall in volumetric efficiency of the engine, and there is a tendency for the amount of fuel delivered by the injection pump at a fixed rack setting to increase slightly with the pump speed. Torque control devices automatically adjust the maximum fuel delivery when the throttle is fully open to suit the engine rpm.

Run-up speed—This is the name often given to the increase in engine speed between full load and no load.

Excess fuel—To aid starting under cold conditions, injection is allowed to continue for a longer than normal period when the 'cold start' device is operated. This ensures that injection persists through the hottest period of compression, giving easy starting, and results in an excessive quantity of fuel being injected. Instead of being referred to as 'extended injection', it is usually known as 'excess fuel'.

Mechanical or centrifugal governors

*I*n mechanical governors, the increase in centrifugal force with speed of rotation is utilised to provide the governing control. Mechanical governors may be of the constant speed, all speed, or limiting speed types.

Simple constant speed governor

Constant speed governors are fitted to engines that are required to run at a set or constant speed, and are governed to this set speed. Applications include engines that power alternator sets, water pumps, conveyors, etc.

The simple constant speed governor (Fig 19.1) consists of two pivoted bob-weights or flyweights, fixed to a pivot plate, which rotates with the pump camshaft, a sliding control sleeve, a pivoted fork and a governor spring. Spring force acts against the sleeve, forcing it against the lever arm of the bob-weights, which are forced in towards the shaft. As the shaft rotates, centrifugal force causes the bob-weights to move outwards from the shaft, the lever arm thrusting against the sleeve. Thus the sleeve is balanced between spring force on the one end and the force exerted by the bob-weights on the other.

The governor mechanism connects to the pump rack via the pivoted fork, one end of which engages in a groove in the sleeve and the other end connects to the rack via a link.

Should the engine speed drop due to an increase in engine load the centrifugal force acting on the weights will decrease, allowing the spring to push the sleeve along the shaft. This movement will move the rack, via the pivoted fork, to increase the fuel supply to the engine.

On the other hand, should the engine speed increase due to a lightening of the load, the subsequent increase in centrifugal force will

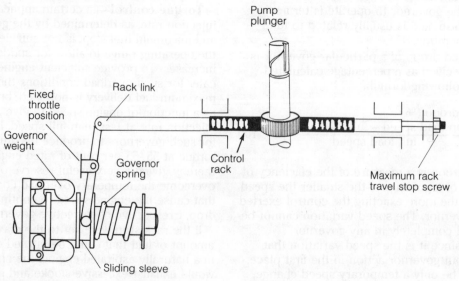

Fig 19.1 A simple constant speed governor

fling the bob-weights outwards and the lever arms will force the sleeve along the shaft against the spring. Movement in this direction will move the rack to reduce the fuel delivery from the pump.

Thus any change in the engine speed will cause an immediate change in the quantity of fuel injected, which will compensate for or at least tend to compensate for the speed change.

Simple variable speed governor

In applications where engines may be required to operate at any selected speed, variable speed governors are used. These governors govern the engine at any set engine speed, from idle to maximum. Governors of this type are used extensively in engines for earthmoving equipment and farm tractors.

The simple constant speed governor can readily be adapted to illustrate the principle of the variable speed governor. The governor shown in Fig 19.2, like the constant speed governor, features pivoted bob-weights, a sliding sleeve and a pivoted fork, but utilises a floating control fork pivot, the position of which is determined by the throttle setting.

When the throttle is moved to increase engine speed, the floating pivot is moved to the right (Fig 19.2), and the fork pivots about its

lower end located in the sliding sleeve. Quite a small amount of throttle control movement will move the rack to the full-fuel position, and provision is provided for further throttle movement by the spring in the linkage, which simply compresses.

As the engine (and governor) speed increases, the increasing centrifugal force will cause the bob-weights to move outwards and force the sliding sleeve along the shaft against the governor spring. This will move the lower end of the control fork to the right and the spring in the throttle linkage will extend to its full length. After this point in the operation, any further movement of the sleeve to the right will cause the control fork to turn on its floating pivot and move the control rack towards the no-fuel position, reducing the fuel delivery to the engine.

Once the throttle has been set, the governor will function as a constant speed governor, the engine's speed being controlled by the position of the floating pivot.

Simple idling and maximum speed governor

Idle and maximum speed governors are used on engines that need governing at idle and

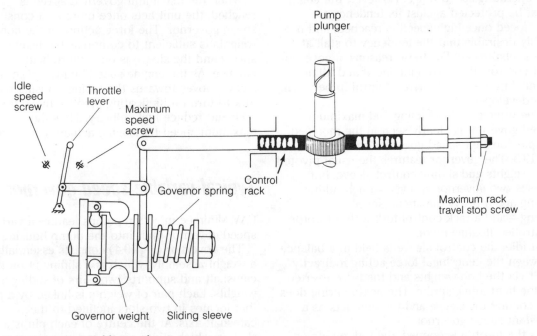

Fig 19.2 A simple variable speed governor

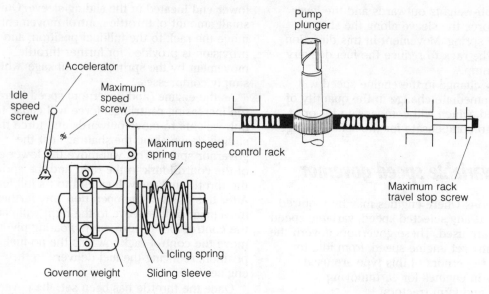

Fig 19.3 A simple idling and maximum speed governor

maximum speeds only. The speed range between these two extremities is entirely controlled and governed by the operator. Governors of this type are used solely for road transport vehicles, and are often referred to as automotive governors.

When a diesel engine is used for automotive applications, it is desirable that the driver retain complete control of the engine fuel supply over the operating speed range. However, the engine must be protected against its tendency to overspeed once high speed is reached, and it is highly desirable that the tendency to stall at idle is eliminated. For these reasons, a governor that controls the maximum speed and idling speed, but allows full driver control in between, was developed.

The principle of the idling and maximum speed governor is clearly seen in the modified simple constant speed governor shown in Fig 19.3. The governor features the usual pivoted bob-weights and sliding control sleeve, but utilises two governor springs—a light idling spring and a heavy maximum speed spring—and a rack control fork with a throttle-controlled floating pivot.

At idle, the control sleeve is held in a balance between the centrifugal force acting indirectly on it via the bob-weights and the force exerted by the light idling spring. The heavy spring does not contact the sleeve and the unit acts as a constant speed governor.

As the throttle is opened, the fork pivots on

the sleeve and moves the rack towards the maximum fuel position. As soon as the engine speed increases, the sleeve is moved against the heavy maximum speed spring by the bob-weights. However, until the governed maximum speed is reached, the centrifugal force acting on the bob-weights is not sufficient to compress the heavy spring, and the rack remains in the position dictated by the throttle.

When the maximum governed speed is reached, the unit acts once more as a constant speed governor. The force acting on the bob-weights is sufficient to compress the heavy spring and the sleeve is once more held in balance. As the engine speed builds up, so the sleeve moves towards the spring, causing the fork to turn on its floating pivot to move the rack and reduce fuel delivery. Thus the maximum speed the engine achieves is limited.

Typical variable speed governor

CAV Minimec injection pumps feature a variable speed governor built into the pump housing.

The governor (Fig 19.4) consists essentially of a weight assembly, which is a sliding fit on the camshaft and supports two pairs of sliding weights. Each pair of weights is linked by a pin that locates in two slots inclined to the camshaft axis. At the centre of each pin is a slipper, which slides on an inclined plane

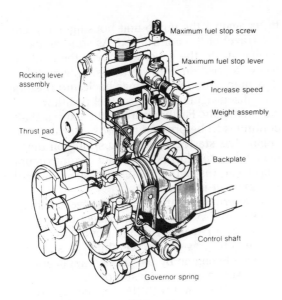

Fig 19.4 *Minimec centrifugal governor*

parallel to the slots. The complete weight assembly is forced to rotate by the backplate, which is bolted to the camshaft and which surrounds the weights. A governor spring, controlled by the control shaft and control lever, acts through a thrust pad and presses the weight assembly against the backplate. A groove in the thrust plate engages with two pins of a

rocking lever fitted over the control shaft, and which transmits the axial movement of the thrust pads and weight assembly to the control rod.

When the camshaft rotates, the weights of the governor will tend to be thrown outwards, and the centrifugal force will be transmitted as an axial force to the thrust pad by the action of the slippers on the inclined planes. The resultant deflection of the governor spring is used to control the unit. If the engine is initially running at idling speed, for example, and the control lever is turned to the maximum speed position, the governor spring will be wound up against the thrust pad and the weight assembly will be pressed hard against the backplate as shown unshaded in Fig 19.5. The resultant forward motion of the rocking lever assembly is transmitted to the control rod, and the control rod moves to the maximum fuel position. The engines speed will now increase until the centrifugal force is great enough for the weights to push the weight assembly back to an equilibrium position in which the amount of fuel supplied is just sufficient to maintain the maximum speed. Similarly, if the control lever is set to the idling position the pressure of the governor spring will be light, a low speed will be sufficient for the action of the weights to push the weight assembly away from the backplate against the spring, and the governor will take up the position shown shaded in Fig 19.5.

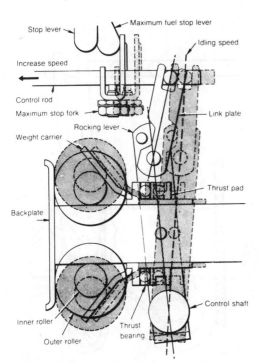

Fig 19.5 *Operation of the Minimec governor*

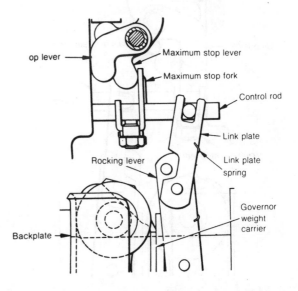

Fig 19.6 *Maximum fuel position of the control rod*

At any other speed between idling and maximum, the governor will take up an appropriate intermediate position.

The travel of the control rod in the maximum fuel direction is limited by a maximum stop fork on the control rod, which engages with a maximum fuel stop lever mounted on the excess shaft supported by the pump unit housing (Fig 19.6). The maximum fuel stop lever is basically a bell crank lever, one arm of which is positioned by the maximum fuel stop screw while the other limits the forward travel of the control rod by contacting the stop fork attached to it. The maximum fuel stop screw projects through the top of the pump housing and is adjusted externally.

The external stop control lever is mounted on a hub that fits over the excess shaft but is not attached to it, and terminates in an internal stop lever butting against the maximum stop fork. When the stop lever is turned towards the stop position, the control rod is pushed back and the fuel is cut off.

To allow excess fuel to be supplied to the engine for cold starting, the excess shaft on which the maximum stop lever is mounted is spring loaded and can be pressed inwards. When this is done the maximum stop lever moves inwards away from the maximum stop fork, allowing the control rod to move to a position of greater than normal fuel supply if the control lever is turned to the maximum speed position (Fig 19.7). The stop lever then butts against the pump unit housing and limits the travel of the control rod. As soon as the engine starts, the governor moves the control rod to a position of lower fuel supply and the maximum stop lever springs back into position, thus disengaging the excess fuel position automatically. Furthermore, if the excess shaft continues to be pressed inwards after the engine has started, a baulking spring on the maximum stop fork will contact the maximum stop lever, restricting the engine speed until the excess shaft is released.

An alternative variable speed governor may be used in CAV Minimec fuel injection pumps. Known as a **reverse linkage governor,** it is designed to prevent the possibility of stalling the vehicle engine by the sudden application of the brakes in emergency stopping.

If a vehicle fitted with the normal type of Minimec governor (Fig 19.4) is stopped abruptly by severe application of the brakes, the control rod can be carried by its own momentum (in the direction of vehicle motion) towards the no-fuel position. If there is no pressure on the governor spring (ie no pressure on the accelerator pedal) and the momentum is sufficiently great, this movement of the control rod could shut off the fuel supply and stop the engine.

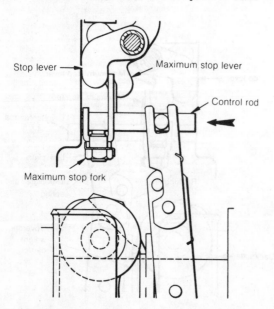

Fig 19.7 Excess fuel position of the control rod

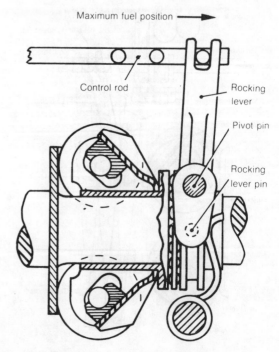

Fig 19.8 Governor with reverse linkage

In the reverse linkage arrangement of the governor, the rocking lever is pivoted as shown in Fig 19.8 so that the control rod operates in the opposite direction for the minimum and maximum fuel positions. In this manner the control rod linkage is counterbalanced. Thus, in the event of sudden application of the brakes in similar circumstances, the control rod is not affected by momentum and the fuel supply is maintained under control by the governor.

A typical reverse linkage arrangement is shown in Figs 19.9 and Fig 19.10. The governor and the governor spring (coiled type or leaf type) are the same as used with the normal form of linkage but the rocking lever is of different design.

The rocking lever is a straight form of lever without a link plate, and pivots on a pin mounted on the side of the pump unit housing. The rocking lever and pivot pin are shown in Fig 19.11.

Adjustments—The Minimec governors feature standard externally adjusted idle and maximum speed stops, but make use of a

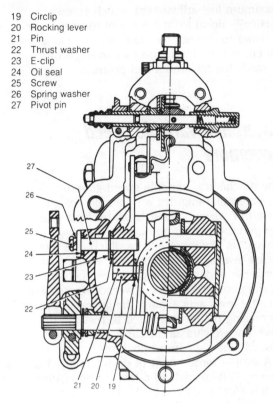

19	Circlip
20	Rocking lever
21	Pin
22	Thrust washer
23	E-clip
24	Oil seal
25	Screw
26	Spring washer
27	Pivot pin

Fig 19.10 Sectioned end view of the rocking lever assembly

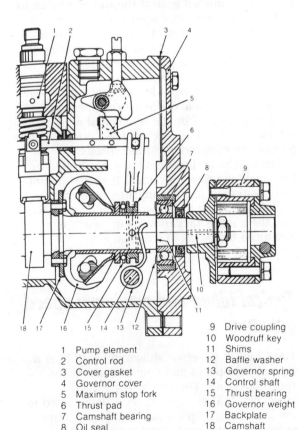

1	Pump element	9	Drive coupling
2	Control rod	10	Woodruff key
3	Cover gasket	11	Shims
4	Governor cover	12	Baffle washer
5	Maximum stop fork	13	Governor spring
6	Thrust pad	14	Control shaft
7	Camshaft bearing	15	Thrust bearing
8	Oil seal	16	Governor weight
		17	Backplate
		18	Camshaft

Fig 19.9 Sectioned view of the governor assembly

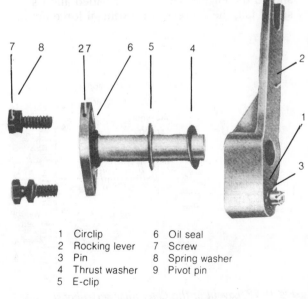

1	Circlip	6	Oil seal
2	Rocking lever	7	Screw
3	Pin	8	Spring washer
4	Thrust washer	9	Pivot pin
5	E-clip		

Fig 19.11 Rocking lever and pivot pin

maximum fuel adjustment, which screws vertically down from the top of the governor housing to contact the maximum fuel stop lever, which, in turn, limits the rack movement towards the maximum fuel position.

Caterpillar variable speed governors

The mechanical variable speed governor used on Caterpillar engines is of simple design. The governor weight carrier is driven by a gear on the pump camshaft and carries the two flyweights, the inner end of which acts against a thrust bearing. As the governor weights move outwards under centrifugal force, the thrust bearing compresses the governor spring. The throttle lever acts against the governor spring, increasing the force applied to restrain the governor weights as the throttle is moved towards the high-speed position.

A simple torque control device is fitted to this governor. It consists of a leaf spring against which a collar affixed to the rack bears. Under normal full throttle running conditions, the force exerted by the governor weights is sufficient to oppose the force applied by the throttle, and the control rack moves towards the full-fuel position only until the collar on the rack bears against the torque spring. However, should the engine become overloaded and its speed fall, the decreased centrifugal force acting on the governor weights will not restrain the governor spring, and the rack will move further in the maximum fuel direction, the collar deflecting the torque spring until it rests against the stop bar behind the torque spring. Thus at full throttle overloaded conditions, more fuel is delivered to the engine than at full throttle normal load conditions, giving increased torque when required.

Obviously, considerable force would be required to increase the engine speed by increasing governor spring force. In applications where the throttle is set this provides no problem, but in applications where the throttle needed to be continually altered, operator fatigue would result. To overcome the problem, the throttle is servo-assisted.

The servo-boost unit utilises a rod connected to the spring seat. The rod rides inside a valve contained in a piston. When the accelerator (or throttle control) is moved towards the maximum fuel position, it also moves the valve. This opens the inlet port and pressurised lubricating oil flows behind the piston. The piston then pushes against the rod and helps to compress the governor spring. When the accelerator movement stops, the oil pressure forces the piston slightly further, permitting the pressure behind the piston to escape.

Adjustments—Caterpillar governors have two adjustments–high idle and low idle (maximum speed and idle speed); these are readily accessible by the removal of a retainer/cover. These are adjusted on the engine to give the required governed speeds.

The maximum fuel delivery is controlled by the maximum rack movement towards the full-fuel position. It may be adjusted on or off the engine according to the workshop manual, to give the correct rack movement, which is measured by means of a dial gauge.

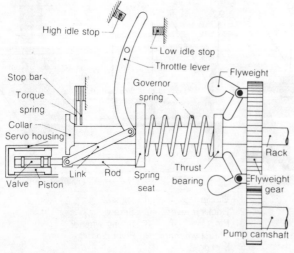

Fig 19.12 Layout of the Caterpillar servo-boost governor

Typical idling and maximum speed governor

The CAV LW governor shown in Fig 19.13 is a typical example of an idling and maximum speed governor. The two spring loaded flyweights (15) are carried on a sleeve fitted to the injection pump camshaft (14), and move outwards under centrifugal force as the camshaft rotates. Bell crank levers (17) transmit

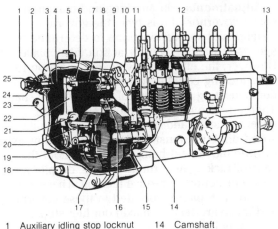

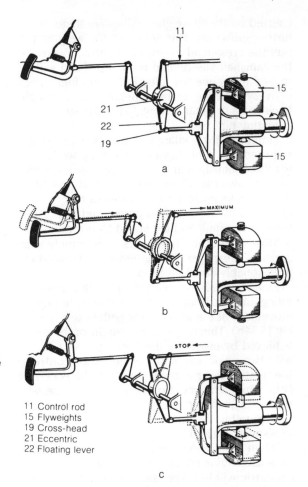

1 Auxiliary idling stop locknut	14 Camshaft
2 Cover	15 Flyweights
3 Auxiliary idling stop valve	16 Spring adjusting nut
4 Link screw	17 Bell crank levers
5 Air breather	18 Drain plug
6 Lever link	19 Cross-head
7 Link springs	20 Shaft
8 Link block	21 Eccentric
9 Stop pawl	22 Floating lever
10 Stopping control lever	23 Control lever
11 Control rod	24 Auxiliary idling stop spring
12 Fuel inlet	25 Auxiliary idling stop sleeve
13 Control rod stop	

Fig 19.13 CAV injection pump with an LW governor

the centrifugal action of the weights to the link pin and crosshead assembly (19) as a longitudinal movement. The crosshead turns the floating lever (22) about the eccentric (21), and the movement of the upper end of the lever is conveyed to the injection pump control rod (11) by the link screw (4) through the link block (8). The position of the control rod is thus adjusted to increase or decrease the quantity of fuel delivered by the injection pump elements.

The accelerator pedal is connected to the control lever (23), which is clamped to the shaft (20). When the shaft is turned by movement of the accelerator pedal, the eccentric—an integral part of the shaft—moves the floating lever and the control rod independently of the weight mechanism.

Operation of the idling and maximum speed governor is shown diagrammatically in Fig 19.14 (components numbered as for Fig 19.13). While the mechanism is at rest, the flyweights (15) lie close to the camshaft sleeve, and the injection pump control rod (11) is held towards the maximum fuel position.

Before the starter motor is engaged with the engine flywheel ring, the accelerator pedal should be fully depressed. After the engine fires and runs up to speed, the accelerator pedal

11 Control rod
15 Flyweights
19 Cross-head
21 Eccentric
22 Floating lever

Fig 19.14 Governor operation

should be steadily released and the governor will then come into operation to control the engine at the predetermined idling speed (Fig 19.14a).

Movement of the flyweights is governed by two sets of springs (see Fig 19.15). At idling speed (Fig 19.15a), the weights move outwards a comparatively short distance and compress only the outer springs (2). Should the engine speed tend to increase, the greater centrifugal force

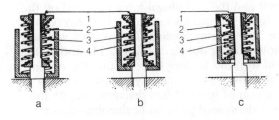

Fig 19.15 Governor springs

exerted on the flyweights will move them further outwards. This action of the weights will pull the crosshead assembly inwards, towards the camshaft sleeve, and the control rod will be moved to reduce the injection pump fuel delivery until the engine idling speed is brought back to its original setting. (1) is the spring adjusting nut.

Similarly, if the idling speed falls below that required, the centrifugal force on the weights will be diminished and the spring will force the weights inwards. The control rod will now be moved to increase the quantities of fuel delivered by the injection pump, and will restore the engine idling speed to normal.

Movement of the flyweights is extremely small and the idling speed is therefore maintained at a reasonably constant level.

Engine speeds between the predetermined idling and maximum limits are controlled by movement of the accelerator pedal (see Fig 19.14b). This ungoverned condition is achieved by providing the floating lever (22) with two pivot points. Under governor control, the floating lever is moved around the eccentric (21), while movement of the accelerator pedal turns the shaft and the eccentric as a whole, thus pivoting the floating lever about the crosshead arms (19).

When the accelerator pedal is depressed to increase engine speed, the shaft and the eccentric will turn the floating lever and move the control rod towards the maximum fuel position, irrespective of the position of the governor weights.

Throughout the intermediate speed range of the engine, the governor flyweights compress the outer idling springs (2 in Fig 19.15), but are prevented from further outward movement by the heavier inner springs (3 and 4). The weights therefore remain in the position shown in Fig 19.15b until maximum speed is reached.

Should the accelerator pedal be depressed so that the injection pump delivers more fuel than the engine requires for the load, the engine speed will tend to exceed the predetermined limit and the centrifugal force imposed on the weights will increase sufficiently to overcome the inner spring loading. The weights will now move further outwards (Figs 19.14c and 19.15c) and turn the floating lever around the eccentric. The control rod is thereby drawn back to a position of reduced fuel delivery, despite the over-depressed position of the accelerator pedal.

Adjustments—In an idling and maximum speed governor of this type, the governor weights and springs combine to determine the speeds at which governor action takes place. However, the effect of any governor action to alter fuelling could be cancelled out by a compensating movement of the eccentric shaft. Therefore the maximum movement of the eccentric shaft and control lever in the full speed direction must be limited to prevent extra throttle movement being applied to hold the control rack against the rack stop, despite any governor action to reduce fuelling. There are two such stops incorporated in these governors, and these are termed maximum fuel stops.

Equally it is true that excessive eccentric movement in the idle speed direction could overcome corrective governor action at idle speed, so causing the engine to stop. An idle fuel stop is included to limit this movement.

By reference to Fig 19.16, it can readily be seen that the governor carries two maximum fuel stops, one (7) covered to prevent unauthorised tampering, which is set on the test bench, and a second (6), which can readily be altered on the engine. An adjustable idle fuel stop (8) and an idle buffer (9) are also incorporated in the governor housing, while a rack stop (including perhaps, an excess fuel device) will certainly be incorporated on the

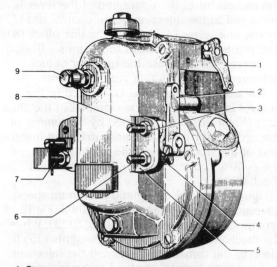

1 Stop control lever	6 Unsealed maximum fuel stop
2 Control lever	7 Sealed maximum fuel stop
3 Idling stop locknut	8 Idling speed stop
4 Pawl on control shaft	9 Auxiliary idling stop
5 Flange	

Fig 19.16 Arrangement of governor stops

other end of any injection pump to which the governor is fitted.

Between the governor speed limits, no difficulty would arise during calibration in holding the control rack at the desired position. However, if the pump were to be run at or near either governed speed, the governor would function and attempt to alter the rack position. All interference can be eliminated by removing the governor cross-shaft, and it is recommended that this be done before calibrating the pump. During calibration, and until it is to be set, the rack stop should be removed or adjusted well clear of the rack.

Once the pump has been calibrated, the governor shaft should be refitted, the stops adjusted, and the governor action checked. The master stop (7) should be adjusted while running the pump at a specified speed with the control lever held against the stop by a spring to give either a specific fuel delivery or rack position. (This will only be correctly set if the governor link has been adjusted to its correct length during reassembly.)

The maximum speed governor action can be checked by running the pump at a specified speed at or near maximum speed and checking the fuel delivery (or rack position) against the specifications, then increasing the pump speed to a second point and rechecking the fuel delivery (or rack position). At the second (higher) speed, the fuel delivery will have dropped to a lower specified figure (or the rack will have moved back to a new specified position) if the governor is functioning correctly. Some slight adjustment is allowable at the governor springs but will not be necessary if the springs were set up correctly during assembly.

The idling stop (8) is adjusted and the governor checked at idle speed in a similar way to the maximum fuel stop (7), but with the control lever held back in the engine idle position. Again two speeds and associated fuel deliveries (or rack positions) will be given in the test sheet for the pump.

The idle stop (8) and the second maximum fuel stop (6) may both be adjusted to give a certain fuel delivery, but this may have to be adjusted on the engine to give the required engine idle and the maximum power with a satisfactory clean exhaust.

The rack stop must be adjusted after setting the maximum fuel stop (7) to allow the rack to move at least as far as the maximum fuel stop

allows. If the rack stop restricted rack movement before stop (7) limited the control lever movement, straining of the governor linkages would result. The rack stop should be adjusted according to the test sheet.

The idle buffer (9) is best adjusted on the engine to prevent stalling when the engine speed is suddenly cut to idle, but not far enough in to start the engine surging or the idle speed increasing.

Pneumatic governors

P neumatic governors utilise the changing inlet manifold depression to control the fuel pump delivery. They consist of two separate sections—the governor unit, which is fitted to the injection pump, and the venturi or throttle unit, which is fitted to the inlet manifold—which are connected by a pipe or pipes.

A simple pneumatic variable speed governor

The simple pneumatic governor (Fig 19.17) shows the basic construction of the two sections. The governor unit consists of its housing, a flexible diaphragm and a spring. The diaphragm connects to the pump rack, and the spring forces the diaphragm and rack towards the maximum fuel delivery position. The throttle unit consists of a butterfly, which is controlled by the operator's throttle control. The unit is fitted to the inlet manifold so that all air passing to the engine must pass through it. The two sections are connected by a pipe.

At idle the butterfly is almost closed, and the depression on the engine side of the butterfly is high. The governor unit is directly connected to the engine side of the butterfly by the pipe, so that the pressure on the spring side of the diaphragm will be equal to the inlet manifold pressure. Atmospheric pressure on the rack side of the diaphragm will force the diaphragm towards the spring until a point of equilibrium is reached, moving the rack to the low-fuel delivery position.

As the throttle is opened, the manifold depression will become progressively less. This

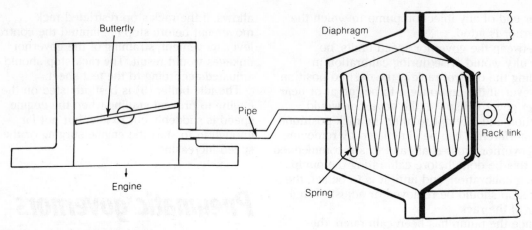

Fig 19.17 A simple pneumatic governor

will cause the pressure on the spring side of the diaphragm to approach atmospheric pressure and the pressure difference across the diaphragm will become less. The spring will then be able to force the rack in the higher fuel delivery direction.

Once the throttle is set and the diaphragm moves the rack to the required position, the engine will run at a steady speed. Should extra load be applied so that the engine speed falls, the manifold depression will also fall—the pressure will rise closer to atmospheric—and the spring will force the diaphragm and rack towards the maximum fuel direction, where it will stay until the manifold depression returns to its original point.

If the engine load becomes lessened and the engine speed increases, the manifold depression will climb correspondingly—the pressure will become less. The atmospheric pressure will be more effective with less pressure opposing it and the diaphragm will move to compress the spring slightly, cutting down the fuel delivery. The fuel delivery will remain at this reduced level until the manifold depression returns to its original level.

A typical pneumatic governor

While the throttle unit shown in Fig 19.17 is sufficient to illustrate the basic principle of the pneumatic governor, in most cases this component is much more correctly termed a **venturi unit.**

A venturi is a tube tapered internally from both ends, the arrangement giving a minimum internal diameter somewhere along the tube. Should a fluid be caused to flow through a venturi, the pressure will change along its length, the point of lowest pressure being the point of minimum diameter. How low the pressure drops is dependent on the fluid's speed through the venturi—the greater the speed, the lower the pressure.

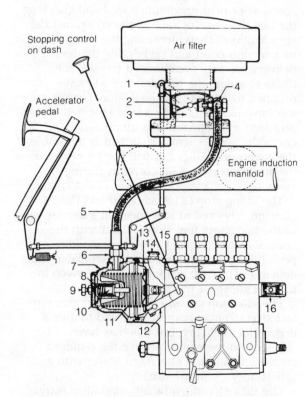

Fig 19.18 CAV fuel pump with pneumatic governor and single-pitot venturi unit

The venturi unit shown in Fig 19.18 is a typical example. A small auxiliary venturi is situated within the main venturi, the butterfly being cut out where necessary. Depending on the position of the butterfly (ie how far it is open), varying proportions of the engine's air supply must pass through the auxiliary venturi. Hence when the engine is running, the velocity of the air passing through the auxiliary venturi (and the subsequent pressure at the smallest diameter) is dependent on the butterfly position and the engine's speed. It is to the auxiliary venturi that the vacuum pipe connects.

The CAV pneumatic governor shown in Fig 19.18 consists of a venturi flow control (venturi unit) located between the engine intake manifold and the air cleaner, and a diaphragm unit mounted directly on the fuel injection pump.

Reference to the diagram will show that the body of the venturi unit is flanged at one end so that it can be secured to the intake manifold, and is spigoted at the other to accommodate the air cleaner. The diameter of the throat must be selected to suit the capacity of the engine to which it is fitted.

Airflow through the throat of the venturi is regulated by a butterfly (2), which is mounted on a spindle carried in bushes pressed into the venturi body. The control lever (1) is secured to the butterfly spindle and is connected by a suitable control linkage to the accelerator pedal. Maximum speed and idling stops (Fig 19.19) provide means of adjusting the limits of movement of the butterfly.

The small auxiliary venturi situated within the main venturi may be secured to or cast integral with the body of the venturi unit. Projecting into the auxiliary venturi at right angles to the airflow is a pitot tube which is connected to the diaphragm unit by the flexible pipe (5).

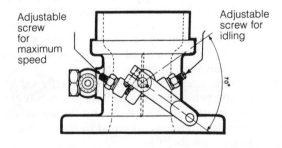

Fig 19.19 Single-pitot venturi unit

Adjustable screw for maximum speed

Adjustable screw for idling

70°

The flexible leather diaphragm (13 in Fig 19.18) is clamped between the governor housing (15) and the governor cover (7). It is connected to the control rod of the fuel injection pump and is spring loaded by the governor spring (11). Pressure exerted on the diaphragm by the spring tends to move the diaphragm and control rod towards the maximum fuel stop (16).

Control rod movement is limited by adjustable stops. Auxiliary idling stops (one type is shown in Fig 19.18) may be fitted to governors with single-pitot venturi units. That illustrated in Fig 19.18 consists of a plunger (10), which is spring loaded by the spring (8). The adjusting screw (9) permits adjustment of the spring force acting on the plunger.

Maximum fuel stops are normally fitted at the end of the pump housing remote from the governor. They provide the means of adjusting the maximum fuelling and may be sealed after setting to prevent unauthorised adjustment.

The operation of a governor fitted with a single-pitot venturi unit may be followed by reference to Fig 19.18. As the throttle is advanced and the butterfly opens, the rate of airflow through the auxiliary venturi changes from a maximum at closed throttle to a minimum at full throttle. Should the throttle be abruptly closed while the engine is running at high speed the air velocity through the venturi will immediatley become very high, but will drop as the engine slows down. Again, should an engine's speed fall while the throttle is fixed, the air velocity will also fall due to the engine requiring less air. Associated with these changes in air velocity are corresponding changes in the depression within the venturi.

A pitot tube projects into the auxiliary venturi, and is connected to a flexible pipe (5). The flexible pipe and the pitot tube interconnect the venturi and the airtight chamber in the diaphragm unit so that the depression in the venturi and the airtight chamber are maintained at the same level. Thus while atmospheric pressure acts on the pump side of the diaphragm (13), air only at the same pressure as that in the venturi acts on the other. The resultant force tends to move the diaphragm away from the pump housing. Because the diaphragm is connected to the control rod, this movement of the diaphragm reduces the fuel injection rate.

The governor spring (11) exerts pressure on the diaphragm in opposition to the force created by the pressure differential and, in consequence, the diaphragm will assume a position where the two forces acting on it are in equilibrium. It follows, therefore, that any change in the value of the depression will cause the diaphragm to move, and, since the diaphragm is connected to the control rod of the fuel injection pump, a change of fuelling will occur.

Force exerted on the diaphragm due to the pressure differential tends to move the control rod towards the auxiliary idling stop, while that exerted by the governor spring tends to move the control rod towards the maximum fuel stop (16). When the engine is running at a fixed throttle setting the control rod assumes a position where the forces acting on the diaphragm are in equilibrium. Fluctuations in engine speed resulting from changes of engine loading will result in corresponding changes of the depression. The increase in the depression that occurs when the engine exceeds the selected speed will cause the control rod to move towards the auxiliary idling stop. The output of the fuel injection pump will then be reduced and engine speed will fall until the selected engine speed is restored. Similarly, a fall in engine speed resulting from increased engine loading will cause the control rod to move towards the maximum fuel stop under influence of the spring (11), thus increasing the output of the fuel injection pump and increasing the engine speed until the selected running speed has been regained.

Any selected engine speed will therefore be maintained within close limits, since any speed change is accompanied by a compensating change in the output of the fuel injection pump.

When the engine is stationary, the control rod is held against the maximum fuel stop by the force exerted on the diaphragm by the governor spring. Although the control rod is in the correct position for starting the engine, the accelerator pedal is usually depressed during this operation. After the engine has started, it may be idled by releasing the accelerator and thus closing the butterfly valve in the flow control unit. Restriction of the main venturi will cause an immediate increase in the velocity of the airflow and the resulting increase in the depression will move the control rod to the idling position. Speed control is maintained at idling in the same manner as at intermediate throttle settings. Violent movement of the control rod is prevented, however, by the spring

loaded auxiliary idling stop. An increased depression caused by any increase in engine speed will move the diaphragm against the stop and overcome the force exerted on the plunger by the spring (8).

When difficulty is experienced in obtaining even running at idling without exceeding the maximum idling speed or the maximum no-load over-run, the use of a pneumatic governor incorporating a cam-operated idling stop is recommended (Fig 19.20). The cam is rotated by movement of the control lever, which is connected by suitable control linkage to the accelerator pedal. The linkage is so arranged that the spring is compressed when the accelerator pedal is released.

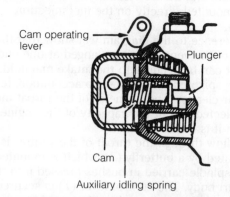

Fig 19.20 Cam-operated idling spring

The operation of governors fitted with double-pitot venturi units (Fig 19.21) is basically the same as that of governors with single-pitot venturi units, movement of the diaphragm and control rod being governed by the velocity of the airflow through the auxiliary venturi (3).

A pitot tube projecting into the second auxiliary venturi (1) is connected by a flexible pipe to an air valve (5), which is fitted in place of an auxiliary idling stop. The butterfly valve is not cut away to permit the passage of air through the auxiliary venturi (1) when the valve is closed, so that at idling speeds there is no airflow through the second auxiliary venturi and if the air valve (5) opens, it opens to atmospheric pressure. When the butterfly valve is open the velocity of the airflow through both auxiliary venturis is the same, and the air valve (5) opens to a depression equal to that in the diaphragm chamber (4).

When the engine is idling, the metal centre of the diaphragm is in contact with the air valve

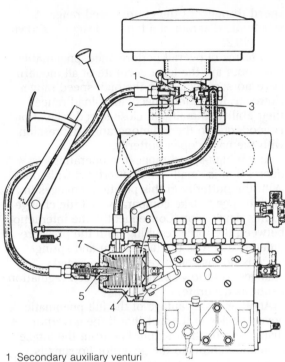

1 Secondary auxiliary venturi
2 Butterfly valve
3 Auxiliary venturi
4 Diaphragm chamber
5 Air valve
6 Diaphragm
7 Governor spring

Fig 19.21 CAV pump with pneumatic governor and double-pitot venturi unit

(5). An increase in engine speed will cause an increase in the depression in the diaphragm chamber (4) and the diaphragm and control rod will move towards the no-fuel position. Movement in this direction beyond the normal idling position will reduce the output of the fuel injection pump and at the same time open the air valve (5). Since the air valve opens to atmospheric pressure when the butterfly valve is closed, the depression in the diaphragm chamber (4) will be relieved and the diaphragm and control rod will move towards the maximum fuel position. This movement will be arrested, however, when the diaphragm reaches the normal idling position, at which point the air valve will be closed and the depression in the diaphragm chamber restored. This interference with normal governor action serves to prevent violent movement of the control rod at idling speeds, but does not interfere at high speeds when the velocity of the airflow through both auxiliary venturis is equal.

Adjustments—Adjustments are provided for maximum fuel delivery, absolute maximum engine speed and idle. The maximum fuel stop is normally adjusted during calibration to specifications and then sealed to prevent unauthorised adjustment. Any adjustment that may be deemed necessary during service may be effected by removing the cap or cover, removing or releasing the locking device, and screwing the adjusting screw outwards to increase the maximum fuel setting, and inwards to reduce it.

Alteration of the setting of the maximum fuel stop will change the power output of the engine and the fuel consumption. An excessive increase of the maximum fuelling will result in a smoky exhaust and increased fuel consumption. When adjustment is made a compromise should be sought between power output, exhaust colour and fuel consumption.

Absolute maximum speed is governed by the throat diameter of the venturi. The maximum speed of the engine can be adjusted to speeds below this figure by limiting the movement of the butterfly valve in the venturi unit. Adjustment is made by movement of the stop screw shown in Fig 19.19.

When single-pitot venturi units are fitted, the idling speed is adjusted by movement of the idling stop (Fig 19.19). Adjustment of the auxiliary idling stop in the governor housing should only be made when hunting occurs at the correct idling speed. If hunting does occur, ensure that it is not attributable to engine faults (such as faulty injectors) before making adjustments.

When double-pitot venturi units are fitted, the positions of the idling stop on the venturi unit and of the air valve assembly in the diaphragm unit must be adjusted at the same time. Adjustment of either will affect the idling speed, and it is only by adjusting both settings simultaneously that a setting can be achieved where even running is obtained at the correct idling speed.

Combined pneumatic–mechanical governor

The combined governor consists of a pneumatic and a mechanical governing assembly, combined together to act as a unit, and designed to give accurate control of engine

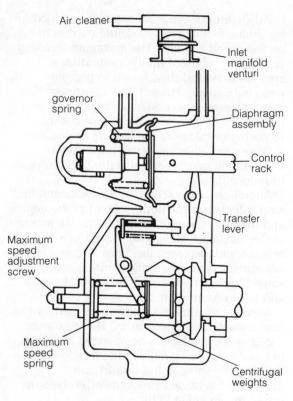

Idle speed—pneumatic governing

speed throughout the entire speed range. A schematic diagram of a typical design is shown in Fig 19.22.

In terms of overall performance, pneumatic governors are the least accurate of all modern governors, especially in the high-speed range. However, in the low-speed, light-load range, their ability to maintain close speed regulation is good, due to the large operating depression with the partly open butterfly.

In the high-speed range, the operating depression is small—the butterfly is wide open. With the butterfly in this position, changes in engine speed make comparatively little change to this depression. Consequently, the interaction between the force resulting from the pressure difference across the diaphragm and spring force is not as sensitive to engine speed changes as at low speeds, thus speed regulation is not as accurate.

Another disadvantage of the full pneumatic governor lies in the fact that if the governor diaphragm or the vacuum line from the intake manifold venturi to the governor housing is punctured, the spring chamber will be in direct connection with the atmosphere. With

Fig 19.22 Operation of the combined pneumatic/mechanical governor

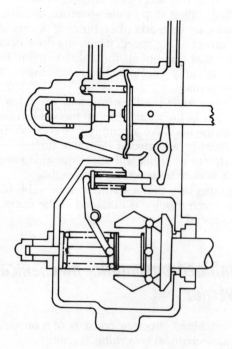

Medium speed—pneumatic
governing

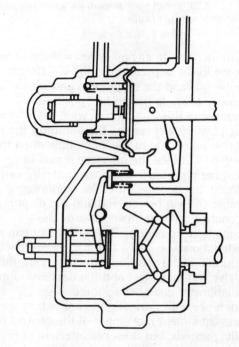

Maximum speed—combined pneumatic
and mechanical governing

atmospheric pressure on both sides of the diaphragm, the governor spring will push the control rack to the maximum fuel position. The governor will be unable to exert any control over the engine speed and the engine will overspeed, possibly causing serious engine damage. Under closed throttle, insufficient air entering the engine with the butterfly closed will limit the engine speed, but excessive exhaust smoke will result.

On the other hand, the centrifugal governor is unaffected by changes in altitude and physical damage to external connections, and is capable of maintaining close speed regulation in the high-speed range.

Therefore, by combining the pneumatic and mechanical governors in a single assembly, each component can be used to its best advantage in maintaining close speed regulation over the full operating speed range. Fig 19.22 illustrates the three phases of operation of the combined pneumatic–mechanical governor covering the operating speed range of the engine.

In conclusion, the pneumatic governor section operates like a conventional pneumatic governor in the low-to-medium speed range,

and at the high-speed range the mechanical governor section operates together with the pneumatic governor section thereby eliminating the influence of altitude changes to the governing effect.

Hydraulic governors

*H*ydraulic governors utilise changes in the pressure of fuel being discharged from a constant displacement pump incorporated in the governor housing, and driven by the injection pump camshaft. As the speed of the pump increases the output rate increases in proportion, and if free flow from the pump is not available, the pressure will also increase. It is the change in fuel pressure with change in speed that is used to exercise speed control in hydraulic governors.

Before pursuing the operation of hydraulic governors further, reference should be made to Fig 19.23, which shows the main components in the layout of a typical hydraulic governor. However, it must be stressed that all

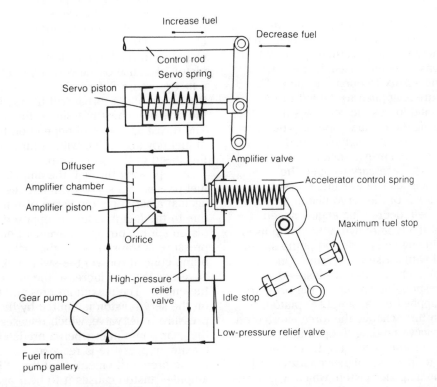

Fig 19.23 Simplified layout of a hydraulic governor

components are not shown in this diagram, and it is included to illustrate the operating principle only.

Reference to the diagram will show that fuel under pressure from the gear pump is admitted to the amplifier chamber, passes through an orifice in the amplifier piston and is conveyed from the rear of the amplifier piston to the front of the servo piston. The servo piston is connected by linkages to the injection pump control rod, and the fuel pressure acting on the front of the piston causes it to move to the right, thus moving the injection pump control rod towards the maximum fuel position.

When fuel from the amplifier chamber passes through the orifice in the amplifier piston, a pressure drop occurs and the amplifier piston is forced to the right because the pressure on the left is greater than that opposing it from the right-hand side of the piston. This movement of the amplifier piston overcomes the force exerted by the accelerator control valve spring, forces the amplifier valve off its seat and allows fuel to move up behind the servo piston, where the pressure of this fuel helps the servo spring to move the servo piston to the left. As the servo piston moves towards the left, the injection pump control rod moves towards the no-fuel position.

The speed control lever varies the tension of the accelerator control spring, increasing the tension to increase the governed speed. When the spring tension is increased, higher fuel pressure is necessary to open the valve. This higher pressure is applied to the left of the servo piston and forces the servo piston to the right, increasing the engine's speed. When it is desired to decrease the engine's speed, accelerator control spring tension is decreased and the pressure necessary to open the amplifier valve is less. Lower fuel pressure is applied to the left of the servo piston, which moves to the left, decreasing engine speed.

Because of the type of hydraulic pump used, an increase in engine speed (and pump speed) is associated with a rise in pressure. Once the throttle is set, a change in engine speed will cause a corresponding change in the pressure that will be applied to the amplifier piston, which will, in turn, change the force exerted on the amplifier valve tending to open it, If, for example, the throttle is set and the engine speed starts to increase, higher pressure will be applied to the amplifier piston, which will open the amplifier valve slightly, allowing the

pressure of the fuel between the amplifier piston and the amplifier valve to drop, while the pressure of the fuel having passed through the valve will increase slightly. These two pressures are applied to the servo piston, and the decreased pressure on the left and the increased pressure on the right will cause the piston to move to the left, decreasing the fuel supply to the engine slightly and slowing the engine.

The high- and low-pressure relief valves are used to control the fuel pressure acting on either side of the servo piston and keep it within limits.

CAV type H hydraulic governor

Construction and layout—The CAV type H hydraulic governor is a self-contained sealed unit carried in a separate housing, which bolts directly to the fuel injection pump housing. The gear pump, which pressurises the fuel to operate the governor, is contained within the housing where it is driven from the injection pump camshaft via a muff coupling.

Fuel passes from the pump gallery to the gear pump where its pressure is raised. It then passes through a diffuser into the amplifier chamber. The fuel escapes from the amplifier chamber through the orifice in the amplifier piston. Because of the restriction of the orifice, a fuel pressure differential will be set up across the piston and a thrust will be applied to it from the side of highest pressure—the diffuser side. The thrust applied will depend on the quantity of fuel flowing, which will, in turn, depend on the speed of the gear pump.

After passing through the amplifier piston, the fuel passes to a servo cylinder where it acts against a servo piston. Servo piston movement due to the fuel pressure is opposed by the servo piston spring, but any movement due to the fuel pressure is transferred to the fuel injection pump control rod via the swing link and governor lever to increase the fuel delivery to the engine. The maximum pressure that can act on the servo piston is limited by the high-pressure relief valve, which releases fuel back to the gear pump inlet when a predetermined maximum pressure is reached.

The previously mentioned thrust on the amplifier piston causes it to bear against the stem of the amplifier valve, tending to open this

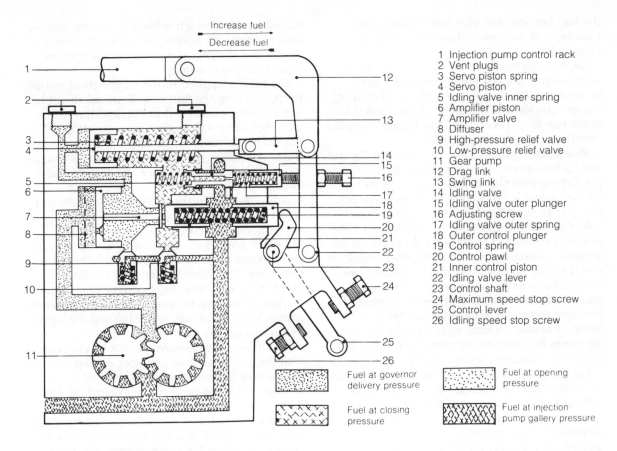

Increase fuel
Decrease fuel

1 Injection pump control rack
2 Vent plugs
3 Servo piston spring
4 Servo piston
5 Idling valve inner spring
6 Amplifier piston
7 Amplifier valve
8 Diffuser
9 High-pressure relief valve
10 Low-pressure relief valve
11 Gear pump
12 Drag link
13 Swing link
14 Idling valve
15 Idling valve outer plunger
16 Adjusting screw
17 Idling valve outer spring
18 Outer control plunger
19 Control spring
20 Control pawl
21 Inner control piston
22 Idling valve lever
23 Control shaft
24 Maximum speed stop screw
25 Control lever
26 Idling speed stop screw

Fuel at governor delivery pressure

Fuel at opening pressure

Fuel at closing pressure

Fuel at injection pump gallery pressure

Fig 19.24 CAV hydraulic governor layout

valve. Opening of the amplifier valve is opposed by the closing force exerted by the accelerator control valve spring, which is, in turn, controlled by the pawl mounted on the control lever shaft. Movement of the control lever towards the maximum speed position forces the outer plunger towards the inner plunger, thus compressing the spring to increase the load on the amplifier valve. Hence the pressure at which the amplifier valve opens is directly controlled by the speed control lever.

After passing through the amplifier valve, the fuel is led to the spring side of the servo piston where it aids the servo spring in opposing servo piston movement. Thus the position of the servo piston is established by a state of balance between fuel at **opening pressure** on the one side, and fuel at **closing pressure** plus the force exerted by the spring on the other. A change in opening pressure or closing pressure will obviously move the servo piston to change the fuel injection rate.

The maximum closing pressure is limited by

the low-pressure relief valve, which returns fuel back to the gear pump inlet when the maximum allowable pressure is exceeded.

The idling valve in the closing pressure system is included to give greater sensitivity under idling conditions than can be obtained by the use of the amplifier alone. The valve can allow fuel to escape to the gear pump inlet through slots in the idling valve body, which are opened or closed by a collar on the valve piston.

The idling valve consists of two plungers carried in a bore in the governor housing, with two springs. The outer spring lies between the two plungers, keeping them apart, while the other lies between the inner plunger and the governor housing, so biasing both plungers outwards towards the governor control lever, where the outer plunger bears against the internal idling stop.

Fuel at closing pressure fills the space between the plungers through a small orifice in the inner plunger but because of the orifice size, the fuel between the plungers cannot move rapidly in or out, and a dash pot action is obtained.

Rapid movements of the control rod, such as occur during uneven idling, are restricted because the fuel between the plungers cannot escape quickly and the plungers are forced to move together. This has the necessary damping effect to steady the engine idle.

When starting the engine, the throttle control should be moved to the maximum speed position to hold the amplifier valve on its seat. As the engine is cranked over (at slow speed) the gear pump delivers a limited quantity of fuel. Because the fuel flow through the amplifier piston is small, so the pressure drop across the amplifier piston is also small and only negligible end thrust is produced. Without sufficient end thrust, the amplifier valve remains seated preventing the admission of fuel at closing pressure to the servo cylinder. As the gear pump continues to turn, opening pressure increases and acts on the servo piston to open the pump rack and fuel the engine.

On firing, the engine speed increases, causing increased fuel flow, which sets up a pressure differential across the amplifier piston. The resultant force moves the piston to open the amplifier valve, admitting fuel to the closing side of the servo piston, which moves to reduce the fuel supply to the engine and so prevent overspeeding.

When the throttle control is returned to the idle position, the spring force tending to close the amplifier valve will be reduced and it will remain open, admitting fuel to the closing side of the servo piston to reduce fuelling, until the piston thrust on the amplifier valve diminishes sufficiently. This will occur as the engine returns to idling speed.

When the engine is running, the speed is directly controlled by the force applied to close the amplifier valve. Hence when the throttle is opened, the amplifier valve is closed (either partially or completely), cutting off the fuel to the closing side of the servo piston. At the same time, fuel pressure on the opening side of the servo piston rises as this fuel can no longer escape, except through the high-pressure valve. The servo piston immediately moves to increase fuelling until the engine reaches the required speed.

As the speed climbs towards the required level, so the piston's thrust on the amplifier valve increases to progressively open the amplifier valve, which admits fuel to the closing side of the servo piston, and allows the escape of fuel in the opening system until a point of equilibrium is reached.

If a load is applied to the engine its speed will drop, reducing the fuel flow. Reduced thrust on the amplifier piston will follow, and the amplifier valve will close. This restriction causes an increase in the opening pressure, and the servo piston moves the rack via the linkages to increase the fuel delivery and restore the engine speed.

Adjustments to the governor—In addition to the control lever stops, there are other adjustments and checks to be made to hydraulic governors. The setting of the pressure relief valves must be checked, and adjusted if necessary. The valve settings are readily checked by removing the two vent plugs from the top of the governor housing, fitting pressure gauges at these points, and adjusting the valves as necessary.

The high-pressure valve should be checked with the pump running at a specified speed with the control lever in the maximum speed position, all stops removed or backed off. Under these conditions, all the fuel delivery from the gear pump must pass through the high-pressure relief valve, which will not open until the fuel pressure reaches the valve setting. The valve setting is adjusted by inserting or removing shims.

To check the low-pressure relief valve, the pump is run at a specified speed with the control lever in the minimum-fuel position and the governor lever held fully open by means of a special tool available for the purpose. All the fuel delivery from the gear pump then passes through the low-pressure relief valve back to the gear pump inlet. The setting of this valve is also adjusted with shims.

The length of the governor link must also be adjusted by means of an adjusting screw provided for the purpose. This may be done by running the pump at a high specified speed with the control lever in the engine idle position, under which conditions no fuel should be delivered. The adjustment is made by altering the length until just a trace of fuel is delivered by the injectors, then adjusting the link length a specified amount to completely cut off fuel delivery.

The idling settings of the pump must be correct to give both correct governed idle speed and correct fuelling at idle. These settings are made after setting the pump to run at a specified speed with both the idling valve stop screw and the idling valve stop fully unscrewed.

The idle stop should first be screwed in until the pump begins to inject fuel, then the idling valve stop screw screwed in to bring the closing pressure to the correct (specified) setting. The idle stop should then be adjusted to give the correct idle delivery, after which the idling valve stop screw may be adjusted to correct the closing pressure. Only by simultaneous adjustment of the two screws can correct pressure and fuelling be set.

Two maximum speed stops are incorporated in this type of governor—one internal and one external. The internal stop should be set first to give a specified fuel delivery reduction when the pump speed is increased from a lower specified figure to a higher specified figure. The second stop may be set to contact the control lever pawl at the same speed as that at which the

internal stop is set, or it may be set to limit the speed to a lower figure for certain applications.

Electronic engine controls

Advanced technology in the internal combustion engine field has brought about electronic control of the operation and performance of today's high-speed diesel engines. More than just a governor, the benefits gained through the use of electronic controls include decreased fuel consumption, reduced exhaust emissions, improved drive-ability, self-diagnosis and engine protection.

In the Detroit Diesel Allison control system, the application of microprocessor control of the

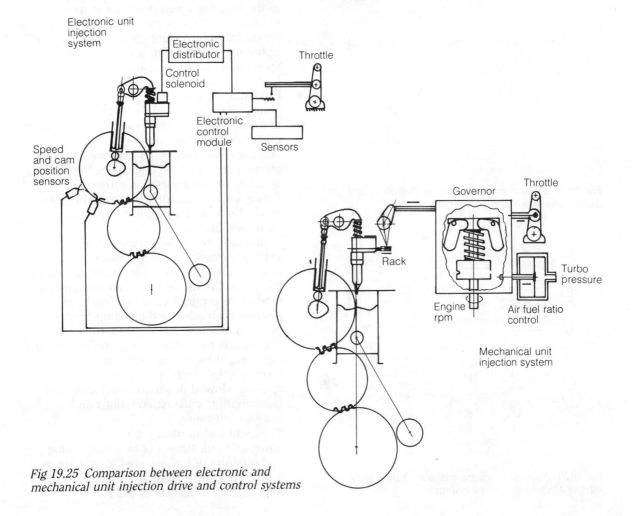

Fig 19.25 Comparison between electronic and mechanical unit injection drive and control systems

fuel injection system provides both flexibility and precision in fuel injection timing, metering sequences and duration of injection through the use of solenoid-operated unit injectors.

System components

The main components in the Detroit Diesel electronic controls system are the electronic control module (ECM), the electronic distributor unit (EDU) and the electronic unit injector (EUI), as shown in Fig 19.26. The complete electronic control system is shown schematically in Fig 19.27.

The electronic control module is the 'brain' behind the complete system, and during engine operation, it continually monitors injection timing, throttle position, oil temperature, oil

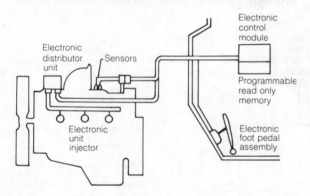

Fig 19.26 Main components of the Detroit Diesel Allison electronic control system

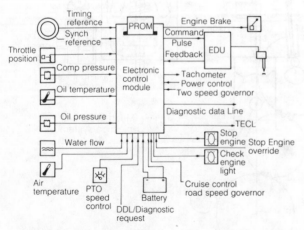

Fig 19.27 Schematic diagram of the Detroit Diesel Allison electronic control system

pressure, air temperature, coolant flow and turbocharger discharge pressure by means of electronic sensors, and analyses the data.

Within the electronic control module there is a programmable memory. The control functions regulated by this memory include torque shaping, idle speed variations and engine governing, power output, cold starting logic, transient fuel control and self diagnosis, with the optional addition of an engine protection system.

An explanation of some of the more unusual control functions mentioned above is as follows:

- Torque shaping is basically the scheduling of (or controlling) fuel injection quantities relative to engine speed, at full throttle position.
- Idle speed variation due to accessory operation can now be controlled at a set constant speed irrespective of changing accessory loads. Also an improved feature of the limiting speed governor is the calibration of speed droop. Speed droop can be calibrated from zero to 150 rpm to gain optimum coordination between engine speed and vehicle gearing, to obtain the best performance for the application.
- Cold starting logic is the term used for the ability of the electronic system to control and time fuel injection relative to the engine oil temperature as opposed to engine coolant temperature. To improve the cold starting ability of the engine, the timing and quantity of fuel injected is programmed during engine cranking. A typical program would be retarded injection timing and increased fuel delivery.

  However, once the engine started, the exhaust emissions (white smoke) and engine vibrations would be increased under these conditions. So the electronic control module re-programs the fuel injection system to an advanced injection timing position at a faster idle speed, thus eliminating the white smoke and providing a quick engine warm up. After the engine heats up, the timing is retarded and idle speed is slowed down, thus reducing fuel consumption, exhaust emissions and combustion noise.
- Transient fuel control refers to the automatic adjustment of injection timing and quantity of fuel injected to suit engine intake air flow. This occurs in the time between one injection and the next.

Consequently, under reduced air flow conditions, the injection timing is advanced, thereby increasing the allowable engine fuelling within the parameters of smoke emissions. Engine air flow is calculated from engine speed and turbocharger boost pressure, monitored by speed and pressure sensors. This constant monitoring and changing of timing and fuel delivery significantly improves acceleration, as engine fuelling is now relative to overall monitored engine operation, instead of the previously fixed monitored engine operation whereby the throttle position and subsequent governor interaction were the only determining factors in engine fuelling.

- The electronic controls system has a built in self-diagnosis system, which automatically checks the microprocessors and continually monitors all other electronic componentry.

  The electronic system is diagnosed on initial startup and throughout the entire engine running period. The system itself is pre-set to detect various faults and records them in memory bank for recall by a diagnosis reader during servicing.

- Engine protection is designed into the electronic circuitry to prevent engine damage resulting from loss of engine coolant, low oil pressure or high oil temperature. Whenever one of the monitored variables falls outside the acceptable limits of the system, the electronic control system will shut the engine down, unless the fault corrects itself within a pre-set limited time period. A function that allows for the engine shutdown to be over-ridden in critical engine operating situations is also programmed into the system.

By using the precise accuracy and the readily programmable abilities of the microprocessor-controlled electronic diesel engine controls, the engine manufacturer can now ensure automatic monitoring and control of the engine throughout all phases of its operation.

20 Distributor-type injection systems

The idea of using the same pumping element for all engine cylinders, with a distributor to supply the fuel to the injectors in turn, is very old—in fact it was patented in 1917. While many manufacturers are now supplying injection equipment operating on

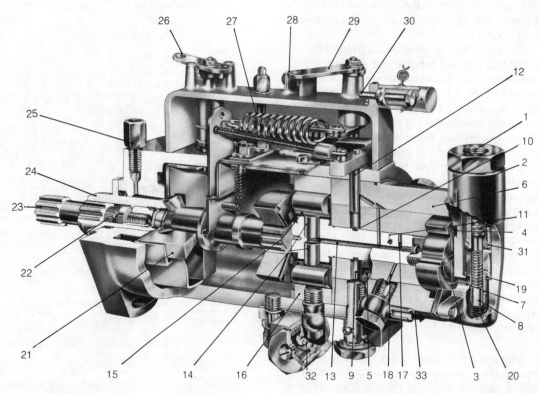

1 Fuel inlet	8 Regulating piston	17 Distributor port	25 Back-leak connection
2 End plate assembly	9 Hydraulic head	18 Outlet port	26 Shut-off lever
3 Transfer pump	drilling	19 Regulating valve	27 Governor spring
4 Transfer pump	10 Groove on rotor	sleeve	28 Idling stop
inlet port	11 Rotor	20 Priming spring	29 Control lever
5 Hydraulic head	12 Metering valve	21 Governor weights	30 Maximum speed stop
drilling	13 Rotor axial drilling	22 Drive hub	31 Nylon filter
6 Hydraulic head	14 Plungers	securing screw	32 Advance device
7 Outlet to	15 Cam follower shoes	23 Quill shaft	33 To injector
regulating valve	16 Cam ring	24 Drive hub	

Fig 20.1 DPA pump with a mechanical governor and automatic advance

this principle, the most extensively used system in Australia is undoubtedly the CAV DPA fuel injection pump.

The CAV DPA fuel injection pump

*I*n the CAV DPA (distributor pump, size A) fuel injection pump the fuel is pumped by a single element, and the fuel charges distributed in the correct firing order and at the required timing interval to each cylinder in turn by means of a rotary distributor, integral with the pump. In consequence, equality of delivery to each

injector is an inherent feature of the pump and deliveries are not subject to maladjustment.

Similarly, since the timing interval between injection strokes is determined by the accurate spacing of distribution ports and high-precision operating cams that are not subject to adjustment, accurate phasing is also an inherent feature.

The pump is a compact, oil-tight unit, lubricated throughout by fuel and requiring no separate lubrication system. It contains no ball or roller bearings, gears or highly stressed springs, and the number of parts and overall size of the pump remain the same irrespective of the number of engine cylinders it is required to serve.

Sensitive speed control is maintained by a governor, either mechanically or hydraulically operated, embodied in the pump.

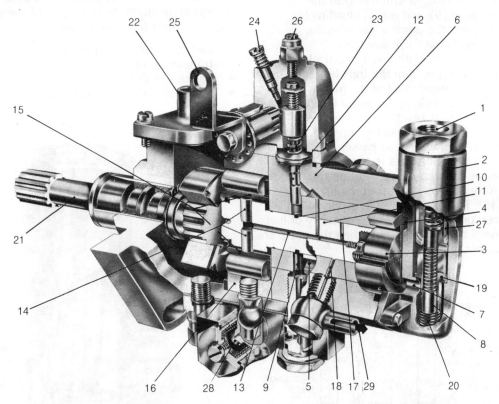

1 Fuel inlet	7 Outlet to regulating	13 Rotor axial drilling	20 Priming spring
2 End plate assembly	valve	14 Plungers	21 Drive shaft
3 Transfer pump	8 Regulating piston	15 Cam follower shoes	22 Back-leak connection
4 Transfer pump	9 Hydraulic head	16 Cam ring	23 Governor spring
inlet port	drilling	17 Distributor port	24 Idling stop
5 Hydraulic head	10 Groove on rotor	18 Outlet port	25 Control lever
drilling	11 Rotor	19 Regulating valve	26 Vent screw
6 Hydraulic head	12 Metering valve	sleeve	

Fig 20.2 DPA pump with a hydraulic governor and automatic advance

Variation of injection timing, which is required on some applications, can be obtained on models of the pump fitted with an advance device. Except where a manually operated start retard is incorporated, the advance device is fully automatic and requires no attention from the operator.

Construction and operation

Briefly, the pump operates as follows (refer to Figs 20.1 and 20.2). Fuel is pumped to the inlet connection (1) incorporated in the end plate assembly (2). The fuel passes through an inlet port (4) to the top of the transfer pump (3) where its pressure is raised to an intermediate level. The fuel leaves the transfer pump through two outlets—one leading forward from the transfer pump along a drilling (5) in the hydraulic head (6), and one (7) leading backward from the transfer pump to the pressure regulating valve assembly.

The pressure regulating valve controls the pressure of the fuel from the transfer pump by allowing a certain amount of the fuel to bypass back to the inlet side of the pump when a certain pressure is reached.

The fuel from the transfer pump that passes through the hydraulic head passes to a second drilling (9) at right angles to the first. The fuel then passes round a groove (10) on the outside of the rotor (11) to pass up to the metering valve (12). The metering valve is controlled by the governor, and regulates the quantity of fuel that passes to the rotor (11), when the ports are in alignment as shown in Fig 20.3a.

The fuel passes along the axial drilling (13) in the rotor to fill the space between the plungers (14). The rotor turns, carrying the trapped fuel with it until the cam followers (15) strike lobes on the stationary cam ring (16). The lobes force the cam followers inwards, taking the plungers with them. The fuel between the plungers is forced back through the axial drilling to pass to the injector lines through the distributor port (17) in the rotor, and one of the outlet ports (18) in the hydraulic head.

Detailed construction

The foregoing explanation gives only an outline of the construction and operation of a CAV DPA fuel injection pump. In order to fully understand the unit, it will be necessary to examine each section in detail, there being the following major sections for consideration:

- end plate and regulating valve
- transfer pump
- metering valve
- pumping and distributing section

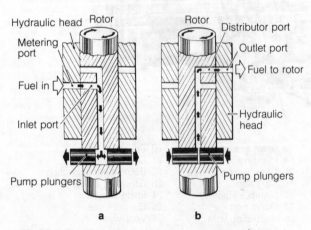

Fig 20.3 Diagram of pumping section, showing **a** charging, and **b** pumping

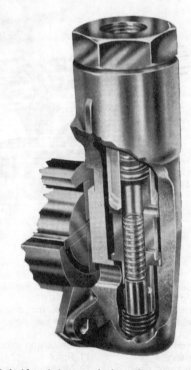

Fig 20.4 Aluminium end plate showing regulating valve

End plate and regulating valve

Two designs of end plate assembly have been used—an early steel type and the current aluminium type. Though different in appearance and layout (the regulating valve assembly lies at right angles to the fuel inlet connection in the steel type), they are identical in function and the current type only will be examined.

The end plate houses a fine-mesh nylon filter held in place by the fuel inlet adaptor, the regulating valve assembly, the transfer pump inlet port and one of the transfer pump outlets. It is secured to the end of the hydraulic head by four studs, thereby enclosing the transfer pump, which is fitted in a recess at the rear of the hydraulic head.

The regulating valve performs two separate functions. Firstly, it controls the level to which the transfer pump raises the fuel pressure, maintaining a predetermined relationship between the pressure and the speed of rotation. Transfer pressure—the pressure of the fuel from the transfer pump—ranges from 80 kPa to a pressure in excess of 825 kPa in some cases, depending on the engine (and pump) speed. Secondly, because of its design, fuel cannot be made to flow through the transfer pump when it is stationary. The pressure regulating valve provides a bypass so that fuel can be pumped through the fuel passages in the hydraulic head and rotor to prime the system.

During normal operation (Fig 20.6) fuel enters the end plate at feed pump pressure, passes through the nylon filter and into the transfer pump through the transfer pump inlet. Some of the fuel from the transfer pump outlet acts on the underside of the regulating valve piston, and forces the piston upwards against the regulating spring.

With an increase in the engine speed the transfer pressure rises and the regulating piston is forced upwards, compressing the regulating

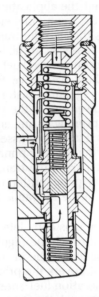

Fig 20.6 Regulating valve, operating position

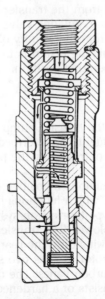

Fig 20.7 Regulating valve, priming position

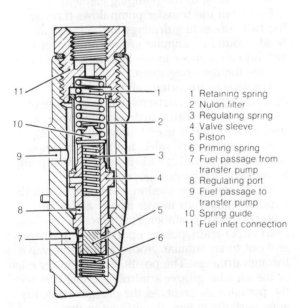

Fig 20.5 Section view of the end plate and regulating valve at rest

1 Retaining spring
2 Nylon filter
3 Regulating spring
4 Valve sleeve
5 Piston
6 Priming spring
7 Fuel passage from transfer pump
8 Regulating port
9 Fuel passage to transfer pump
10 Spring guide
11 Fuel inlet connection

spring. Such movement of the piston progressively uncovers the regulating port, and regulates transfer pressure by permitting the escape of a metered quantity of fuel back to the inlet side of the transfer pump. The effective area of the regulating port is increased as engine speed is raised, and is reduced as engine speed falls. On certain pumps, the maximum movement of the piston is restricted by a screw in order to increase the rate at which transfer pressure rises. The screw, which is referred to as a **transfer pressure adjuster**, is set during manufacture to suit the application concerned.

When priming the pump, fuel entering the end plate cannot pass through the transfer pump and into the fuel passages in the hydraulic head and rotor in the normal way. Fuel at priming pressure enters the valve sleeve and acts on the upper face of the regulating piston. The piston is forced to the lower end of the valve sleeve, compressing the priming spring and uncovering the priming ports. Fuel then passes through the priming ports, to the transfer pump outlet port, through the lower part of the transfer pump to the second transfer pump outlet, which leads along the drilling in the hydraulic head (Fig 20.7).

As soon as priming is completed and fuel is no longer forced through the regulating valve, the priming spring forces the piston slightly upwards, so that it covers the priming ports. During normal operation fuel enters the regulating valve through the priming ports, but with these covered, it cannot. When the starter is operated, fuel from the transfer pump passes down outside the regulating valve sleeve and up into the sleeve from the vicinity of the priming spring. The regulating piston is thus forced back to its normal operating position.

Transfer pump

The purpose of the transfer pump is to raise the pressure of the fuel to an intermediate level, the pressure being controlled by the pressure regulating valve.

The transfer pump is a positive displacement, vane-type pump, which functions in the same way as a vane-type lubrication system oil pump or a vane-type blower. The transfer pump rotor, a steel unit screwing onto the rear of the injection pump rotor, carries two slots in which lie two sliding carbon vanes. The outer section of the pump consists of a hardened steel liner with inner and outer surfaces not concentric.

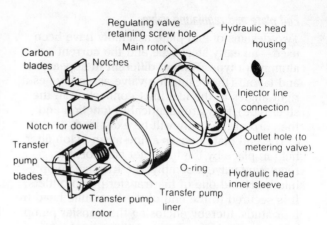

Fig 20.8 Exploded view of the transfer pump

This liner is prevented from turning by a locating dowel in the end plate, which engages with a corresponding slot in the liner.

The transfer pump inlet lies in the end plate, while one outlet lies in the end plate and one in the hydraulic head. A fuel-tight seal between the end plate and hydraulic head is provided by an 'O'-ring in a recess around the transfer pump liner.

Metering valve

The purpose of the metering valve is to provide a means of governor control over the quantity of fuel passing to the pumping element.

Fuel from the transfer pump flows through the two connecting drillings in the hydraulic head, around an annular groove on the rotor and into a chamber in the hydraulic head that houses the metering valve.

The type of metering valve used in a particular pump is determined by the type of governor fitted; with a hydraulic governor, a piston-like valve is used (shown in Fig 20.2), while a semi-rotary valve, not unlike a simple tap and shown in Fig 20.1, is employed in pumps featuring a mechanical governor.

The piston valve used in conjunction with the hydraulic governor moves axially along the metering valve chamber. Fuel flows to the valve from underneath, passes up a vertical passage and out to an annular groove around the valve through drillings. The position of the lower edge of the annular groove controls the fuel flow to the pumping element. As the plunger rises, the edge partially covers the drilling in the hydraulic head to reduce the flow; as the plunger descends and the groove edge progressively uncovers the drilling, the flow increases.

The semi-rotary valve used in pumps fitted with mechanical governors is caused to rotate through a limited arc by the governor. A groove, similar to a very small keyway, runs axially along one side of the round valve body and carries fuel upwards to the drilling in the hydraulic head leading to the pumping element. Rotation of the valve in one direction aligns the groove in the valve body with the hydraulic head drilling, while rotation in the other direction causes the edge of the groove to progressively cover the drilling, reducing the fuel supply to the pumping components.

Pumping and distributing section

The function of this section of the pump is to raise the fuel pressure from transfer pressure to injection pressure, to distribute the fuel charges to the correct engine cylinders in turn, and to reduce fuel pressure in the injector lines after injection to preclude the possibility of the injector nozzle dribbling.

The section is composed of six subsections of components, the:

- hydraulic head
- rotor and plunger assembly
- cam followers
- cam ring
- maximum fuel adjusting plates
- drive plate.

Hydraulic head—The hydraulic head provides a housing in which the rotor turns. It combines with the rotor to distribute the fuel charges to the injector pipes. In addition to these primary functions, the hydraulic head carries the metering valve, conveys fuel from the metering valve to the rotor, conveys fuel from the transfer pump to a groove on the rotor from whence it passes to the metering valve and provides the injection line connections.

The hydraulic head is constructed of two separate components—an inner sleeve and an outer barrel (Fig 20.9). The sleeve is an interference fit in the barrel, which is heated in manufacture to allow the sleeve to be fitted.

Because the sleeve is shorter than the barrel, and because they are flush at the rotor drive end, the necessary recess to accommodate the transfer pump is provided without any special machining.

The inner sleeve carries the horizontal fuel

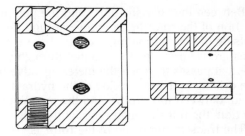

Fig 20.9 Hydraulic head

feed drilling from the transfer pump, the large feed drillings at right angles to the horizontal drilling that carry the fuel to the metering valve, and inlet and outlet drillings. The barrel is drilled and tapped to provide accommodation for the fuel outlet unions etc, and carries the metering valve drilling and an angled drilling that conveys fuel from the metering valve to the fuel inlet in the sleeve. The transfer pump end is counter-bored to accommodate the 'O'-ring seal, which prevents leakage at the transfer pump. The outside of the large diameter end that fits into the pump housing is machined to accommodate a second 'O'-ring, which provides a seal between the hydraulic head and the housing.

Rotor and plunger assembly—The rotor and plunger assembly is driven by the drive shaft, which connects to the drive plate bolted to the rotor, and rotates in the axial bore of the hydraulic head. The rotor comprises two sections—a pumping section and a distributing section. The distributing section is a precision, minimum-clearance fit in the hydraulic head, while the pumping section is of larger diameter and has two diametrically opposed slots machined in it. A transverse bore runs between the slots and carries the two opposed pumping plungers.

An axial drilling running the length of the distributing section connects to the transverse bore. Along the distributing section, and connecting to the axial drilling, are a number of radial drillings—a single-outlet port remote from the pumping section and a group of radial inlet or charging ports nearer to the pumping section. (The number of ports is governed by and is equal to the number of engine cylinders the pump is to serve.) The charging ports are so positioned that they align in turn with the inlet port in the hydraulic head sleeve, while the outlet port in the rotor aligns with the outlet ports in the hydraulic head, in turn.

Between the charging ports and the pumping section of the rotor lies a groove around the outside of the distributing section of the rotor. This groove provides the channel through which the fuel passes to reach the metering valve from the drilling in the lower side of the hydraulic head. Between this groove and the pumping section, the rotor diameter is slightly reduced.

The transfer pump end of the rotor is counterbored and internally threaded to accommodate the threaded spigot of the transfer pump rotor—thus the rotor provides the means of conveying the drive to the transfer pump. A screwed plug seals the end of the axial drilling in the rotor, and is centrally situated within the counterbore.

Cam followers—The cam followers are made in two parts—a roller and a shoe. The followers are situated in the pumping section of the rotor, and ears or lugs on the ends of the shoes protrude slightly past the surfaces of the rotor pumping section. The cam followers rotate with the rotor and force the pumping plungers together when the rollers strike the lobes of the stationary cam ring situated around the rotor pumping section.

Cam ring—The cam ring (Fig 20.10) is located around the pumping section of the rotor between the end of the hydraulic head and a circlip in the housing. This circlip carries timing reference marks and is known as a timing ring.) Cylindrical in general terms, the cam ring has one small flat on its outer surface with a tapped hole in the centre, and a number of lobes on its inner surface. Normally the cam ring has as many lobes as the engine for which the pump is

designed has cylinders, the lobes being diametrically opposed to operate both pumping plungers together. The exception is the case where the pump has to be fitted to an engine with an uneven number of cylinders, in which case the cam ring will feature twice as many lobes as the engine has cylinders.

The cam lobes are designed to fulfil two functions—to force the plungers together via the cam followers and raise the fuel pressure to injection pressure, and to reduce the pressure of fuel in the injector line to slightly below injector opening pressure to prevent dribbling at the nozzle. Each cam lobe is so shaped that the plungers move outwards slightly after injection while the outlet ports are still in register. That part of the lobe that allows the plunger movement is known as a **retraction curve** (Fig 20.11). This outward movement increases the volume enclosed between the plungers, thus lowering the pressure slightly, and since the injector line is still connected to this space, the line pressure is allowed to drop.

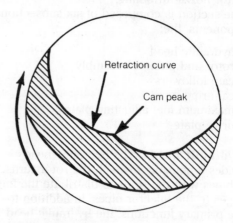

Fig 20.11 Enlarged view of the cam lobe

Note   As a further means of preventing injector dribble, a large number of DPA fuel pumps are now being fitted with conventional delivery valves, located in the fuel outlet fittings

The cam ring is either locked in position by means of a stud that screws into the cam ring from the bottom of the pump, or it is provided with a ball coupling that screws into the cam ring and engages with a piston in an automatic advance cylinder housed underneath the pump. The automatic advance system partially turns the cam ring in the housing to advance or retard the injection point.

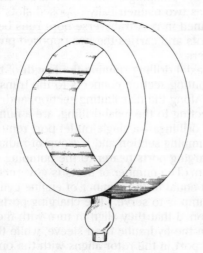

Fig 20.10 Cam ring

Maximum fuel adjusting plates—The maximum amount of fuel the pump is capable of delivering is governed by the distance the plungers can move outwards before being forced together. Therefore the maximum fuel delivery must be controlled at this point. To provide a maximum fuel stroke that can be adjusted to suit the requirements of different engines, two adjusting plates are fitted to the rotor pumping section, one on each side (Fig 20.12). These plates are provided with eccentric slots or cut-outs, which engage with the ears or lugs on the cam follower shoes. The ears' protrusion through the slot provides the stop that limits the plunger's outward movement, and since these slots are eccentric, partial rotation of the plates varies the maximum outward movement. The two plates move together because of two lugs on the forward plate, which engage with slots in the rear one. These plates are adjusted in manufacture and should only need readjusting after the pump has been dismantled.

Drive plate—Bolted to the front of the rotor, and clamping the forward maximum fuel adjusting plate in position, is the drive plate. This ground plate carries a number of lines on its outer surface, each with an associated letter. These lines are used for timing purposes. The centre of the drive plate is splined to carry the drive from the drive shaft to the rotor, a master spline being provided to ensure correct timing.

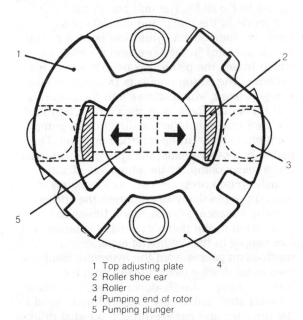

1 Top adjusting plate
2 Roller shoe ear
3 Roller
4 Pumping end of rotor
5 Pumping plunger

Fig 20.12 Maximum fuel control

Detailed pump operation

Fuel is supplied to the inlet connection incorporated in the end plate, from the fuel tank, by the fuel feed pump via (usually) two fuel filters. The fuel passes through a fine (usually) nylon filter to the transfer pump inlet port. The transfer pump, being a positive displacement unit, increases its fuel delivery with its speed of rotation. Since fuel flow through the system is somewhat restricted, the transfer pump pressure rises as the speed of pump rotation increases. The resulting pressure varies between 80 kPa and an upper maximum, which may be 825 kPa or more in some applications, depending on pump speed, the relationship being controlled by the pressure regulating valve.

From the transfer pump, the fuel passes either to the pressure regulating valve through the fuel outlet in the end plate or along the drilling in the hydraulic head sleeve. The fuel passing into the regulating valve acts against the end of the regulating valve piston, which is moved against the regulating valve spring. When the fuel pressure becomes high enough, the piston compresses the spring sufficiently to allow the fuel to pass through the regulating port to the transfer pump inlet.

The fuel passing along the drilling in the hydraulic head sleeve flows around the rotor groove and into the metering valve drilling. The rate at which the fuel passes through the metering valve is dependent on two factors, the position of the metering valve as determined by the governor, and the fuel pressure as controlled by pump speed per medium of the transfer pump. Hence at high engine speed and large throttle opening, although the inlet ports are in register for only a very short time, the high transfer pressure and large opening at the metering valve allow the required large quantity of fuel to enter the rotor. Conversely, at low speeds with only a small throttle opening, the combination of low fuel pressure and small metering valve opening results in only a small quantity of fuel passing to the rotor.

After passing through the metering valve, the fuel passes through an angled drilling in the hydraulic head barrel, a radial drilling in the hydraulic head sleeve, through one of the radial inlet ports in the rotor (when they are in alignment), to the axial drilling in the rotor. The fuel enters the space between the plungers, forcing them apart. The distance the plungers are forced apart is dependent on the quantity of fuel passing through the metering valve and is limited, as previously discussed, by side plates on the pumping section of the rotor, to a predetermined maximum. Hence at low speed, the plungers are forced apart a small distance only and the subsequent injection is of a small fuel quantity, while at large throttle opening the plungers are pushed apart to their limit, giving the maximum injection quantity. That part of the pump's operation during which the fuel flows to the rotor and forces the plungers apart is known as the **charging stroke** and is shown in Fig 20.3a.

Further rotation of the rotor causes the rotor inlet port to move out of alignment with the hydraulic head charging port, and as the rotor outlet port moves into alignment with one of the delivery drillings in the hydraulic head, the cam followers strike the lobes on the cam ring and force the plungers together. This is the **pumping stroke** and is shown in Fig 20.3b.

After passing the peaks of the cam lobes, the cam followers move outwards slightly due to the cam ring retraction curve. This outward movement of the plungers occurs while the rotor outlet is still in register with the hydraulic head outlet drilling, and the pressure in the injector lines is reduced slightly due to the plungers' outward movement. Once the ports are no longer in register, due to further rotation of the rotor, the cam followers move outwards to the stops due to centrifugal force.

As soon as the next rotor inlet port aligns with the hydraulic head charging port, the flow

of fuel into the rotor pushes the plungers apart and the cycle begins again.

A line diagram of the DPA fuel circuit is shown in Fig 20.13.

Automatic injection advance

Under light load conditions, the quantity of fuel injected (and hence the plungers' stroke), is very small. The injection stroke does not begin until the cam followers strike the plungers, after having travelled up the cam lobe to a point near the peak.

On the other hand under full load conditions the injection stroke begins almost immediately the cam rollers strike the lobes. Thus as the quantity of fuel required becomes greater, the plungers start their stroke when the rollers are at a lower point on the cam lobe, that is, earlier in rotor rotation.

It follows, then, that injection timing becomes progressively advanced as fuelling is increased, since contact between the plungers and followers occurs progressively earlier in the pump's rotation.

In many applications, some form of automatic advance device is also included to increase the amount of advance. This extra automatic advance of the fuel injection timing is accomplished by partially rotating the cam ring in the pump housing. This movement of the cam ring is accomplished by means of an automatic advance unit, a typical example being shown in Fig 20.14. The unit fits on the underside of the pump and consists of a housing containing a cylindrical bore, a piston, which operates in the bore, and a spring(s), which biases the piston to one extreme of the cylinder. The piston connects to a ball-headed screw or stud, which screws into the cam ring, and the spring force is in such a direction that the cam ring is held in the direction of pump rotation, namely in the retarded position. The piston is moved against the spring by fuel from the transfer pump. As the pump speed and transfer pump pressure rise, so the piston is moved progressively to compress the spring, thereby advancing the injection timing in proportion to the pump's speed of rotation. The fuel supply to the automatic advance mechanism passes from the hydraulic head horizontal drilling (which leads from the transfer pump), travels down through a hollow locating stud that secures the hydraulic head in the housing, and passes through a radial drilling in the locating stud to flow into the automatic

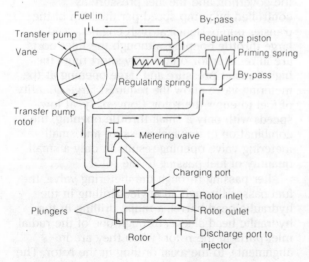

Fig 20.13 Line diagram of the DPA fuel circuit

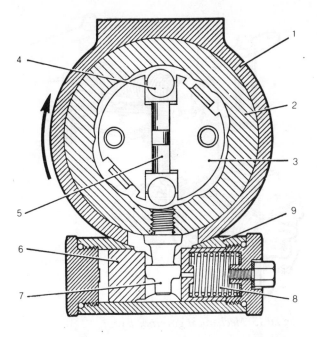

1 Pump housing
2 Cam ring
3 Pumping and distributing rotor
4 Roller
5 Pumping plunger
6 Automatic advance piston
7 Cam advance screw
8 Piston spring
9 Automatic advance housing

Fig 20.14 Automatic advance mechanism

advance cylinder through a drilling in the advance unit housing.

In some instances, a means of cutting off the fuel supply to the automatic advance unit when starting from cold is incorporated in the locating stud; this may be either a cock or a cable-actuated valve. If automatic advance cannot be prevented, starting under cold conditions may become virtually impossible, since transfer pressure may be sufficient at cranking speed to carry the cam ring (per medium of the piston) towards the advance position, and cylinder air temperature at cranking speed is insufficient before TDC to efficiently ignite the fuel. As soon as the engine starts, the device is .opened and the piston (and cam ring) move to the correct (more advanced) position.

Any tendency for the cam ring to be dragged towards the retarded position by the impact of the cam follower rollers on the cam ring lobes is prevented by a ball check valve in the fuel supply stud, which causes fuel to be trapped in the cylinder. When the engine is stopped or its speed is reduced, the advance mechanism is allowed to return to the retarded position under

the influence of the spring due to a slight but controlled-rate leakage of fuel between the piston and the cylinder.

Mounting and driving arrangements
Both types of DPA pump—mechanically or hydraulically governed types— are equipped with a triangular mounting flange for attachment to the engine's timing case or cylinder block. Because the pump is filled completely with fuel during operation for lubrication purposes, it can be mounted horizontally, vertically or in any intermediate attitude that may be necessary for a particular engine installation.

Both types of pump are driven by a splined shaft provided with a master spline. The master spline is engaged with a corresponding master spline on the engine drive during fitting, and ensures correct location of the pump in relation to the drive.

On mechanically driven pumps a short detachable splined shaft, known as a quill shaft, engages with splines in the drive hub that protrudes from the mounting flange. Carrying the drive from the drive hub to the rotor drive plate and secured to the drive hub by an axial stud, is a second drive shaft. Hence the drive passes from quill shaft, to drive hub, to drive shaft, to drive plate, via two clamping studs to the rotor, and to the transfer pump rotor. On some applications, the quill shaft is replaced by a special keyed drive hub that bolts to the engine coupling—some early models of mechanically governed pumps have a quill shaft with a small locking screw at one end. It is important that this screw is engaged in the socket head of the drive hub securing screw before tightening the pump to the engine.

On hydraulically governed pumps the detachable quill shaft is not employed, and a one-piece drive shaft engages at one end with the engine drive and at the other with the rotor drive plate. On certain hydraulically governed pumps a torsion bar is fitted to iron out backlash in the drive shaft. The device consists of a flat bar engaged between the end of the rotor and the pump drive coupling on the engine. The bar runs through the centre of the drive shaft and is made to twist during engagement of the pump to the engine.

Lubrication
Reference has already been made to the fact that the pump housing is filled with fuel. Fuel

reaches the annular groove on the rotor at transfer pump pressure, and a small quantity leaks from this groove around the rotor shank (which at this point is of slightly reduced diameter) to fill the pump housing. The rest of the rotor and hydraulic head is lubricated by slight fuel leakage from the various ports. Once the housing is full, the fuel is returned to the secondary fuel filter for recirculation to the pump. When an automatic advance unit is fitted, additional fuel enters the body after leaking past the automatic advance piston.

DPA pump governors

*T*he need for an effective governing device on a diesel engine was clearly shown in Chapter 19, and engines equipped with CAV DPA fuel injection pumps are naturally no exception. Because the governors are integral with the pump and directly control the metering valve, they are included in this work with the pumps rather than in the chapter on governors.

As the preceding comments on these pumps have clearly shown, there are two entirely different types of governor fitted— the mechanical governor and the hydraulic governor. While the type of governor has a great deal of bearing on the appearance of the pump, the pumping sections are identical except for the metering valve. The mechanically governed pump is somewhat longer than the other, due to the governor weight assembly, which is situated within the pump housing.

The mechanical governor

The mechanically governed DPA fuel injection pump is shown in Fig 20.1. Reference to this figure will show that the flyweight assembly is carried on the splined drive shaft within the pump housing, and that the governor-controlled linkage is enclosed by a cover fitted to the upper face of the housing.

Governor operation can be readily understood if reference is made to Fig 20.15, which shows the governor weights and control sleeve diagrammatically.

The governor weights are housed in pockets in the weight carrier, which is rigidly clamped

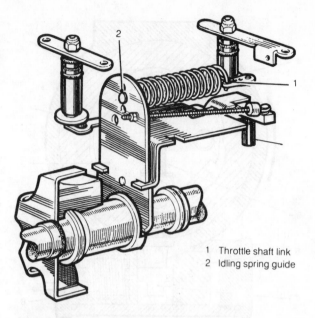

1 Throttle shaft link
2 Idling spring guide

Fig 20.15 Mechanical governor layout

between the end of the drive hub and a step on the drive shaft. Drive hub, governor weight assembly and drive shaft thus rotate as a single unit when the pump is operating. The weights are so shaped that they pivot about one corner under the influence of centrifugal force. Such movement causes a thrust sleeve, with which they are engaged, to slide along the drive shaft.

Movement of the thrust sleeve is transmitted to the metering valve by the pivoted governor arm and the spring loaded hook lever.

Outward movement of the weights tends to close the metering orifice by rotating the metering valve, thus reducing the quantity of fuel reaching the engine cylinders at each injection.

The governor arm is spring loaded by the governor spring; the tension of this spring acts in opposition to centrifugal force, tending to oppose outward movement of the weights and to hold the metering valve in the maximum fuel position. Spring tension can be varied by moving the control lever to which the spring is connected, tension being increased as the throttle control is moved towards the maximum speed setting.

When the control lever is moved to a position that calls for engine acceleration, increased spring tension is applied to the governor arm. This increased spring tension overcomes centrifugal force acting on the governor weights, and the metering valve is rotated to the maximum fuel position. Engine speed builds up until the centrifugal force acting on the weights is sufficient to overcome the increased spring tension. The weights then move outwards and reduce the fuelling by rotating the metering valve until the two opposed forces acting on the thrust sleeve are in equilibrium.

While running at a selected speed, the spring tension opposing movement of the governor weights remains constant. When speed fluctuations occur, the resulting change of centrifugal force causes movement of the weights and brings about a compensating change of fuelling—increased fuelling when engine speed falls below the selected speed, and decreased fuelling when it is exceeded.

It will be noted that the governor spring is not connected directly to the governor arm. It is coupled to the idling spring guide, which passes through a hole in the governor arm and is, in most cases, spring loaded by a light idling spring. At speeds outside the idling range the tension applied to the governor spring is sufficient to compress the idling spring completely, thus rendering it ineffective. At idling speeds, when the tension of the governor spring is reduced to a minimum, the idling spring comes into action. Movement of the control arm by governor action is opposed by the light idling spring only, so that stable idling can be maintained when changes of centrifugal force are small.

Idling and maximum speed adjustment screws are provided on both hydraulically and mechanically governed pumps. Maximum speed and idling stops are adjusted while the pump is on the engine and should be set in accordance with the engine manufacturer's instructions.

Fuel shut-off is achieved by rotating the metering valve to a position where the metering orifice is completely closed. An eccentric on the shut-off shaft engages a bar, which is brought into contact with the arm on the metering valve when the control is operated. The hook lever connecting the metering valve to the governor arm is spring loaded so that the control can be operated at any engine speed without need for over-riding the governor.

The hydraulic governor

A CAV DPA fuel injection pump featuring a hydraulic governor is shown in Fig 20.2.

The components of the hydraulic governor are contained in a small casting secured to the upper face of the injection pump housing, a speed control lever and a shut-off being externally mounted.

Within the governor is a pinion that engages with a rack carried on the stem of the metering valve. The rack is not secured to the valve stem, but is usually located between two springs, a lower governor spring and an upper light idling spring. In some instances the governor spring only is fitted as is the case with the hydraulic governor shown in Fig 20.16.

Movement of the piston-type metering valve controls the effective area of the metering orifice, thus controlling the metering pressure at the inlet port in the rotor, and regulating the quantity of fuel admitted to the pumping element at each charging stroke.

Two forces act on the metering valve, the force exerted by the governor spring, and the force exerted by the fuel at transfer pressure acting on the underside of the valve. The two forces are opposed, spring force tending to hold the valve in the maximum fuel position and transfer pressure tending to hold it in the idling position. During operation the valve assumes a

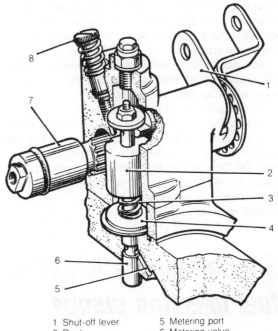

1 Shut-off lever	5 Metering port
2 Rack	6 Metering valve
3 Governor spring	7 Pinion shaft
4 Damping valve	8 Idling stop screw

Fig 20.16 Section through a hydraulic governor

position where the two forces acting on it are in equilibrium.

When the control lever is moved towards the full throttle setting, the metering valve moves to the maximum fuel position and the governor spring is compressed. Engine speed increases in response to increased fuelling, thus causing a rise in transfer pressure. As transfer pressure becomes sufficiently high to overcome spring force, the metering valve moves towards the idling position until the two forces are in equilibrium. The engine will then run at a speed corresponding to the setting of the speed control lever, and the selected speed will be maintained by governor action.

Any change in engine speed resulting from a change of loading is accompanied by a corresponding change of transfer pressure. This will cause movement of the metering valve and bring about a compensating change of fuelling—increased fuelling when engine speed falls below the selected speed and decreased fuelling when the selected speed is exceeded.

Within the idling range, the idling spring is compressed. The force exerted by it on the metering valve is in the same direction as the force exerted by transfer pressure. Thus at idling speeds a reduced force exerted by the governor spring is opposed by transfer pressure plus the force exerted by the idling spring. This enables close governing to be maintained at low speeds where transfer pressure changes are relatively small.

The dished damper plate is immersed in fuel oil, and prevents violent metering valve movement by dashpot action.

A cam, controlled by the shut-off lever, contacts the underside of a washer fitted at the upper end of the metering valve stem. When the control is operated, the valve is lifted upwards and the metering orifice completely closed. The control can be operated at any engine speed, thus enabling the operator to shut off the engine in an emergency.

Special application DPA fuel injection pumps

Some DPA fuel injection pumps are used in automotive applications. For driver convenience and improved performance, certain

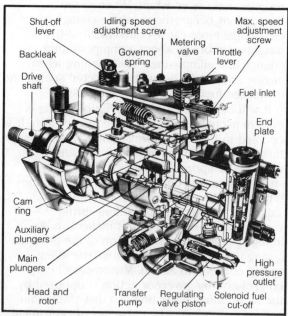

Fig 20.17 Special application DPA fuel injection pump

modifications are carried out on many of these pumps. The modification options available include:

- A manual advance lever to allow injection timing to be advanced by the driver for easier cold weather starting. This is the reverse of standard DPA pumps because of the considerably modified pump timing and advance systems.
- An electric solenoid shut-off control valve to allow the engine to be stopped by means of the operator key.
- A light load advance system to ensure smooth running during light loaded operation. In this modification, fuel supply to the advance unit passes through a helical groove on the metering valve and a drilling in the hydraulic head, instead of directly from the transfer pump.

As the metering valve moves under

governor action, this helix covers and uncovers the port leading to the automatic advance mechanism. Because there is a constant leakage past the automatic advance piston, the restriction in the supply circuit causes a pressure drop, the amount of pressure loss being dependent on the amount of restriction.

Thus the amount of advance is determined by the position of the metering valve helix, which controls the fuel pressure to the automatic advance. Because of the relative positions of the helix and the normal fuel passage in the metering valve, under light load conditions (moderate to high engine speed) with the metering valve partly opened, the advance port is fully opened allowing full fuel pressure to act on the advance unit and fully advance the timing.

Under heavy load conditions when the metering valve is fully opened, the advance port is partly closed reducing the fuel pressure acting on the automatic advance unit and therefore retarding injection timing slightly. Light load advance pumps are distinguished by a circular groove machined in the outside of the hydraulic head adjacent to the transfer pump housing. A modified rotor assembly featuring four plungers to provide excess fuel above normal maximum delivery for ease of starting. The main pair of plungers operate at all times the pump is running, and behave as the plungers in the standard rotor. The auxiliary, smaller diameter plungers operate only when starting the engine and are located in tandem beside the existing main plungers in the rotor. At cranking speed all four plungers operate, but once the engine has started, fuel at transfer pressure acts against a piston valve, which closes off the fuel drillings to the auxiliary plungers, leaving only the larger plungers operational.

DPA pump service

*O*nce again, the need for absolutely clean working conditions, specialised equipment, special service tools, detailed service information and experience must be stressed

before any thoughts of injection pump service can be entertained. Provided that these prerequisites are available, the DPA pump does not present any special service difficulties.

To ensure that the correct service information is used, the pump type and model should be identified from the pump typeplate on the housing, and reference should then be made to the appropriate pump test sheet. However, a setting code below the type/model data also provides specific information on assembly and setting details.

For example, in the setting code EX52/600/4/2850, the '52/600' indicates that the pump will deliver 52 millilitres of fuel per 1000 shots when it is running at 600 rpm on an injection pump test bench. The '4' indicates the governor spring location code on mechanically governed pumps; a zero appears in this position for hydraulically governed pumps. The '2850' is the maximum no-load speed of the engine.

Both calibrating and phasing, as they apply to jerk pumps, are eliminated with the DPA pump— calibrating because all fuel deliveries come from the same element and cannot be adjusted individually, and phasing because the firing intervals are set in manufacture when the cam ring lobes are ground at the correct positions.

The condition of the hydraulic head assembly can be ascertained when the pump is assembled by setting it up on the test bench and running it at a specified speed. The fuel escaping from the pump is made up of that escaping from between the plungers and their bore and between the rotor and hydraulic head, as well as a small quantity escaping past the automatic advance piston. Hence, the quantity of fuel escaping is a direct indication of the clearances between the pumping and distributing components of the pump. If this fuel is passed into a calibrated glass and measured during a specified number of injections, a good indication of the hydraulic head condition can be gained, since a maximum allowable leakage rate is quoted in the manufacturer's manual.

However, it only takes a very small quantity of dirt entering the inlet connection when the pump is removed from the engine to destroy the pumping components if the pump is run. Again, a speck of dirt in an outlet union will be pumped into the test injectors and will damage them if the pump is run on a test bench. Because of these dangers, many servicemen prefer to dismantle the pump without performing any preliminary test.

Dismantling a mechanically governed DPA pump

A number of special tools are necessary before a DPA pump can be dismantled. Every engine manufacturer, as well as service equipment manufacturers, make these tools available. The following dismantling sequence by courtesy of British Leyland Motor Corporation of Australia Limited, Austin-Morris Division, may well be applied to almost any mechanically governed DPA pump, the special tools necessary being listed under this manufacturer's part numbers.

To dismantle the pump, proceed as follows (part numbers refer to Fig 20.18).

1 Remove the cover-plate (106) from the side of the pump housing, and drain the fuel from the injection pump.
2 Withdraw the quill shaft (105) from the drive hub (100) and check the drive hub end float by inserting a feeler gauge between the drive hub and the pump body (92). The end-float should not exceed 0.254 mm (0.01 in.). Excessive end-float can be corrected by renewing the pump body and the governor weight retainer (45).
3 Mount the pump on an assembly base (a mounting plate for service work, part no 18G633) secured in a vice and remove any

high-pressure connections (27) from the hydraulic head.
4 Unscrew the nuts and remove the shut-off lever (79) and throttle arm (73) from their shafts. Withdraw the dust cover from each shaft and remove the two nuts (66) and washers (67) securing the control cover (87).
5 Press the throttle shaft downwards and withdraw the control cover complete with the shut-off shaft (74). Discard the control cover gasket.
6 Detach the governor spring (52) from the governor arm (50) and the shut-off bar (54) from the control bracket (47). Remove the two control cover studs (64) and the small set screw (48) securing the control bracket. Detach the keep plate (63) and lift the control bracket assembly from the pump. This assembly will include the control bracket (47), the governor arm (50), the spring for the governor arm (51), the spring guide (53) and its associated idle spring, governor spring (52), linkage hook (56) (complete with spring retainer (57), linkage spring (58), linkage washer (59), pivot ball washer (60), backing washer (61) and linkage nut (62), and the metering valve (55).
7 Disconnect the metering valve from the linkage hook and place it in a clean

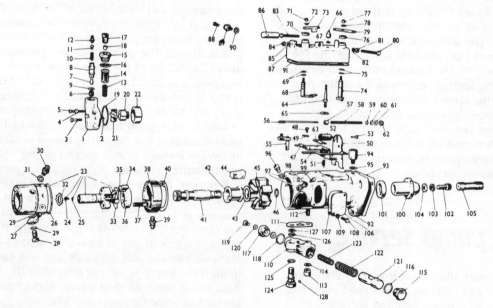

Fig 20.18 Exploded view of CAV DPA mechanically- governed fuel injection pump

container of test oil to protect its precision-ground surface.

8 Disconnect the governor spring from the idle spring guide, and remove the idle spring and guide from the governor arm.

9 Disconnect the linkage hook from the governor arm. Detach the governor arm spring and separate the governor arm from the control bracket.

10 Slacken both the spring cap (117) and end plug (115) in the automatic advance unit (if fitted). Remove the hydraulic head locating bolt (124) complete with its outer 'O'-ring (125), and take care not to lose the non-return valve ball (128) located in the side of the head locating bolt.

11 Remove the cap-nut (113) and washer (114) and withdraw the advance unit from the pump. Detach the inner 'O'-ring (126) and washer (127) from the head locating bolt hole, and discard the advance unit gasket.

12 Unscrew the spring cap (117) (complete with 'O'-ring (118)) from the advance unit, and withdraw the spring(s) and the piston. Note that there is a shim inside the spring cap.

13 Remove the end plug (115) and 'O'-ring (116) from the advance unit.

14 Unscrew the cam advance screw (39) from the cam ring (38) with the special spanner (part no 18G646).

15 Slacken the fuel inlet connection (15) and remove the screws and studs (3, 4) securing the end plate to the hydraulic head (23). Lift out the transfer pump vanes (21), and withdraw the transfer pump liner (22).

16 Unscrew the fuel inlet connection (15) withdraw the regulating valve components in the following order: sleeve retaining spring (13), nylon filter (14), transfer pressure adjuster (12), regulating spring (10) and peg (11), regulating sleeve (8) with piston (7) and joint washer (9), and lastly the piston retaining spring (priming spring) (6). If necessary, remove the transfer pump liner locating pin from the inner face of the end plate. The pin can be in one of two positions, marked 'A' and 'C' on the end plate outer face, depending whether the pump's rotation is anticlockwise or clockwise.

17 Hold the drive hub (100) with the special tool for the purpose (part no 18G659) and, using the special tool (part no 18G634) slacken the transfer pump rotor (20) by

turning it in the direction of pump rotation as shown on the pump nameplate.

18 Remove the two hydraulic head locking screws (96, 98), one of which incorporates an air vent screw (99), and withdraw the hydraulic head and rotor assembly (23) from the pump. Remove the 'O'-ring (26) from the groove in the periphery of the hydraulic head.

19 Unscrew the transfer pump rotor (20), but do not allow the pumping and distributing rotor assembly to fall out of the hydraulic head.

20 Stand the hydraulic head on the bench with the drive plate (36) uppermost. Hold the drive plate with the special tool (part no 18G641) and unscrew the two drive plate screws (37). (See Fig 20.19.) Remove the drive plate and top adjusting plate (33), and withdraw the rollers (34) and shoes (35) from the pumping and distributing rotor.

21 Withdraw the rotor from the hydraulic head, remove the bottom adjusting plate (32), and refit the rotor to the hydraulic head. Immerse the assembly in test oil to protect the working surfaces.

22 Withdraw the cam ring from the pump housing, noting the arrow etched on the visible face of the cam ring. This arrow indicates the direction of rotation as shown on the pump's nameplate.

23 Remove the cam ring locating circlip (40) from inside the pump, using suitable circlip pliers.

24 Hold the drive hub with the special tool

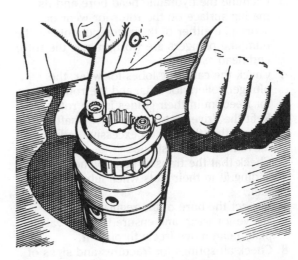

Fig 20.19 Holding the drive plate while removing the drive plate screws

(part no 18G659) and, with a suitable Allen key or special tool, unscrew the drive shaft screw (102) from inside the drive hub. Withdraw the drive shaft (41) and governor weight assembly from inside the pump housing as an assembly.

25 Remove the 'O'-ring (46), the weight retainer (45), weights (44), thrust washer (43), and sleeve (42) from the drive shaft.
26 Withdraw the drive hub from the pump and remove the spring washer (103) and support washer (104) from inside the drive hub.
27 Remove the drive hub oil seal (101) from the pump housing using a suitable tool (such as part no 18G658).

Inspection of components

All components should be washed in clean test oil, after which the following procedure should be followed:

1 Remove the rotor plug (24) and washer (25). Hold the pumping plungers in their bores and blow out the passages in the rotor with compressed air. Coat the threads of the plug with Araldite and refit the plug.
2 Remove the pumping plungers one at a time and examine them and their bore in the rotor for wear and abrasion. The end of each plunger will be polished where it contacts the roller shoe, and the plungers should be replaced in their original positions.
3 Examine the hydraulic head bore and its mating surface on the rotor for wear or scoring. If either of these components or the pumping plungers are worn, renew the rotor and head as a unit.
4 Check the cam ring lobes for wear, the plunger rollers for flats, and the roller shoes for freedom in their guides in the rotor.
5 Refit the drive plate to the drive shaft and ensure that there is no excessive radial movement on the splines.
6 Check that the transfer pump vanes are a sliding fit in their slots when lubricated with fuel.
7 Inspect the bore of the regulating valve sleeve for wear, and ensure that the valve piston can move freely through it.
8 Check all springs for fractures and signs of weakness, and the governor weight retainer, thrust washer, and sleeve for signs of wear.

9 Inspect the bore in which the drive hub runs. If it is worn or scored, the housing must be renewed.

Note   One method that may be used to gauge the amount of wear sustained by the hydraulic head components is as follows:

1 Machine a cylindrical sleeve of fair wall thickness to fit over the roller and plunger assembly in the position the cam ring normally occupies. The inner diameter of the sleeve should be such that it will just fit over the rollers when the plungers are forced together.
2 Connect the hydraulic head assembly to an injector tester, as in Fig 20.22. Operate the injector tester and turn the rotor until the rotor outlet port registers with one of the hydraulic head outlet ports connected to the tester. At this point, the rollers will be forced out against the sleeve by the plungers.
3 Note the quantity of fuel leaking from the assembly. Experience will provide the necessary judgement as to how great this leakage may be.

Reassembly of the mechanically-governed DPA pump

All components should be rinsed in clean test oil and assembled wet in the following order (all component numbers again refer to Fig 20.18.):

1 Fit a new hub oil seal (101) to the pump housing (using tool part no 18G663). Insert the correct inspection plug (part no 660) into the oil seal; a continuous black line should be visible through the plug.
2 Fit the two washers (104, 103) into the drive hub (100) and insert the hub into the hub oil seal.
3 Using the special tools (part nos 18G661, 18G662), assemble the governor weights (44), thrust washer (43), and sleeve (42), to the weight retainer (45). (See Fig 20.20.) The stepped flange of the sleeve must go away from the thrust washer.
4 Slide the governor weight assembly onto the drive shaft (41), fit a protection cap (part no 18G657) over the drive shaft splines and a

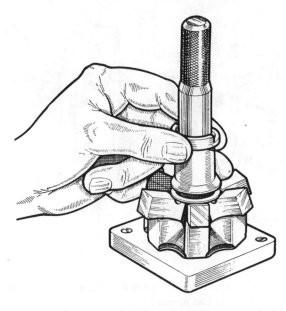

Fig 20.20 *Assembling the governor weights*

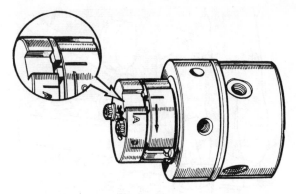

Fig 20.21 Hydraulic head and rotor assembly, showing the top adjusting plate position relative to the rotor

new 'O'-ring (46) in the groove on the drive shaft.

5 Insert the drive shaft and weight assembly into the pump housing and engage the drive shaft splines with the splines in the drive hub. Fit the drive shaft screw (102) and, using a drive hub holding tool (part no 18G659) hold the drive hub while tightening the screw with an adaptor (part no 18G664) and torque wrench to the figure laid down in the specifications. Check the drive hub end float as described in 'Dismantling a mechanically governed DPA pump' (2).

6 Using circlip pliers, fit the cam ring locating circlip (40) against the shoulder in the pump housing. If the circlip is the type that has a timing line scribed on one face, ensure that the timing line faces forward to the inspection cover.

7 Place the cam ring (38) in position against the circlip, and ensure that the direction of the arrow on the visible face of the cam ring conforms with the direction of the arrow on the pump name plate. Fit the cam advance screw (39) finger tight and check the cam ring for freedom of rotation. If the hydraulic head and rotor are being renewed, ensure that the direction of the arrow on the pumping end of the rotor conforms with the direction of the arrow on the pump nameplate.

8 Withdraw the rotor from the hydraulic head and fit the top adjusting plate (33) so that the slot in its periphery is in line with the mark on the rotor (see Fig 20.21).

9 Fit the drive plate (36) to the rotor with its relieved face next to the top adjusting plate, and the slot in the periphery of the drive plate in line with the mark on the rotor (see Fig 20.21). Tighten the drive plate screws (37) lightly and insert the roller (34) and shoe (35) assemblies into their guides in the rotor. Make sure that the contour of the roller shoe ears conforms with the contour of the eccentric slots in the top adjusting plate.

10 Fit the bottom adjusting plate (32), engaging its slots with the lugs on the top adjusting plate, and ensuring that the contour of the eccentric slots matches the contour of the roller shoe ears.

11 Insert the rotor assembly into the hydraulic head, then fit and lightly tighten the transfer pump rotor (20).

12 Stand the hydraulic head and rotor assembly on the bench, drive plate uppermost. Fit the relief valve timing adaptor (part no 18G653), preset to 15 atmospheres, to two opposite high-pressure outlets on the hydraulic head and connect this assembly to the injector tester (see Fig 20.22).

Note The relief valve timing adaptor connects to the injector tester and provides two connections for attachment to the hydraulic head. The unit contains a relief valve, which can be adjusted to any recommended pressure. The hydraulic head outlets are each stamped with a letter for identification and the choice of outlets to which the relief valve timing adaptor is

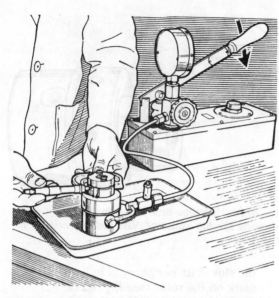

Fig 20.22 Setting the roller-to-roller dimension

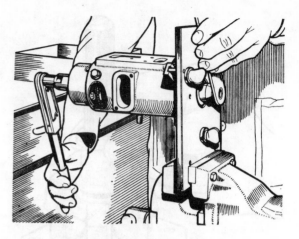

Fig 20.23 Tightening the transfer pump rotor

connected is specified by the manufacturer, and varies from one pump application to the next.

13 Operate the pumping lever of the injector tester and turn the rotor in the normal direction of rotation until the pumping plungers are forced outwards as far as the eccentric slots in the adjusting plates will allow; this is the maximum fuel position. Using a maximum fuel adjusting probe (part no 18G656), rotate the adjusting plates as necessary to set the roller-to-roller dimension at the figure given in the specifications (see Fig 20.22). Hold the drive plate with the correct tool (part no 18G641) and tighten the drive plate screws to the torque figure given in the specifications. Recheck the roller-to-roller dimension. Disconnect the adaptor from the hydraulic head.

14 Rotate the drive shaft in the pump housing to position the master spline at 12 o'clock. Fit a new 'O'-ring (26) to the groove in the periphery of the hydraulic head and align the master spline in the drive plate with the metering valve bore in the hydraulic head. Lubricate the periphery of the hydraulic head and the bore of the pump housing liberally with clean test oil, and assemble the hydraulic head to the pump body. Fit the two hydraulic head locking screws (96, 98) finger tight, positioning the screw with the vent valve in the position where it

will be readily accessible when the pump is fitted to the engine.

15 Hold the drive hub with the correct tool (18G659) and, using the spanner designed for the purpose (part no 18G634), tighten the transfer pump rotor to the torque figure given in the specifications (see Fig 20.23). Fit the transfer pump liner (22), and insert the transfer pump vanes (21) in their slots.

16 Carefully drive the transfer pump liner locating peg into the position in the pump end plate that corresponds to the pump's direction of rotation—position A—pump rotation anticlockwise, position C—pump rotation clockwise. Seat the piston retaining spring (or priming spring) (6) in the bottom of the regulating valve bore.

17 Fit a new seal washer (9) to the small diameter end of the regulating valve sleeve (8) and fit the piston (7) into the sleeve. Insert the regulating spring (10) and peg (11) into the large diameter end of the sleeve, and place the transfer pressure adjuster (12) on the top of the sleeve. Fit the sleeve retaining spring (13) onto the pressure adjuster and pass the filter (14), small end leading, over the spring and onto the shoulder of the valve sleeve. Insert this assembly, valve sleeve first, into the bore of the end plate and fit the fuel inlet connection (15) and washer (16).

18 Place a new sealing ring (19) in its recess in the hydraulic head face, and fit the end plate to the head engaging the locating peg with the slot in the transfer pump liner. Tighten the end plate screws (3) and studs (4) to the torque figure given in the

specifications, then tighten the fuel inlet connection (15) to the torque figure given.

19 Using the correct spanner (part no 18G646), tighten the cam advance screw (39) to the torque figure given in the specifications and check the cam ring for freedom of rotation.

20 Fit new 'O'-rings (116, 118) to the advance unit end plug (115) and spring cap (117), using a protection cap (part no 18G640) to pass the rings over the threads. Screw the end plug finger tight into the end of the advance unit where the fuel duct enters the bore. Fit the piston (121) into the advance unit with its counter bored end at the open end of the housing, place the two springs (122, 123) in position in the piston. Place the shim washer inside the spring cap (117) and screw the cap finger tight into the housing. If the spring cap or end plug are renewed, ensure that the new part has the same identification letter as the component it replaces. Unmarked components should be used to replace unmarked components.

21 Fit a new 'O'-ring (125) under the head of the hydraulic head locating bolt (124), using a protection cap (18G639). Position the non-return valve ball (128) in the side of the head locating bolt and fit the bolt to the advance unit. Using an assembly cap (part no 18G647), fit a new inner 'O'-ring (126) to the shank of the locating bolt, and place the plain washer (127) on top of the 'O'-ring.

22 Place a new advance unit joint gasket (111) on the pump housing with the straight side of the 'D'-shaped hole at the drive end of the pump; to ensure sealing, this joint washer should be fitted dry. Position the advance unit on the pump, fit a new aluminium and rubber washer to the stud, and fit the cap nut (113).

23 Tighten the two hydraulic head locking screws, the hydraulic head locating bolt and the advance unit cap nut to the torque figures given in the specifications. Tighten the advance unit end plug and spring cap to the torque figures given in the specifications.

24 Insert the metering valve (55) into its bore in the hydraulic head.

25 Assemble the governor arm (50), control bracket (47) and governor arm spring (51), then fit the assembly to the pump housing, ensuring that the lower end of the governor arm engages the stepped plate of the thrust sleeve flange. Fit the keep plate (63) with its

open end towards the shut-off bar (54), and fit new tab washers (65) with their pointed tabs towards the governor arm. Screw in the two control cover studs (64) to the torque figure given in the specifications, and secure them with the pointed tabs.

26 Fit the small screw (48) and tab washer (49) to the metering valve end of the control bracket. Tighten the screw to the torque figure given in the specifications and lock it with the tab washer.

27 Assemble the spring retainer (57), spring (58), and linkage washer (59) onto the linkage hook (56). Pass the threaded end of the hook through the governor arm, fit the pivot ball washer (60) and backing washer (61), and screw on the linkage nut about three turns.

28 Press back the spring retainer and attach the linkage hook to the metering valve so that the hook end is turned towards the metering valve.

29 Press the governor arm lightly in the direction of the metering valve and, using a vernier gauge held parallel to the pump axis, set the governor link length (see Fig 20.24) to the dimension given in the specifications. This adjustment is made by slackening or tightening the linkage hook nut.

30 Locate the spring guide (53) in the correct hole in the governor arm and connect the governor spring to the guide. In many cases, a very light spring is fitted over the spring guide. Refer to Fig 20.25 for the correct governor spring location in conjunction with the pump setting code discussed previously.

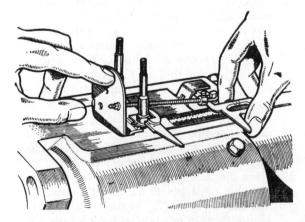

Fig 20.24 Setting the governor link length

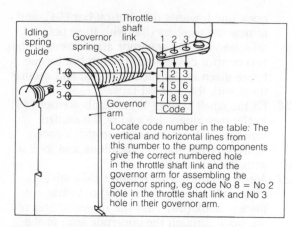

Fig 20.25 Application of the setting code

31 Insert the plain end of the shut-off bar (54) into the slot in the control bracket and position the shut-off bar under the tab of the control cover locking washer.

32 Using a protection cap (part no 18G654), fit new lower 'O'-rings to the shut-off and throttle shafts (68, 74). Fit a new upper 'O'-ring to each shaft, using a protection cap (part no 18G665) and pack the groove between the 'O'-rings on each shaft with Shell Alvania no. 2 grease.

33 Press the shut-off shaft into its bore in the control cover, positioning the eccentric peg close to the edge of the control cover and projecting slightly from the joint face.

34 Place a new control joint gasket (91) in position on the pump housing, engaging the tabs with the slots under the keep plate (63). To ensure sealing, this joint washer should be soaked in test oil before assembly.

35 Connect the free end of the governor spring to the appropriate hole in the throttle shaft link (as per Fig 20.25), and press the throttle shaft into its bore in the control cover. Place the control cover in position on the studs, ensuring that the shut-off peg engages the shut-off bar. Pull the shut-off shaft fully home as the control cover is lowered onto the pump housing.

36 Fit new sealing washers (67) to the control cover studs, and screw on the stud nuts to the torque figure in the specifications.

37 Place the dust caps (70, 76) on the throttle and shut-off shafts, fit the throttle arm (73) and shut-off lever (79) to their shafts, and secure them in position with their nuts and washers.

38 Refit the inspection cover (106) to the side of the pump housing using a new gasket, and refit the quill shaft (105).

Testing and adjusting DPA pumps

After an overhaul, a DPA fuel injection pump must be tested and adjusted to specifications. While there are some slight variations, the tests and adjustments made fall into eight general categories:

- transfer pump vacuum test
- transfer pump pressure test
- automatic advance test
- back-leakage test
- maximum fuel delivery test
- fuel cut-off test
- governor setting adjustment
- timing setting.

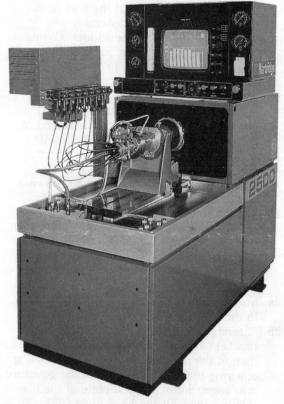

Fig 20.26 Hartridge series 2500 fuel pump test stand

To make the first seven tests, the pump should be set up on a power-driven test bench. A mounting plate to carry the pump is usually available for any particular test bench and bolts to the test bench table. The pump can then be driven directly by the test bench drive coupling.

Before setting up the pump on the test bench, it is essential to establish that the test bench is capable of providing an adequate fuel supply. Usually, a fuel supply of not less than one litre per minute is specified.

The pump should be set up on the test bench as follows:

1 Mount the pump on the test bench, ensure that the test bench is set to run in the direction of rotation indicated on the pump nameplate, and connect the pump drive.
2 Connect the hydraulic head outlet unions to the test bench test injectors by means of a set of matched injector lines. The usual specification is to use test injectors set to 175 atmospheres and injector lines 6 mm × 2 mm × 863.5 mm (34 ins) long.
3 Using transparent (where possible) flexible pipes, connect the injection pump to the test equipment as follows:
   i Fit a suitable adaptor in place of the hydraulic head locking screw which does not have a vent screw, and connect a pressure gauge. The gauge should be capable of registering a maximum pressure slightly above the maximum transfer pressure. (The maximum transfer pressure should be quoted in the pump specifications.)
   ii Fit an end plate adjuster to the fuel inlet union. The adjuster provides both a means of connecting the supply pipe and adjusting the transfer pressure. Use a 'T'-coupling in conjunction with the end plate adjuster and connect both the fuel supply and vacuum gauge.
   iii Connect the pump drain union to a measuring glass, and the measuring glass drain cock to the test bench return connection.
4 Once the pump has been connected to the test bench and gauges etc connect an automatic advance gauge to the advance unit spring cap in place of the small set screw and set the degree scale to give a zero reading.
5 Unscrew the idling and maximum speed stop screws to ensure that the throttle lever has its full range of movement.

6 Unscrew the pressure adjuster in the pump end plate (by means of the end plate adjuster) to the maximum extent, then screw it in approximately 1.5 turns.
7 Fill and prime the pump as follows:
   i Turn on the fuel supply to the injection pump and open the pump air vent (or bleeder) screws.
   ii When test oil, free from air bubbles, flows from the hydraulic head vent screw, close the vent.
   iii When test oil, free from air bubbles, flows from the housing vent screw, close this vent also.
   iv Turn the pump drive in the direction of rotation for 90:, and again open the hydraulic head vent. Close again when there are no air bubbles in the escaping test oil.
   v Turn the pump drive a further 90° and again check both vents.
   vi Check that test oil is flowing from the pump drain connection.
   vi With the throttle fully open, run the test bench at 100 rpm. Loosen the injector lines at the injectors and retighten them when test oil, free from air bubbles, issues from the connections.

During testing, it is essential that the pump handles sufficient fuel to ensure lubrication of the rotor and hydraulic head assembly. For this reason, the pump must not be run for long periods at high speed with low fuel output, or run for long periods with the shut-off control in the closed position. Further to this, all test procedures should be carried out with the throttle and shut-off controls in the fully open position except where specifically stated otherwise.

Check the oil-tightness of all joint washers, oil seals and pipe connections while the pump is running and stationary.

Once items (1) to (8) have been completed, the pump may be adjusted. Unless otherwise stated, the test must be made with the throttle and shut-off controls in the full open position.

Transfer pump vacuum test—This test is designed to show the efficiency of the transfer pump and the presence of any faults on its inlet side.

The test is made by running the test bench at 100 rpm and turning off the test oil supply valve for a maximum time of 60 seconds. The vacuum gauge should reach 405 mm Hg in 60 seconds maximum. Because the test creates a shortage of test oil to lubricate the rotor, the pump must not be run for more than 60 seconds with the valve closed. After completion of the test, the pump must be bled again at the hydraulic head vent screw while the pump is running at 100 rpm.

Transfer pump pressure test—Because the transfer pressure exerts very considerable influence on the fuel delivery and controls the automatic advance, it is essential that the transfer pressure is kept within specifications through the pump operating speed range. Because of this it is usual to check the transfer pressure at three or four different pump speeds.

Because the transfer pump connects directly to the hydraulic head locating screw holes, the pressure gauge connected to one of these holes will directly read the transfer pressure. Provided the correct regulating spring and regulating sleeve have been fitted to the end plate, the transfer pressure should lie within limits at all specified speeds, the maximum pressure being adjusted by means of the end plate adjuster which is screwed in to increase the pressure.

Automatic advance test—The rate of automatic injection advance and the total advance vary from one pump application to the next, and are governed by two factors—the change in the transfer pressure and the spring(s) fitted to the automatic advance unit.

Like the transfer pump pressure test, readings are taken at a number of different pump speeds. Adjustment is made by shimming the automatic advance spring(s) to change the spring force opposing the force exerted on the advance piston by transfer pressure, but should only be attempted if the amount of advance is consistently above or below specifications. If the reading is low at one point and high at another, then some mechanical fault would seem to be indicated.

Back-leakage test—This test is designed to establish whether the amount of test oil leaking between the rotor and hydraulic head, past the pumping plungers and past the automatic advance piston is within desirable limits. Excessive back-leakage is a sign of worn components or a leak into the pump housing, while too little shows that there is probably insufficient test oil leaking past the pumping components for efficient lubrication.

The test specifications vary from one pump application to the next, but the back-leakage quantity is usually measured during 100 injections. The procedure is to run the pump at

specified rpm, and use the measuring glass to measure the quantity of fuel that flows, during 100 injections, from the drain connection.

Maximum fuel delivery test—The specified maximum fuel pump output can be found by reference to the setting code on the pump typeplate as detailed in the beginning of 'DPA pump servicing', or to the pump test sheet.

The maximum fuel delivery is controlled by the maximum plunger movement and is adjusted by partial rotation of the adjusting plates. Although the roller-to-roller dimension is set during pump assembly, it is likely that a final adjustment will have to be made after checking the maximum fuel delivery on the test bench.

The normal procedure is to run the pump at specified rpm and measure the deliveries from the test injectors during 200 injections. There should be very little variation between injectors, since the same pumping element supplies all hydraulic head outlets, and the deliveries should lie within close specified limits. Before taking the readings, the test oil should be allowed to settle for 15 seconds in the measuring glasses, and the glasses should be allowed to drain for 30 seconds before taking a second test.

If the fuel delivery is incorrect, it must be adjusted at the adjusting plates. The fuel supply should first be turned off, the cover plate removed, and the drive plate screws loosened. Adjustment is made by means of a probe engaged in the slot in the periphery of the top adjusting plate (see Fig 20.27), and the pump drive must be turned until the slot is accessible. Very little movement should be necessary, since the adjusting plates were adjusted closely when setting the roller-to-roller dimension; clockwise

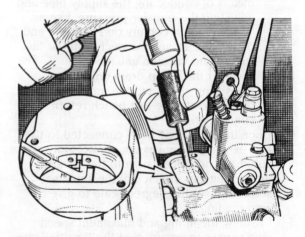

Fig 20.27 Adjusting the maximum fuel setting

rotation (from the drive end) will increase fuel delivery.

When an adjustment has been made, the drive plate screws must be retightened to the correct torque, using a torque wrench and adaptor. It is essential that the wrench and adaptor are kept in a straight line. Finally the cover plate must be replaced, the fuel supply reconnected and air bled from the pump housing and the maximum fuel delivery checked again.

After adjusting the fuel delivery at the high specified pump speed, a second check is usually made at a lower speed. If this second test is less than the first by more than the specified figure, it indicates either an incorrect relationship between pump speed and transfer pump pressure or that pumping efficiency is falling with speed to a greater degree than it should.

Fuel cut-off test—One very important factor that must be checked during pump testing is that the engine can be stopped by means of the fuel shut-off control. Not only should the shut-off control be able to prevent all but a slight injection, but also the throttle should be able to reduce the injection rate to such a point that the engine stops.

Both of these can be checked by running the pump at low rpm with one control closed and the other fully open. A maximum delivery of 200 injections is usually quoted for each case.

Governor setting adjustment—The maximum engine speed is controlled by the governor, which should decrease fuel delivery once the corresponding maximum pump speed is exceeded. The pump is run on the test bench at specific rpm (above normal maximum), and the maximum speed screw adjusted to give the correct delivery from 200 shots, which is less than the maximum fuel setting adjusted previously.

Timing setting—This setting is made after all the foregoing tests and adjustments have been completed, with the pump removed from the test bench.

Because the plungers move outwards further as the fuel injection rate is increased, the commencement of injection changes with the fuel requirements of the engine and an invariable reference point must be found. The point used is the point at which the cam followers are in contact with both the cam ring and the plungers, with the plungers fully outwards in the maximum fuel position, ie the beginning of injection, for no. 1 cylinder.

To locate the timing point, a special tool known as a timing flange is fitted to the pump drive, and the relief valve timing adaptor (as used for checking the roller-to-roller dimension) is used to connect two specified hydraulic head outlets to the injector tester. The timing flange incorporates a protractor and a scribing guide.

The protractor is fixed to the hub, which fits the pump drive, and the scribing guide can be rotated in relation to hub and protractor. For each pump application, a specified protractor reading must be aligned with a datum line on the scribing guide and the two components locked together.

To ensure that it is no. 1 injection, the relief valve timing adaptor is connected to specified hydraulic head outlets (they are identified by means of letters stamped adjacent to them), and the pump turned in the direction of normal rotation until a specified timing letter on the drive plate becomes visible through the pump inspection hole.

With the timing flange set and mounted on the pump, the correct timing letter visible through the inspection hole and the relief valve timing adaptor connected to the specified outlets, the injector tester is operated until a specified pressure is reached and the pump turned in the direction of rotation until resistance is felt. This resistance occurs when the cam rollers press on the plungers to commence pressurising the fuel and this point must be commencement of injection. A scriber run through the scribing guide will mark the pump flange at the correct timing point (see Fig 20.28).

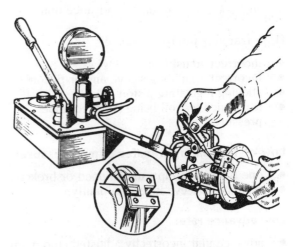

Fig 20.28 Scribing the timing mark

Test bench faultfinding

When a DPA fuel injection pump is being tested and adjusted on a test bench, there are a number of factors that could cause any fault encountered. The following faultfinding guide indicates possible causes of some of the more common problems encountered in service.

Insufficient transfer pump vacuum:

- damaged transfer pump 'O'-ring sel
- worn or damaged transfer pump vanes
- loose or unevenly tightened end plate
- transfer pump liner turning due to absence of locating dowel
- incorrectly located transfer pump liner locating dowel
- loose or damaged inlet or vacuum gauge connections
- priming spring missing or broken
- regulating spring missing or broken
- damaged regulating sleeve gasket
- regulating piston missing.

Low transfer pump pressure:

- incorrect adjustment
- damaged transfer pump 'O'-ring seal
- worn or damaged transfer pump vanes
- loose or unevenly tightened end plate
- regulating spring missing or damaged
- regulating piston missing
- damaged regulating sleeve gasket
- incorrect regulating sleeve or piston
- fuel leakage at hydraulic head vent screws, pressure gauge connection or hydraulic head locating studs
- fuel leakage at automatic advance unit.

High transfer pump pressure:

- incorrect adjustment
- incorrect regulating sleeve and/or piston
- sticking regulating piston
- test bench test oil being supplied under pressure.

Low or fluctuating transfer pump pressure:

- one transfer pump vane chipped or broken
- regulating sleeve inner seal faulty.

Low advance rate:

- advance unit incorrectly adjusted (too many shims fitted)

- incorrect advance spring(s) fitted
- low transfer pump pressure
- sticking advance piston
- excessive clearance between the advance piston and the housing
- sticking cam ring.

High advance rate:

- advance unit incorrectly adjusted (insufficient shims fitted)
- incorrect advance spring(s) fitted
- high transfer pump pressure.

Incorrect maximum fuel delivery:

- air in the system
- throttle not held fully open
- fuel shut-off control not held fully open
- incorrect maximum fuel setting
- cam ring reversed
- cam ring worn
- governor link adjustment incorrect (mechanically governed pump)
- incorrect transfer pump pressure
- sticking plungers or roller shoes
- rotor plug loose or leaking
- pumping components worn excessively
- sticking metering valve.

Low fuel delivery at reduced rpm:

- low transfer pump pressure
- plungers scored or worn
- scoring in the region of the hydraulic head ports
- excessive clearance between rotor and hydraulic head
- scored metering valve
- throttle not fully open
- fuel leak at rotor sealing plug.

Difficulty in setting delivery with maximum speed screw:

- incorrect assembly of governor
- sticking metering valve
- sticking governor thrust sleeve (mechanical governor)
- incorrect transfer pump pressure (hydraulic governor).

Pump installation and timing

Since the procedure for timing and fitting is not standard for all engines, reference must be made to the engine manufacturer's handbook before carrying out this work. In an emergency, however, and if the timing marks are clearly marked on the engine and the pump mounting flange, most pumps can be fitted as follows:

1 Turn the pump drive shaft so that the master spline is in alignment with the master spline on the engine coupling.
2 Enter the drive shaft into the engine coupling as the pump is pushed onto the securing studs.
3 Push the pump hard against the mounting face and secure lightly with the three holding nuts.
4 Rotate the pump on the retaining studs until the timing mark scribed on the edge of the pump mounting flange is accurately aligned with the timing mark on the engine.
5 Tighten the retaining nuts. The pumps should now be correctly timed and all pipes, controls etc may be fitted and the air bled from the system.

Bleeding the DPA fuel system

Air will have to be bled from the DPA fuel system on installation of the pump and if the system should be drained by running out of fuel.

Before loosening any of the bleed screws, ensure that the surrounding area is thoroughly cleaned to prevent dirt and foreign matter getting into the system. Proceed as follows, carrying out operations (1) and (2) below while operating the fuel feed pump priming lever:

1 Slacken the filter outlet connection or the injection pump inlet connection (whichever is the higher) and allow fuel to flow until free of air. Tighten the connection.
   Note Filters of the type that provide two inlets and two outlets, two of which are plugged during service, must be vented at the plugged connections. This must be done irrespective of the height of the filter in the system.
2 Slacken the vent valve fitted on one of the two hydraulic head locking screws, and the vent screw on the governor housing. When fuel free of air flows from the vents, tighten the housing vent screw and then the governor vent screw.
3 Slacken any two injector high-pressure pipe unions at the injector end. Set the accelerator to fully opened position and

ensure that the stop control is in the 'run' position. Turn the engine until fuel, free of air, flows. Tighten the unions.
4 Start the engine.

The Roosa Master rotary injection pump

*T*he Roose Master DM and DB2 rotary distributor fuel injection pumps are similar in construction and operation to the CAV DPA fuel injection pumps. For this reason, only the basic operation and fuel flow will be discussed. However, the Roosa Master pump differs from the DPA pump in its maximum fuel adjustment, its speed adjustment and its injection pump timing to the engine. There are other variations as well that will be mentioned during the description of operation.

Basic pump operation

The transfer pump located at the rear of the injection pump is driven via the distributor pump drive shaft and rotor. It is a positive displacement vane-type pump with four vanes fitted to a rotor turning within a cam ring. Transfer pressure is controlled by a regulating valve adjacent to the vane pump.

The distributor rotor has one or two charging ports for efficient charging of the pump plungers and one discharge port. The distributor rotor also houses the two pumping plungers. High pressure pumping is achieved by the pump plungers being simultaneously forced toward each other by the lobes on the cam ring via the cam rollers, the number of cam lobes being the same for an even number or double for an odd number of engine cylinders. The rotor rotates within very precise tolerances inside the hydraulic head, which houses the metering valve, the circular fuel passage, the discharge ports and discharge fittings. An injection advance device is mounted under the pump and automatically advances or retards injection timing depending on engine speed.

The governor is of the mechanical flyweight design incorporating the fundamentals of spring force and centrifugal force working in balance

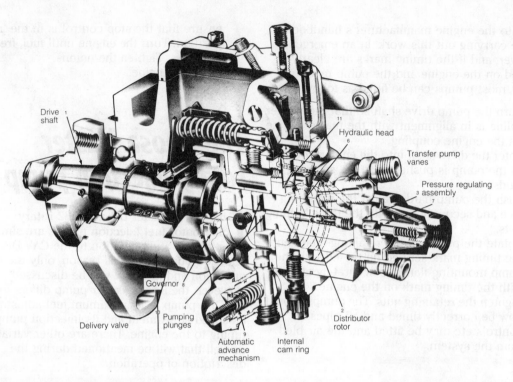

Drive shaft 1

Hydraulic head 6 11

Transfer pump vanes

Pressure regulating assembly 7 3

Governor 8

Delivery valve 10

Pumping plunges 4

Automatic advance mechanism 9

Internal cam ring 5

Distributor rotor 2

Fig 20.29 A Roosa Master DM fuel injection pump

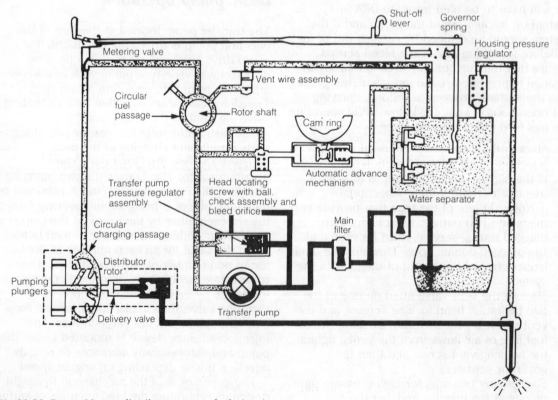

Metering valve

Vent wire assembly

Circular fuel passage

Rotor shaft

Cam ring

Shut-off lever

Governor spring

Housing pressure regulator

Automatic advance mechanism

Water separator

Transfer pump pressure regulator assembly

Head locating screw with ball, check assembly and bleed orifice

Main filter

Circular charging passage

Pumping plungers

Distributor rotor

Delivery valve

Transfer pump

Fig 20.30 Roosa Master distributor pump fuel circuit

with one another to control the metering valve position.

Engine shutdown is achieved by moving the metering valve into a closed position by the operation of an external shut off lever.

Charging and discharging cycles

By reference to Fig 20.30, it can be seen that fuel from the tank passes through the fuel filters and feed pump (if fitted) to the fuel inlet opening and screen to the vane type transfer pump. The fuel pressure from the transfer pump is controlled by the transfer pump pressure regulating valve. Fuel at transfer pressure leaves the vane pump via a passage in the hydraulic head and passes to the automatic advance device and the circular fuel passage surrounding the rotor shaft.

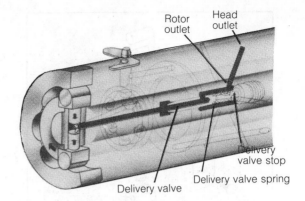

Fig 20.32 Discharge cycle—Roosa Master DB fuel injection pump

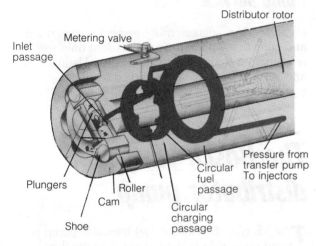

Fig 20.31 Charging cycle—Roosa Master DB fuel injection pump

The fuel then flows through a port in the hydraulic head to the metering valve, which, under the control of the governor, will determine the amount of fuel flowing to the circular charging passage. With the rotation of the rotor, the inlet passage aligns with the circular charging passages in the hydraulic head thus allowing fuel to flow into the pumping chamber, as shown in Fig 20.31. Further rotation of the rotor will move the charging port and inlet passages out of alignment and at the same time the plunger rollers will begin to ride up on the cam ring lobes. At this point the high-pressure discharge passage will come into alignment with one of the outlets in the hydraulic head. Further rotation of the rotor will cause the plungers to be forced together by the cam ring lobes, pressurising the fuel that is delivered to the outlet connections and on to the injectors (Fig 20.32).

During pump operation when the plungers are not being intermittently charged with fuel at transfer pressure, the flow of fuel is diverted via slots in the rotor shaft for circulation through the pump housing for the purposes of lubrication, cooling and purging air that may have entered the system.

The removal of air from the system is via a special air vent passage in the hydraulic head, shown in Fig 20.30, which allows small quantities of fuel (and air) to flow into the governor housing and back to the tank, making the system entirely self-bleeding with no other means of bleeding provided.

Distributor pump adjustments

Maximum speed adjustment

There are two control levers on this type of pump, one on each side of the governor housing. One is the speed control and is fitted with adjustable stops for adjusting the maximum and idle engine speeds; the other is the stop control, which over-rides the governor's control of the metering valve to move the metering valve to the no-fuel delivery position when operated.

Maximum fuel adjustment

This adjustment should only be made when the pump is on a test bench and the quantity of fuel delivered by the pump should always be within

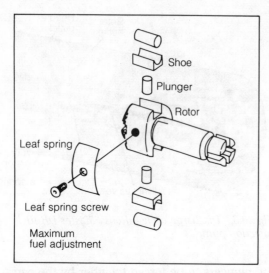

Fig 20.33 Rotor assembly showing leaf spring and screw fuel adjustment

the manufacturer's specifications. The adjustment is carried out by first removing the top from the governor housing and inserting an Allen key down through the pump housing (there is a special hole that allows entry), and altering the setting of the leaf spring adjusting screw (refer to Fig 20.33)

Pump timing

In the absence of a workshop manual giving the specific procedure, the following general method can be applied to almost any installation:

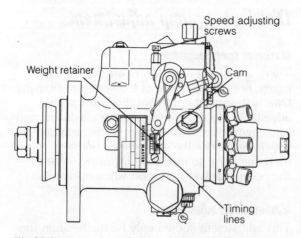

Fig 20.34 Roosa Master injection pump showing timing marks

1   Rotate the engine crankshaft in the normal direction of rotation until no. 1 piston approaches TDC on the compression stroke. Continue to turn the crankshaft until the injection timing marks on the fan pulley or flywheel align.
2   Remove the small timing cover plate on the side of the pump housing and examine the timing marks for alignment (Fig 20.34). If the timing marks don't align, loosen the pump mounting bolts and rotate the pump till the marks align, retighten mounting bolts.
3   Recheck the timing by rotating the engine forward for two revolutions and checking that the crankshaft timing marks and the pump timing marks align.

Pump service

Pump service—disassembly, inspection, overhaul and adjusting—should be undertaken only with reference to Roosa Master Workshop Manuals and Service Data, using approved testing equipment.

The Bosch VE-type distributor pump

The VE type distributor fuel injection pump is used on a wide variety of small to medium power diesel engines used in such applications as cars, commercial vehicles, small boats, earthmoving equipment and agricultural tractors. The VE injection pump is also manufactured under licence by other injection equipment manufacturers and therefore the following text may apply to other brand-name pumps.

The VE type injection pump has a single high-pressure pumping chamber and plunger regardless of the number of engine cylinders it has to serve. Fuel delivered from the high-pressure chamber is directed by means of a distributor groove in the plunger to a number of outlet ports corresponding to the number of engine cylinders. The pump is driven by a drive shaft, which is connected via a driving disc to a cam plate and plunger. The cam plate is designed with the same number of face cams as the

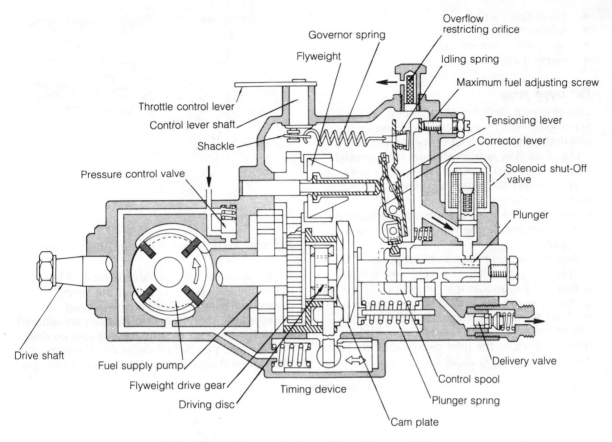

Fig 20.35 Schematic diagram of a Bosch VE fuel injection pump

Note Fuel-supply pump is shown in two views, one being a 90°—rotated view with the pump vanes and rotor in section and the other being a normal side view.
Timing device is shown as turned around by 90°

engine has cylinders. As the cam plate rotates, it rides up and down a set of fixed rollers causing the plunger to rotate and reciprocate to provide both the pumping and distributing action.

Governing of the VE pump is accomplished by means of a mechanical governor incorporated in the pump housing.

To stop an engine fitted with a VE pump, a solenoid shut-off valve is used, which prevents the flow of fuel to charge the chamber, thereby preventing injection from taking place.

The pump also incorporates an automatic advance unit, which advances and retards the timing relative to the changing engine speed and load conditions.

As a starting aid in cold operating conditions, an optional manually or automatically controlled injection advance device can be fitted external to the pump to improve starting capabilities.

The pump is not only compact in size, but weighs about half as much as its in-line counterpart. The internal components are completely lubricated throughout by fuel and no special lubricating oil or system is required. The fuel continually circulates through the pump and back to the fuel tank, thereby continually cooling and purging air from the pump.

Construction and operation

The construction and operation of the pump may be followed by considering it to be composed of six individual working modules, and examining each of those modules in turn:

- fuel supply pump
- pumping and distribution section

- governing section
- solenoid shut-off valve
- automatic advance unit
- starting aids.

Fuel supply pump

Low-pressure charging of the pump housing is accomplished by a vane-type fuel supply pump situated at the drive end of the distributor pump. Fuel flow is from the fuel tank through the fuel filter and into the vane pump; from here, it enters the pump housing at pressures that vary between 360 kPa and 810 kPa. Generally, there is no fuel feed pump fitted to this fuel system as the vane pump serves this purpose.

Being a constant displacement pump, the fuel supply pump can deliver several times the amount of fuel required for injection. Therefore, when the pump housing pressure reaches a predetermined level, excess fuel delivery is relieved via a pressure control valve and returned back to the inlet side of the fuel supply pump as seen in Fig 20.36

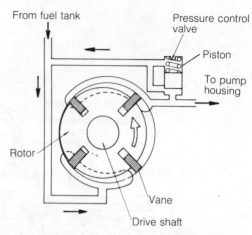

Fig 20.36 Arrangement of the fuel supply pump

The pressure control valve is located beside the fuel supply pump and is of the spring-loaded piston type. This valve is pre-set on manufacture and requires no further adjustment.

For the purpose of self-bleeding and cooling of the entire pump, fuel circulates through an overflow restricting orifice back to the fuel tank.

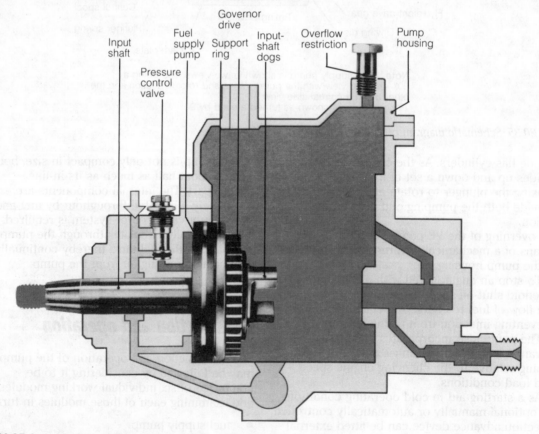

Fig 20.37 Layout of the pressure control valve and overflow restriction

The overflow restricting orifice is 0.6 mm in diameter and is situated in the banjo bolt in the fuel return line on top of the pump housing. While this orifice allows fuel to return to the fuel tank, it offers sufficient restriction to fuel flow from the fuel supply pump to cause the injection pump housing to be pressurised. Further, in conjunction with the fuel supply pump pressure control valve, the orifice is responsible for the pump housing fuel pressure necessary for the charging of the high-pressure chamber and the operation of the automatic injection advance unit.

A hand-operated plunger-type pump is incorporated into the body of the fuel filter housing for priming or bleeding the system. The hand pump is used to prime the system whenever the fuel filter is changed or the fuel pump housing is drained.

Pumping and distribution section

The distributor pump drive shaft simultaneously drives the fuel supply pump, the cam plate and the pump plunger as shown in Fig 20.38. The plunger pumping stroke length is determined by the profile of the cam plate, so that as the cam plate rotates and remains in contact with the four cam rollers, the plunger reciprocates as it rotates. Two plunger return springs are fitted to the plunger via a mounting plate, and ensure that the plunger follows the contour of the cam face plate.

Fuel delivery

As the plunger rotates on its downstroke, the metering slit aligns with the intake passage in

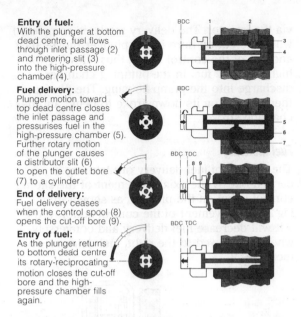

Entry of fuel:
With the plunger at bottom dead centre, fuel flows through inlet passage (2) and metering slit (3) into the high-pressure chamber (4).

Fuel delivery:
Plunger motion toward top dead centre closes the inlet passage and pressurises fuel in the high-pressure chamber (5). Further rotary motion of the plunger causes a distributor slit (6) to open the outlet bore (7) to a cylinder.

End of delivery:
Fuel delivery ceases when the control spool (8) opens the cut-off bore (9).

Entry of fuel:
As the plunger returns to bottom dead centre its rotary-reciprocating motion closes the cut-off bore and the high-pressure chamber fills again.

Fig 20.39 Pumping cycle of the VE fuel injection pump

the distributor head, and fuel, under pressure in the pump housing, flows into the high-pressure chamber and drilling in the plunger body, as shown in Fig 20.39.

As the cam face runs up on the rollers, the plunger direction is reversed and the delivery stroke begins. The plunger rotates as it moves up and the intake passage is closed off early in the pumping stroke so that further upward movement of the plunger will raise the pressure of the fuel. Continued rotation will bring the distributor slit into line with an outlet port in the distributor head, and high-pressure fuel will open the delivery valve and flow into the injector line to be injected into the combustion chamber

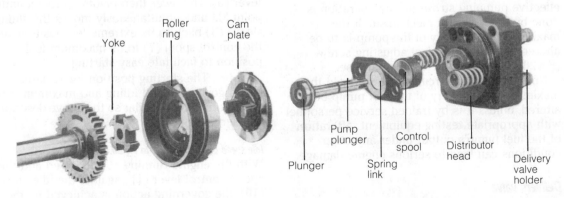

Fig 20.38 Exploded view of the high-pressure pumping assembly

via the injector. The delivery stroke is completed when the cut-off bore in the plunger moves past and out of the control spool, thus allowing the high-pressure fuel in the pumping chamber to discharge into the pump housing. The plunger then returns on its downstroke to begin another charging stroke.

Fuel metering

The control of fuel delivery or metering is carried out by the axial movement of the sliding control spool on the plunger, as shown in Fig 20.40. Movement of the control spool to the left will decrease fuel delivery and conversely when it is moved to the right, fuel delivery will increase.

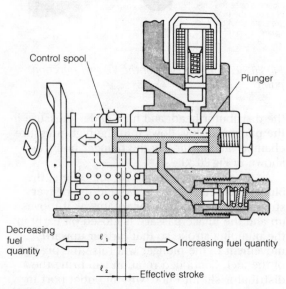

Fig 20.40 Spool control of effective pumping stroke and fuel delivery

Movement of the control spool to alter the effective pumping stroke during operation is done by the governor mechanism. If the maximum fuel delivery of the pump is to be altered, the maximum fuel adjusting screw (Fig 20.35) is provided for this purpose.

Note Under no circumstances should the maximum fuel delivery of the fuel pump be altered, unless it is by trained service personnel with appropriate testing equipment. Alteration of the fuel pump's output under any other conditions can lead to serious engine damage.

Delivery valve

The delivery valves in the VE type injection pump perform the same functions as in other types of injection pumps, and for more detailed explanation, refer to Chapter 18, Jerk-type Injection Pumps.

Governing section

Two types of governor are available for VE-type pumps to suit a range of engine applications. They are all mechanical flyweight governors, classified as follows:

- variable speed
- idling and maximum speed.

The variable speed governor is designed to control the engine rpm at any selected speed, including idle and maximum no-load engine speeds. This governor is generally fitted to injection pumps used on industrial engines and tractors.

The idling and maximum speed governor is a typical automotive governor, controlling the engine rpm at idle and maximum speed only. The speed range between idle and maximum speed is controlled entirely by the operator. This governor is used in automotive applications.

Variable speed governor operation

Starting position (Refer to Fig 20.41)

With the centrifugal weights at rest and in a fully collapsed position, as the speed control lever is moved to the maximum speed position, the tension of the governor spring moves the tensioning lever (4) firmly to the left. As a result, the starting leaf spring (6) presses on the starting lever (5). The lever then pivots on its mounting point M2 and simultaneously moves the sliding sleeve (3) back to its extreme left position, and the control spool (7) to its maximum fuel position to facilitate easy starting.

Note The starting position for variable speed governors and idling and maximum speed governors is very similar so the above description will serve for both.

Idle speed operation (Refer to Fig 20.41)

With the engine running at idle speed and the speed control lever (11) against the idle stop (10), the governing action is achieved by the balance of forces exerted by the idle spring (14)

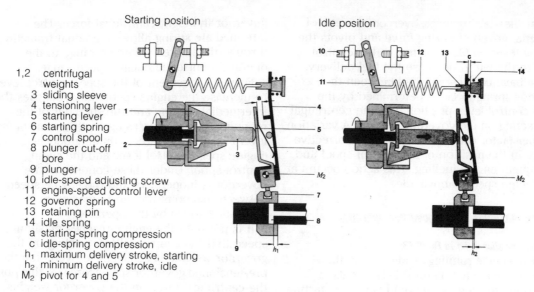

1,2 centrifugal weights
3 sliding sleeve
4 tensioning lever
5 starting lever
6 starting spring
7 control spool
8 plunger cut-off bore
9 plunger
10 idle-speed adjusting screw
11 engine-speed control lever
12 governor spring
13 retaining pin
14 idle spring
a starting-spring compression
c idle-spring compression
h_1 maximum delivery stroke, starting
h_2 minimum delivery stroke, idle
M_2 pivot for 4 and 5

Fig 20.41 Starting and idling positions of the variable speed governor mechanism

and the centrifugal governor flyweights. In this position, the idling spring (14) pushes the tensioning lever (4) to the left, which moves the starting lever (5) against the sliding sleeve (3). Therefore, the balance between the force exerted by the idling spring and centrifugal force acting on the governor weights determines the control spool's (7) position and relative fuel delivery for idle speed.

Higher speed operation (Refer to Fig 20.42)
When the speed control lever is turned from idle toward the maximum speed position, the

governor spring (4) is tensioned, and the idle spring (5) is compressed until the shoulder of the idle spring guide contacts the tensioning lever (7). The tensioning lever then pivots on M2, compressing the starting spring (9) until the stop bears against the starting lever (6), which then also pivots on M2. The lower end of the starting lever moves the control spool (10) to the right, increasing the fuel delivery to the engine.

The engine speed increases due to the additional fuel delivery until the governor weight force, acting through the sliding sleeve

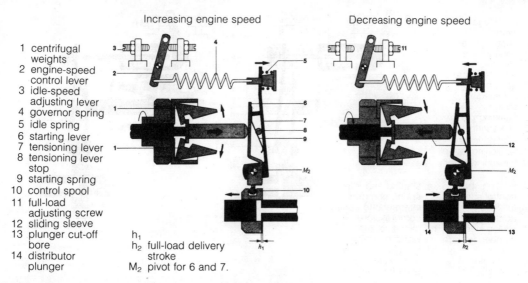

1 centrifugal weights
2 engine-speed control lever
3 idle-speed adjusting lever
4 governor spring
5 idle spring
6 starting lever
7 tensioning lever
8 tensioning lever stop
9 starting spring
10 control spool
11 full-load adjusting screw
12 sliding sleeve
13 plunger cut-off bore
14 distributor plunger
h_1
h_2 full-load delivery stroke
M_2 pivot for 6 and 7.

Fig 20.42 Effect of engine speed changes—variable speed governor

against the starting lever, overcomes some of the initial governor spring force and pivots the starting lever on M2. This moves the control spool to the left to decrease the fuel delivery. Thus a balance is established between the governor spring force as determined by the speed control lever position, and the centrifugal force acting on the governor weights. A variation to either factor causes an immediate corrective change in the position of the control spool and subsequent engine fuelling. This action occurs at all selected speeds above idle.

Idle and maximum speed governor operation

Idle speed operation (Refer to Fig 20.43)

With the engine running at idle speed, the centrifugal force acting on the flyweights is applied via the sliding sleeve (14) to the starting lever (8), which reacts with the idle spring (7). Therefore the pre-set tension of the idle spring reacting and balancing with the governor weights' centrifugal force determines the control spool (12) position and subsequent idle speed.

Idle speed through to maximum speed operation (Refer to Fig 20.43)

When the engine speed is to be increased, the speed control lever (2) is moved to the right, collapsing the starting spring (11) and the idling spring (7), and leaving the force of the intermediate spring (5) balancing with the

governor flyweight centrifugal force. The intermediate spring allows a gradual transition through the limited governed range to the operator-controlled engine speed range.

Further movement of the speed control lever to increase the engine speed fully collapses the intermediate spring and the shoulder of the retaining pin (6) presses against the tensioning lever (9), giving a direct coupling between the engine speed control lever and the sliding control spool. Under these conditions, the governor is inoperative and the engine speed between this point and up to maximum is controlled solely by the operator.

It is not until the engine reaches maximum speed that the centrifugal force acting on the governor weights is sufficient to overcome the pre-tensioned governor spring force. At this point, the centrifugal force on the governor weights, acting through the sliding sleeve, moves the starting lever to the right against the governor spring force, thereby moving the control spool to the left to decrease the fuel delivery and limit the maximum speed.

Solenoid shut-off valve

An electrically operated shut-off valve is used to stop the engine by cutting off the flow of fuel from the fuel supply pump to the high pressure

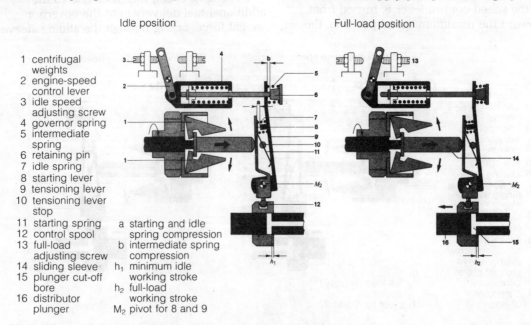

Idle position   Full-load position

1 centrifugal weights
2 engine-speed control lever
3 idle speed adjusting screw
4 governor spring
5 intermediate spring
6 retaining pin
7 idle spring
8 starting lever
9 tensioning lever
10 tensioning lever stop
11 starting spring
12 control spool
13 full-load adjusting screw
14 sliding sleeve
15 plunger cut-off bore
16 distributor plunger

a starting and idle spring compression
b intermediate spring compression
h_1 minimum idle working stroke
h_2 full-load working stroke
M2 pivot for 8 and 9

Fig 20.43 Idle and full-load positions—idle and maximum speed governor

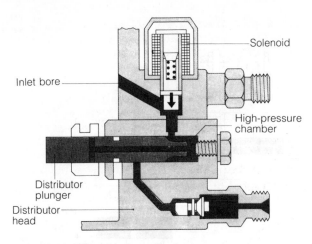

Fig 20.44 Solenoid shut-off valve

Automatic advance unit

The timing advance device (Fig 20.45) is designed to automatically adjust injection timing to suit changes in engine speed.

Operation

At low engine speed, the timer piston is held against its stop (in a fully retarded position) by a timer spring. As the speed of the engine increases, the fuel supply pump pressure also increases, progressively moving the timer piston against the force exerted by the timer spring. The timer piston's movement is transmitted via a pin to the roller ring assembly, causing it to rotate in a direction opposite to that of the pump, thereby advancing injection timing.

Load sensing advance and retard unit

As loads vary on a diesel engine, the injection timing must be advanced or retarded if optimum engine performance is to be achieved.

With an engine operating in its normal working load range, the automatic advance unit would have moved the roller ring to its maximum advanced position. If the load on the engine decreases with the throttle position unchanged, the engine speed will increase.

chamber. Being electrically operated, the device is readily operated by the stop/start key.

When the key is operated to start the engine, the solenoid is energised and lifts the valve from its seat in the fuel intake passage, allowing fuel to flow into the high-pressure chamber. Conversely, when the engine is to be stopped, the power to the solenoid is switched off and a spring in the solenoid pushes the valve down, closing off the fuel intake port and stopping the engine. When it is operating correctly, an audible click is heard when the start switch is turned to the engine 'on' position.

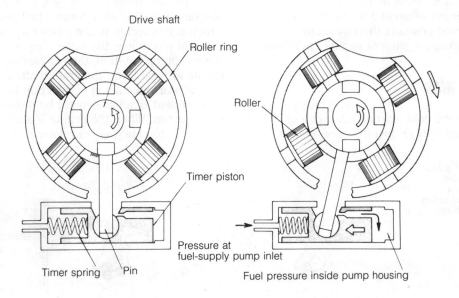

a Beginning of angular advance   **b** Conclusion of angular advance

Fig 20.45 Operation of the automatic advance unit

a	starting (initial) position	2	governor with shaft
b	just before opening	3	transverse bore in sliding sleeve
c	opening; pressure drops in pump cavity.	4 5	sliding sleeve transverse bore in governor shaft
1	longitudinal bore in governor shaft	6 7	governor-shaft port transverse bore in governor shaft

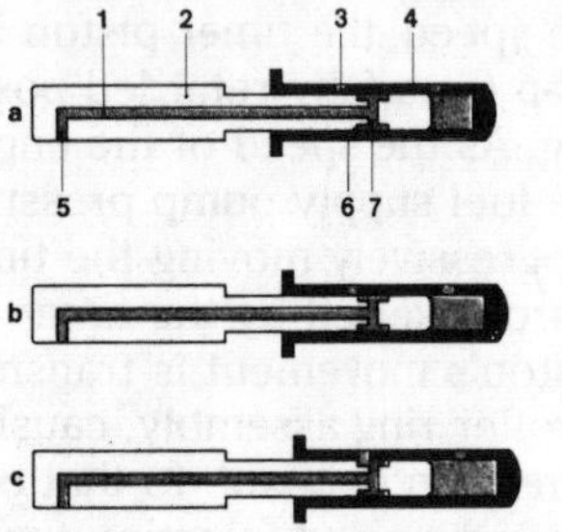

1 governor spring
2 sliding sleeve
3 tensioning lever
4 starting lever
5 control spool
6 distributor plunger
7 governor shaft
8 centrifugal weights
M_2 pivot for 3 and 4

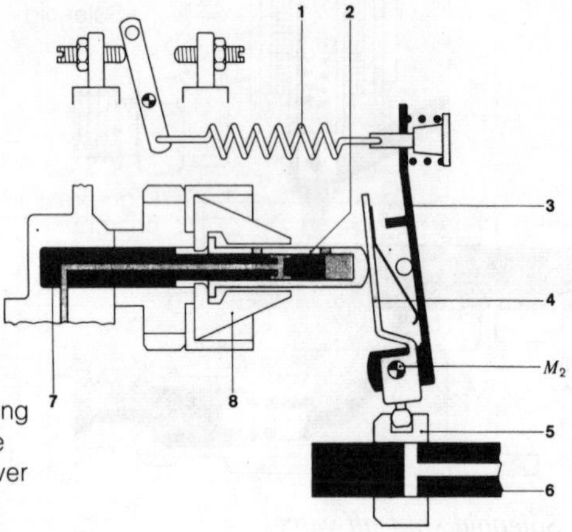

Fig 20.46 Governor assembly with load sensing advance/retard unit

By reference to Fig 20.46b, it can by seen that the resultant movement of the governor weights will move the sliding sleeve (4) to the right, until the port (3) in the sleeve uncovers the drilling (7) in the governor shaft allowing fuel at supply pressure to escape to the low-pressure side of the fuel supply pump. This causes a drop in the pump housing pressure acting on the timer piston, allowing the timer spring to move the timer piston and roller ring assembly into a retarded position, thereby retarding the injection timing for smooth engine operation.

Starting aids

To improve the cold starting efficiency of engines fitted with VE fuel injection pumps, a means of advancing the injection timing during starting is incorporated in the pump. The injection advance operation can be either mechanical or automatic, as shown in Fig 20.47.

Operation (*Refer to Fig 20.48*)

In the manual system, when the cold start cable is pulled, an advance lever (see previous diagram, Fig 20.47 (5)), turns a ball pin (3), which is connected to the roller ring (6). Turning this ball pin moves the roller ring against normal pump rotation, thus advancing injection timing at least 5° on the crankshaft.

Cold start advancing of injection timing is independent of advancement by means of the timer piston. When the engine is started, the normal automatic (speed related) advance

1 set screw
2 cable
3 stop
4 coil spring
5 advance lever

1 Temperature-sensitive control unit

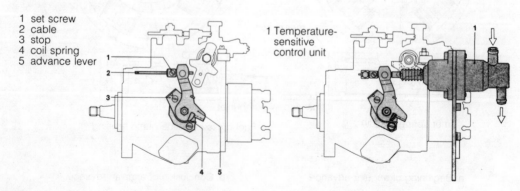

Fig 20.47 Cold start injection advance devices

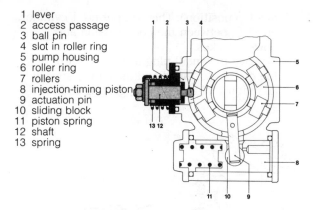

1 lever
2 access passage
3 ball pin
4 slot in roller ring
5 pump housing
6 roller ring
7 rollers
8 injection-timing piston
9 actuation pin
10 sliding block
11 piston spring
12 shaft
13 spring

Fig 20.48 Operation of the mechanically operated cold start device

mechanism takes over without interference from the cold start advance device.

The automatic cold starting system advances injection timing in the same way, except that the advance lever is operated by a temperature-sensitive bellows unit that reacts to changes in the water temperature of the engine. As the engine warms up, the advance unit becomes inoperative.

Air–fuel ratio control

The air–fuel ratio control unit is mounted on top of the fuel pump housing, as shown in Fig 20.49, and is used on fuel pumps fitted to turbocharged engines. When the turbocharger boost pressure is less than maximum, the device reduces fuel delivery accordingly so that the amount of fuel injected can be completely burned. If there were no set ratio between the maximum quantity of fuel injected and the quantity of air available, the turbocharged engine would be over-fuelled on acceleration, with resultant black smoke and poor fuel economy.

Operation

A pressure chamber of the top of the unit is connected to the engine inlet manifold by means of a flexible hose or pipe. Intake manifold boost pressure acts on the diaphragm (1) to oppose the force exerted by the spring (5) of the air–fuel ratio control. Therefore, when the engine and turbocharger increase in speed, boost pressure will also increase to act on the diaphragm and compress the spring.

Under low boost conditions, however, the spring holds the adjusting rod (2) in a higher

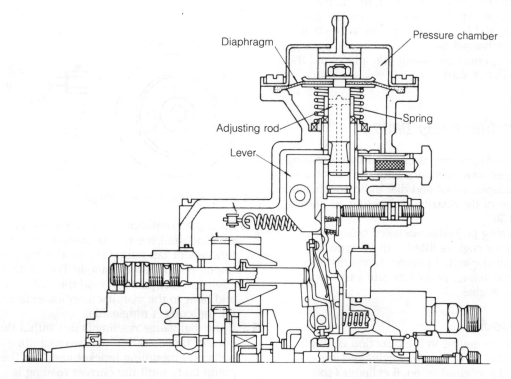

Fig 20.49 Air–fuel ratio control fitted to the Bosch VE fuel injection pump

position so that the larger diameter end of the control contour groove is in contact with the lever (3), which limits the leftward movement of the top of the tensioning lever, so limiting the maximum fuel delivery.

As boost pressure increases and the spring is compressed, the adjusting rod (2) moves down, causing the lever (3) to follow the contour of the adjusting rod, which allows the tensioning lever to move the control spool toward the maximum fuel position.

Anti-reverse operation

Because of its design, the VE type injection pump will not inject fuel when turned in the direction opposite to that of normal rotation. This ensures that the engine will not fire under stall or roll-back conditions.

During normal operation, on the fuel intake stroke the intake port is open to allow fuel to flow into the high-pressure chamber, charging the system. By the start of the delivery stroke, the plunger has rotated so that the distributor slit is aligned with an outlet port in the distributor head. However, if the plunger is rotated in the reverse direction, the distributor slit will align with the intake port on the plunger up-stroke, and fuel in the high-pressure chamber will be discharged back into the intake passage. As a result, injection cannot take place and the engine will not start.

VE injection pump timing

Injection pump timing is checked by measuring the plunger travel distance with the aid of a special adaption tool and dial indicator installed in the end of the distributor housing, as shown in Fig 20.50.

The timing procedure detailed below is common for engines fitted with VE-type fuel pumps but the actual plunger travel distance used in the timing procedure will vary from engine to engine.

Timing procedure

1 Turn the engine in the direction of rotation until the valves in no. 4 cylinder (four-cylinder engine) or no. 6 cylinder (six-cylinder engine) are both rocking (valve

overlap position). This will put no. 1 cylinder on TDC, between the compression and power strokes.
2 Remove the small plug from the centre of the distributor head and install the special adaption tool and dial gauge. (Fig 20.50).

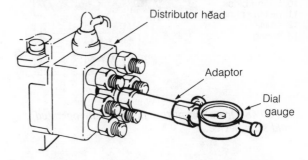

Fig 20.50 The timing dial gauge attached to the pump

3 Rotate the crankshaft backwards against the direction of rotation approximately 30° to place the plunger at the bottom of its stroke. Set the timing dial gauge to zero and move the crankshaft a few degrees in either direction to ensure that the plunger is at the bottom of its stroke.

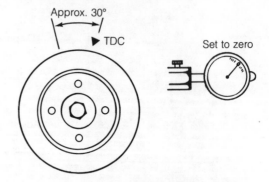

Fig 20.51 Zero setting the dial gauge

4 Turn the crankshaft in the direction of rotation until the timing pointer on the flywheel or fan pulley aligns with the correct timing position, for example TDC. With the engine in this position read the dial indicator and refer to the manufacturer's specifications for the correct dimension.
5 If the dial gauge reading is not within the correct range, loosen the two mounting bolts and mounting bracket and turn the pump body until the correct reading is obtained.

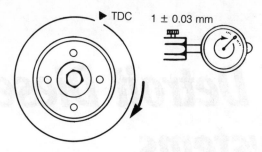

Fig 20.52 Checking the timing dimension

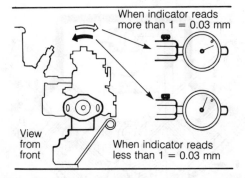

Fig 20.53 Adjusting the injection timing

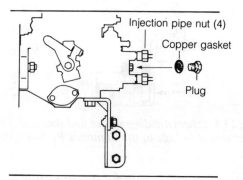

Fig 20.54 Refitting the access plug

6 Tighten the mounting bolts and recheck the reading.

7 Disconnect the special measuring tool and refit the plug.

8 No bleeding is required as the pump is self-bleeding.

9 Start the engine.

21
Cummins and Detroit Diesel Allison fuel systems

Two of the world's major diesel engine manufacturers—the Cummins Engine Company Inc and Detroit Diesel Allison—both utilise unconventional fuel injection systems. Each system features a camshaft-operated unit injector mounted in the cylinder head to raise the fuel pressure to injection pressure, and spray the fuel directly into the combustion chamber. In the case of Cummins, the system may be considered mechanical, since there is no injector valve to be opened hydraulically by fuel pressure. The system used by Detroit Diesel Allison, however, consists essentially of a combined plunger-type injection pump and hydraulically operated injector.

The Cummins PT fuel injection system

The Cummins PT fuel injection system is used on the majority of Cummins engines. However, the 'B' and 'C' series Cummins engines use either CAV or Bosch rotary fuel injection pumps and conventional fuel injectors. The PT (or pressure-time) concept derives its name from two of the primary variables affecting the amount of fuel metered and injected per cycle in the Cummins fuel system. 'P' refers to the pressure of the fuel at the inlet of the injectors. This pressure is controlled by the fuel pump. 'T' refers to the time available for fuel to flow into the injector cup. This time is controlled by engine speed through the camshaft and injector train.

Metering of fuel in the system is based on the principle that the volume of liquid that will pass

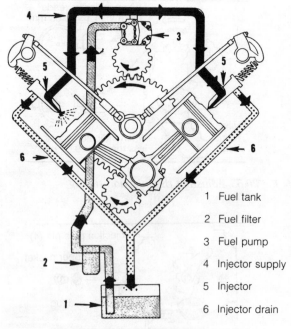

1	Fuel tank
2	Fuel filter
3	Fuel pump
4	Injector supply
5	Injector
6	Injector drain

Fig 21.1 Schematic diagrams of fuel flow and mechanical linkage in the Cummins PT fuel system

through an orifice is proportional to the pressure of the liquid, the time of flow and the orifice size.

In basic terms, the PT fuel system utilises simple mechanically operated injectors through which fuel continually circulates. In each injector, a fixed metering orifice leads fuel into a pressure chamber below a reciprocating plunger. As the plunger descends, fuel in this chamber is forced through multiple spray roles into the combustion chamber. The injector plunger is operated from the engine camshaft via a pushrod and rocker lever. Fuel can flow through the metering orifice into the pressure chamber only during the upper part of the

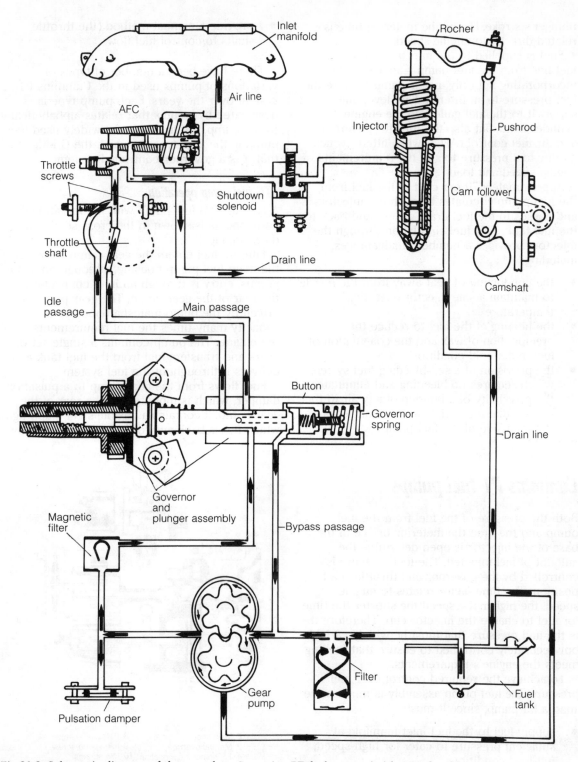

Fig 21.2 Schematic diagram of the complete Cummins PT fuel system (with AFC fitted)

plunger's stroke. Hence the metering time is related directly to engine speed.

Fuel is supplied to the injectors through a fuel line from the fuel pump—an assembly incorporating not only a gear pump to raise the fuel pressure to an intermediate level and supply it to the fuel gallery in the engine cylinder head, but also the throttle, governor and air fuel control (AFC) (when fitted), which modify fuel pressure to suit the requirements of engine speed and load.

Approximately 70 per cent of the fuel from the fuel pump circulates through the injectors and passes to the fuel drain gallery and back to the fuel tank. This fuel circulation through the injectors provides a number of advantages, including:

- the conducting of heat away from the nozzle to maintain a safe injector operating temperature
- the heating of the fuel to reduce the precipitation of wax and the coagulation of fuel under cold conditions
- the provision of a self-bleeding fuel system, which requires no bleeding and eliminates the possibility of a buildup of any air in the system that may originate from air leaks on the inlet side of the fuel pump.

Cummins PT fuel pumps

Both the pressure of the fuel from the fuel pump and the time the metering orifice in the base of the injector is open determine the amount of fuel injected. The fuel pressure is controlled by the governor and throttle shaft position. The time factor relates to engine speed, the higher the speed the shorter the time for fuel to charge the injector cup. Therefore it is the fuel pressure that must be widely variable but accurately controlled to ensure that fuelling meets the engine's requirements.

To achieve the required control over fuel pressure, the fuel pump assembly is much more than a lift pump, since it must:

- supply fuel to the fuel inlet manifold at sufficient pressure to cater for high-speed, full-load conditions
- limit fuel pressure to control maximum engine speed
- regulate fuel pressure between idle and maximum speed

- provide a manual method (the throttle shaft) to control fuel flow.

There have been a number of types or variations of pumps used in the Cummins PT system over the years. Each pump type is designated by a suffix that relates alphabetically to the pump design. The most widely used fuel pump is the PTG type, in which the G indicates that it is a governor-controlled system.

PTG fuel pump operation

A typical PTG fuel pump with its operational components is shown in the cross-sectional diagram, Fig 21.3.

Filtered fuel enters the gear pump through the main fuel pump housing, although on some pumps, entry is through an inlet connection at the rear of the gear pump. The gear pump is driven by the pump mainshaft and has a capacity many times the fuel requirements of the engine. The pump contains a single set of gears and transfers fuel from the fuel tank and delivers it throughout the fuel system.

Fuel flows from the gear pump to a pulsation damper, which is mounted on the rear of the gear pump. It contains a flexible steel diaphragm, which flexes back and forth to absorb fuel flow

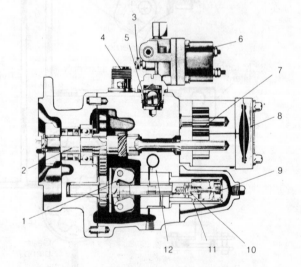

1 Governor weights	7 Gear pump
2 Main shaft	8 Pulsation damper
3 Manual override	9 Idle speed screw
4 Tachometer connection	10 Idle springs
5 Filter screen	11 Maximum speed spring
6 Shut-down valve	12 Throttle shaft

Fig 21.3 The PTG fuel pump with automotive governor

pulsations and so stabilise the fuel pressure. From the gear pump, fuel also flows through a wire gauze magnetic filter and on to the governor assembly.

Essentially, the governor consists of seven parts:

- a governor weight assembly
- a governor plunger
- a ported control sleeve
- an idle spring plunger (or button)
- an idle plunger guide
- an idle spring
- a (main) governor spring.

The governor plunger moves axially in the ported sleeve, under the influence of the centrifugal force on the flyweights from the one end and the force exerted by the governor springs from the other.

The plunger may be likened to a spool valve, with a 'waist' between two control lands. A radial drilling leads from this area to an axial bore emerging at the spring end of the plunger.

The ported sleeve contains four ports:

- a supply passage located to supply fuel to the area between the plunger lands
- a main passage leading to the throttle shaft
- an idle passage leading directly to the injector supply gallery
- a bypass passage leading fuel back to the gear pump inlet.

The pre-tensioned governor spring acts against the idle plunger guide to hold it firmly

against a shoulder in the housing. The idle spring, located inside the idle plunger guide, acts against the button, which is, in turn, held against the end of the governor plunger.

When the engine is cranked, fuel from the gear pump enters the governor plunger through the supply passage and fills both the bore and the waist. Fuel is able to flow through the idle passage, which is fully open to the injectors.

As soon as the engine starts, centrifugal force acting on the flyweights is transferred to the plunger, moving it against the idle spring, so that the control edge of the plunger land partly covers the idle passage, restricting fuel flow and pressure to the injectors. This movement is opposed by the idle spring and a balance is reached between centrifugal force and idle spring force. Engine speed is controlled by the fuel pressure, in turn controlled by the flow through the idle passage. During idle, the throttle shaft remains closed.

At the same time, gear pump supply pressure is controlled as the button is forced from its seat against the end on the governor plunger to dump excess fuel back to the gear pump inlet through the bypass passage.

When the throttle is opened to increase the speed of the engine, fuel passes at supply pressure through the throttle shaft to the injectors to increase the rate of fuel injection. The engine speed increases and the flyweight centrifugal force acts on the governor plunger to force the plunger against the button in the idle spring guide. When the force created by the fuel pressure acting on the button exceeds the

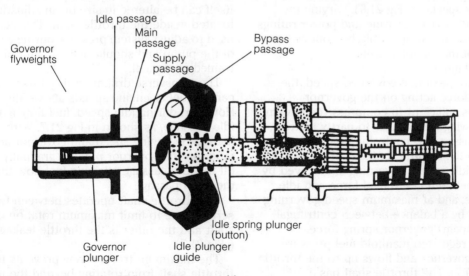

Fig 21.4 Sectional view of the PTG governor assembly

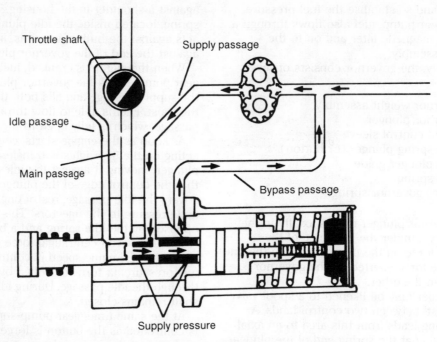

Fig 21.5 Schematic diagram of governor fuel pressure control

force holding the plunger and button together, the button is unseated and fuel is bypassed to the inlet side of the gear pump. This is shown in Fig 21.5. The bypass regulator, therefore, maintains the correct supply by unseating the button at the designated pressure and bypassing the excess fuel.

The size of the recess in the button controls the fuel pressure to the injectors, and a range of buttons is available to enable the fuel pressure to be set as specified (Fig 21.6). Varying the pressure will alter the torque and power ratings of the engine, and should only be done within the limits of the manufacturer's recommendations.

Further, at maximum governed speed, the centrifugal force acting on the governor flyweights is sufficient to compress the governor spring, allowing the plunger to partly restrict the main passage, so controlling the fuel pressure to the throttle shaft.

Thus, at idle speed, governing is achieved by a balance between centrifugal force and idle spring force, and at maximum speed governing is achieved by a balance between centrifugal force and (main) governor spring force.

The now regulated manifold fuel pressure leaves the governor and flows up to the throttle shaft (Fig 21.7). The throttle shaft has a

transverse hole drilled through it, and partially rotates in a ported sleeve pressed into the fuel pump housing. The shaft allows the operator to alter the system fuel pressure between idle and maximum engine speeds thereby controlling the torque and power output during this operating range.

The transverse drilling in the throttle shaft functions as a variable orifice to control the amount of fuel flow to the injectors. The drilling itself can be altered in size by an adjusting screw located inside the throttle shaft. This screw is used to adjust system pressure during calibration of the pump and should only be altered by trained personnel.

The transverse drilling in the throttle shaft carries fuel at engine speeds above idle. However, during idle speed, fuel flow is via an idle passage, as shown in Fig 21.7. With increasing engine speed and governor action, a taper on the governor plunger gradually closes off the idle passage allowing fuel flow through the main passage.

The throttle shaft operates between two stop screws, one to limit maximum rotation of the shaft and the other is the throttle leakage screw (Fig 21.7).

The maximum travel screw prevents the throttle shaft from rotating beyond the full open

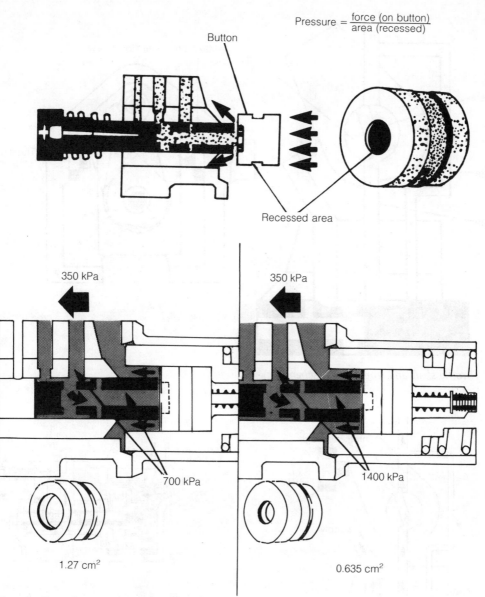

$$\text{Pressure} = \frac{\text{force (on button)}}{\text{area (recessed)}}$$

Button

Recessed area

350 kPa

350 kPa

700 kPa

1400 kPa

1.27 cm²

0.635 cm²

Fig 21.6 Use of a 'button' to vary the system fuel pressure

position, and allowing unrestricted fuel flow through the transverse drilling.

Note   This screw does not adjust the maximum engine speed; in this system, the maximum engine speed is adjusted by the governor spring tension.

The throttle leakage screw is designed so that when the throttle shaft is closed, there is always a small amount of fuel flowing through the throttle shaft, as shown in Fig 21.8. This is defined as throttle leakage and is required to keep the fuel lines filled with fuel to cool and lubricate the injectors when the throttle is closed. Throttle leakage is an important setting on the fuel pump. If it is set too high, it can result in slow acceleration and excessive carboning of the injectors. If it is set too low, it causes a hesitation in engine response when the throttle is reopened after a downhill run, and leads to injector plunger and barrel damage.

It can now be seen that there are two main components controlling system fuel pressure in the PTG pump. They are the governor and the throttle shaft.

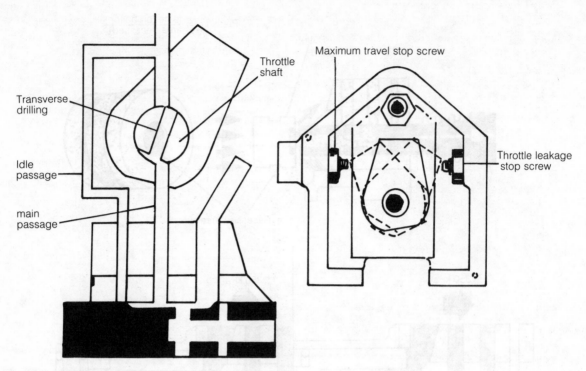

Fig 21.7 Throttle shaft and stop screws (schematic)

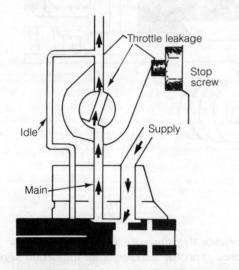

Fig 21.8 Throttle shaft in throttle leakage position (schematic)

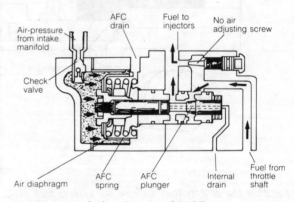

Fig 21.9 Air–fuel ratio control (AFC)

From the throttle shaft, fuel flows to the air fuel control (or AFC) section of the pump, if fitted (refer to Fig 21.9). The AFC assembly is necessary on turbocharged engine applications to limit the amount of fuel injected and so maintain an acceptable maximum fuel–air ratio for complete fuel combustion and a clean exhaust during acceleration. It does this by altering the amount of fuel supplied to the injectors to a level compatible with the air supplied by the turbocharger.

Intake manifold air pressure is applied to the diaphragm and plunger via a connecting air line. As the air pressure increases with turbocharger speed, the force acting on the diaphragm progressively overcomes the AFC spring force, causing the plunger to uncover the fuel passage

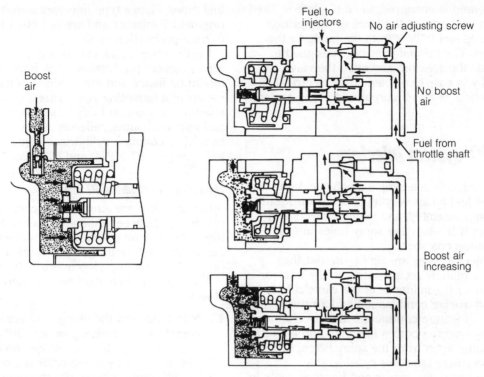

Fig 21.10 Operation of air–fuel control (AFC)

and allow more fuel to flow through the AFC (refer to Fig 21.10). Thus, as the turbocharger boost increases, the AFC plunger opens the passage further to increase the fuel flow to the injectors.

When there is little or no air pressure applied to the AFC (eg idle or start of acceleration), the flow of fuel around the AFC plunger is blocked off. Under these conditions, all fuel flows around the 'no air' adjusting screw, as shown in Fig 21.10.

From the AFC, the fuel finally flows to the shutdown valve (see Fig 21.11), before passing to the fuel gallery and the injectors.

Most shutdown valves are controlled by an electrically operated solenoid. In the shutdown mode, a wave spring forces a disc plate onto a seat, preventing fuel flow from the pump. When

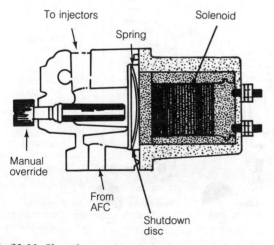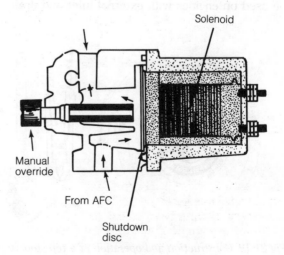

Fig 21.11 Shut-down valve operation

the solenoid is energised, the disc plate is lifted from its seat against the force of the spring, permitting fuel to flow from the pump to the injectors. The manual over-ride, designed to over-ride the solenoid and open the valve manually in case of a solenoid or electrical malfunction, is also incorporated in the system.

The PT system injectors

The PT injector is a simple mechanical unit that receives fuel under varying pressures from the fuel pump assembly, and meters, injects and atomises it through fine spray holes into the combustion chamber.

All injector types are similar in that they feature a plunger reciprocating in the injector body under the influence of the camshaft–pushrod–rocker combination and a return spring. All feature continuous fuel circulation, so every type has both a fuel inlet and an outlet —an outlet apart from the spray holes, that is. However, there are some considerable constructional differences and improvements in the more recently developed types, which are also a little different in operation from earlier designs.

PT injector types

There are two basic types of PT injectors— flanged injectors and cylindrical injectors. Flanged PT injectors are retained in the cylinder head by means of two capscrews passing through the flange, which also provides the fuel line connection points. Injectors of this type would be used on engines with external inlet and drain

fuel pipes. Flange-type injectors were the original PT injector and are not fitted to current-production engines.

On the other hand, cylindrical PT injectors are in current production. They have no mounting flange and are retained in the head by means of a mounting yoke. Three 'O'-rings are fitted to the injector body to seal between the fuel inlet and outlet galleries in the cylinder head. The common cylindrical injectors in use today are the PTC, PTD and PTD 'top stop' types.

Operation of cylindrical PT injectors

Although there are a number of constructional differences between different types of PT injectors, they all operate in the same manner. Fig 21.12 shows both the construction and operation of the top stop injector (type D).

The injector cycle may be considered in three stages:

1   **Metering**—As the plunger moves upward it uncovers the metering orifice (20) and fuel enters the injector cup. At this point, the fuel flow leaving the injector is closed off by the plunger, momentarily stopping circulation of fuel through the injector.

2   **Injection**—As the plunger is driven down by the rocker arm, the metering orifice (20) is covered and fuel can no longer enter the cup, nor can it escape except through the

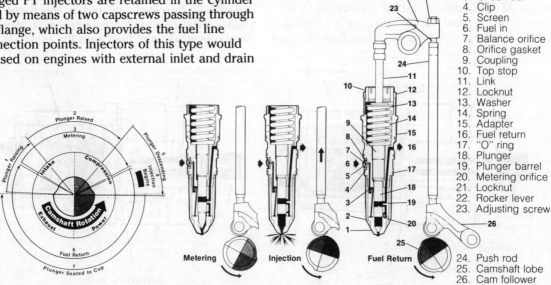

1. Cup
2. Cup retainer
3. Check ball
4. Clip
5. Screen
6. Fuel in
7. Balance orifice
8. Orifice gasket
9. Coupling
10. Top stop
11. Link
12. Locknut
13. Washer
14. Spring
15. Adapter
16. Fuel return
17. "O" ring
18. Plunger
19. Plunger barrel
20. Metering orifice
21. Locknut
22. Rocker lever
23. Adjusting screw
24. Push rod
25. Camshaft lobe
26. Cam follower

Fig 21.12 Construction and operation of a top stop (type D) injector

spray holes. With further downward movement, the plunger contacts the trapped fuel and forces it through the spray holes into the combustion chamber.

3   **Fuel return**—After injection is completed and the plunger is seated in the cup, the plunger remains seated until it begins to rise to start the injection cycle again. Fuel again circulates freely through the injector, providing cooling and lubrication.

Top stop injector

This is the latest type of Cummins injector, so called because it features a splined·locknut and stop mechanism (10) in the top of the injector body, which limits the upward movement of the injector plunger (Fig 21.12). The plunger travel for top stop injectors cannot be set with the injectors in the engine. A special top stop adjustment fixture, shown in Fig 21.13, is required. The injector spring is pre-loaded by means of a weighted handle at a prescribed travel as shown by the dial indicator. The plunger travel is changed by turning the adjusting nut (10, Fig 21.12). Plunger travel is pre-set in manufacture or on reconditioning.

Most of the time during engine operation, the plunger spring force reacts against the stop in the top of the injector and not against the camshaft, which in previous designs, carried this load at all times. Therefore the injector mechanical drive train is greatly relieved of this

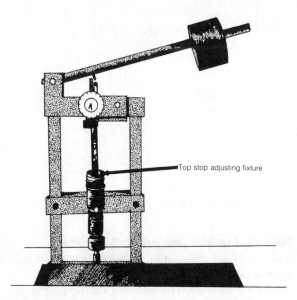

Fig 21.13 The top stop injector plunger travel adjusting fixture

continuous loading, which in turn allows improved lubrication of the camshaft lobes and cam followers, extending the life of the drive train.

Automatic timing control systems

As a means of controlling pollution and increasing engine performance, Cummins have introduced automatic devices to control injection timing. The hydraulic variable timing (HVT) and the mechanical variable timing (MVT) units are quite different in operation from one another but both produce the same results. The MVT unit was designed to operate on in-line engines, but was far too complex and expensive to be incorporated in 'V' engines, and the HVT unit was developed for this application.

Hydraulic variable timing (Refer to Fig 21.14)

The HVT system consists of special PTD injectors with an hydraulic tappet between the injector rocker lever and the plunger, and an oil control valve mounted separately on the engine. Regardless of engine speed, the system allows the engine to operate at advanced injection timing under light-load conditions, and at retarded timing during high-load conditions.

The hydraulic tappet consists of an outer cylindrical body, which reciprocates in a housing at the top of the injector, an inner piston and piston return spring, an inlet ball check valve and an unloader valve. On the upward movement of the rocker lever, the return spring moves the piston upward, extending the length of the tappet. When system fuel pressure is less than 221 kPa, an oil control valve opens and allows oil from the engine lubrication system to charge the hydraulic tappet, the oil being trapped below the piston because of the ball check valve, and maintaining the tappet at its extended length. Therefore, when the rocker lever comes down to drive the injector plunger, injection timing is advanced. At the end of the injection stroke when the plunger bottoms in the cup, the increased oil pressure in the tappet opens the unloader valve and allows the oil to drain from the tappet back to the sump.

When engine fuel pressure exceeds 221 kPa, the oil control valve is closed, preventing oil from charging the hydraulic tappet, so that the piston compresses the return spring and bottoms

Advanced timing

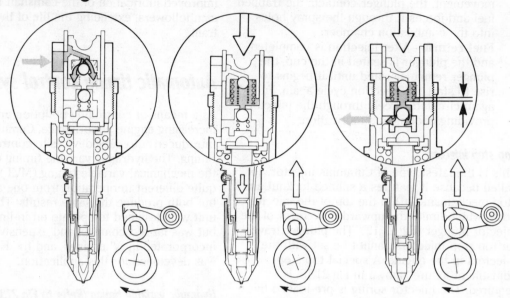

Retarded timing

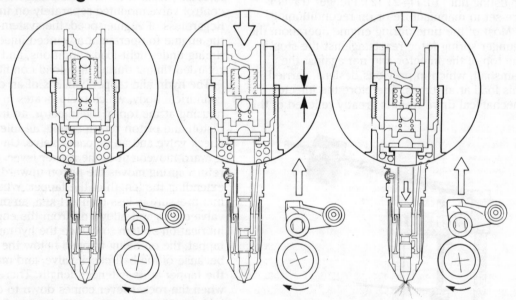

Fig 21.14 The PTD injector fitted with a hydraulic variable timing tappet assembly

on a shoulder in the tappet body, retarding the injection timing.

Several advantages are claimed for the HVT system, including improved cold-weather idling characteristics, reduced white smoke in cold climates, and reduced injector carboning.

Mechanical variable timing (Refer to Fig 21.15)

The mechanical variable timing unit is another means of controlling engine emissions on Cummins engines. The variable timing unit is a two-position injection timing control that will advance or retard injection timing depending on

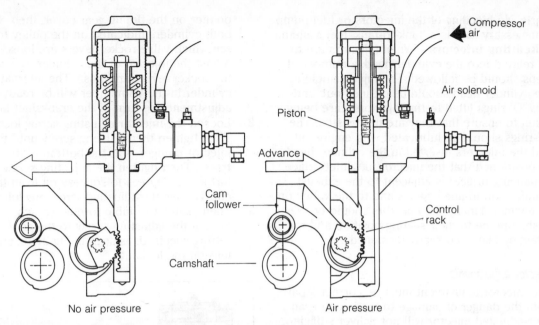

Fig 21.15 Operation of a mechanical variable timing (MVT) unit

engine speed and load conditions. The injection timing has to be advanced whenever the engine is started, to prevent white smoke emissions from becoming excessive. On the other hand, whenever the engine is operated at cruise and full-load conditions, the injection timing must be fully retarded to minimise the otherwise excessively high cylinder pressures, which would damage the engine bearings and piston assemblies.

The MVT unit consists of an air-operated cylinder moving a rack and pinion attached to the pivot of the injector drive train cam followers, as seen in Fig 21.15. Injection advance is achieved by moving the pivot of the lever-type cam follower so that the follower roller is vertically above the axis of the camshaft. The retarded position is achieved by moving the pivot so that the follower roller is moved slightly around the camshaft in the direction of normal rotation.

Air pressure from the vehicle's air reservoir applied to the air cylinder will act on the piston to move it upward, compressing the return spring and moving the control rack upward, as well. This movement of the control rack will partially rotate the eccentric follower pivot via the control pinion, and move the cam followers to an advanced injection position. Cutting off the air supply to the air cylinder will allow the return spring to move the control rack downward, rotating the follower pivot in the

opposite direction and moving the injection timing to a retarded position. Engine fuel pressure, working through an electric solenoid, operates the MVT unit in accordance with engine operating conditions.

Servicing the PT fuel system

Like all fuel systems, the PT system needs periodic maintenance checks and adjustments. These include checking (and perhaps making adjustments to) the fuel system pressures and governor adjustments, and since there is a large range of Cummins engines, each with a number of power ratings and governors to suit the diversity of applications to which these engines are put, these checks and adjustments can only be undertaken if the relevant service data are available. In the absence of relevant service data, the following adjustments may be helpful.

Refitting the fuel pump—No pump timing is required when refitting a PTG fuel pump to the engine, since injection timing is controlled by the position of the engine camshaft, which ultimately drives the injector plunger. Once fitted, the fuel pump housing must be primed with clean diesel fuel. This can be done by removing a plug from the top of the housing and filling the pump housing with diesel fuel. After refitting the plug, the engine should be cranked over with the starter motor until it starts. No bleeding of the injectors or fuel pump is necessary as it is a 'tank-to-tank' fuel system.

Refitting injectors—When injectors are to be refitted into the cylinder head, a number of steps should be followed. The injector hole in the cylinder head should be cleaned out, and new 'O'-rings fitted to the injectors, care being taken to ensure that they are not twisted. The 'O'-rings should be lubricated with engine oil and the injectors inserted in the cylinder head so positioned that the injector fuel inlet orifice (balancing orifice) is adjacent to the exhaust manifold on in-line engines and to the valley on 'V' engines. Finally the injectors should be pushed home to their seating position and the retaining clamp correctly tensioned.

Injector adjustments

The injector adjustment must be correct. Apart from the danger of damage to the injector, an over-adjusted injector will not deliver sufficient fuel, could bend an injector pushrod and will advance injection timing. An injector set too loose will cause increased fuel delivery, retarded injection timing and carboning up of the injector spray holes.

In practice, this is a very simple operation, with two methods being employed—the torque method and the dial indicator method.

The torque method (as applied to a six cylinder engine)

1 Rotate the crankshaft in the direction of engine rotation. On the front accessory drive pulley or crankshaft pulley there are a series of three VS (valve set) marks, such as A or 1-6VS, B or 2-5VS, C or 3-4VS, as shown in Fig 21.16. Align one of these marks with the

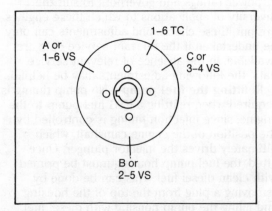

Fig 21.16 Timing marks on the accessory drive pulley

pointer on the timing gear cover, then check both cylinders indicated on the pulley to see which valve rocker levers are loose. Adjust the injector of the cylinder in which the rocker levers are loose. The alternative cylinder in the VS number will be ready for adjustment one turn of the crankshaft later.

2 Loosen the injector adjusting screw locknut. Then tighten the adjusting screw until the injector plunger is at the bottom of its travel. Then tighten the adjusting screw a further 15 degrees to squeeze oil from the cup. Loosen the adjusting screw one full turn.

3 Then, using a suitable torque wrench, tighten the adjusting screw to the cold setting specified, as seen in Fig 21.16, and tension the locknut.

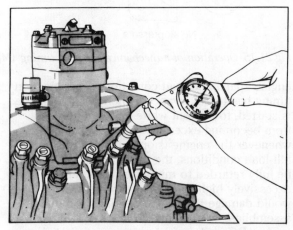

Fig 21.17 Adjusting the injector plunger by the torque method

With the torque method, after the injector has been set, the valve crossheads and valve clearances can be adjusted on the same cylinder. Refer to the manual for the adjustment procedure and clearances.

The dial indicator method (again as applied to a six-cylinder engine)

Cummins have now introduced a more refined method of injector adjustment. With this method, the injector plunger travel is measured by a dial indicator, which requires the injectors and valves to be set on separate cylinders. The injector is set with the piston at the beginning of the compression stroke and valves on the corresponding cylinders at the end of the same stroke.

1 Rotate the engine crankshaft in the direction of rotation. Align the A or 1-6VS mark on

the accessory drive pulley or crankshaft pulley with the pointer on the timing gear cover.

2 When the A or 1-6VS mark is aligned with the pointer, the inlet and exhaust valves for cylinder no. 5 will be in the closed position. Also the injector plunger for cylinder no. 3 will be at the top of its travel. If the valves on no. 5 cylinder are not closed, that is the rockers are loose, refer to the alternative A or 1-6VS in the table and check valves on cylinder no. 2 and injector on cylinder no. 4.

Right-hand rotation engine

Bar in direction	Pulley position	Set cylinder Injector	Valve
Start	A or 1–6 VS	3	5
Adv. To	B or 2–5 VS	6	3
Adv. To	C or 3–4 VS	2	6
Adv. To	A or 1–6 VS	4	2
Adv. To	B or 2–5 VS	1	4
Adv. To	C or 3–4 VS	5	1

Note   This table is to be used for dial indicator setting only

Fig 21.18 Injector–valve setting table

Note   The instructions using cylinder no. 3 to begin injector adjustment are for illustration purposes only. It is possible to start with any cylinder for injector adjustment provided that the crankshaft and camshaft positions are in accordance with the tune-up table (Fig. 21.18).

3 Use the injector adjustment kit to check the travel of the injector plunger. Install the dial

indicator and support so that the extension for the dial indicator is against the injector plunger, as shown in Fig 21.19.

4 Actuate the rocker lever to push the injector plunger to the bottom of its travel, using the rocker lever actuator from the adjustment kit, as illustrated in Fig 21.20. Let the plunger rise to the top of its travel. Actuate the lever again and set the dial indicator to zero, while holding the plunger at the bottom of its travel seated in its cup.

5 Tighten the rocker lever adjusting screw until the injector plunger has the correct travel, as per specifications.

Fig 21.20 Actuating the rocker lever to check plunger travel with the dial indicator

6 Hold the adjusting screw in position and tighten the locknut to the specified torque. Actuate the rocker lever two or three times to check that the adjustment is correct.

7 After the plunger travel has been adjusted the crossheads and valves must be adjusted for the cylinder shown in the table in Fig 21.18 before the crankshaft is rotated to the next valve set mark.

Adjustment of top stop injectors

1 Rotate the crankshaft in the direction of engine rotation and align the A or 1-6VS mark on the accessory drive or crankshaft pulley with the pointer on the timing gear cover. Refer to the table (Fig 21.18) for the location of the injector and valve adjustments.

2 Loosen the locknut for the rocker lever adjusting screw and tighten the adjusting

Fig 21.19 Dial indicator set up to measure plunger travel

screw until all the clearance is removed from between the rocker lever and injector link. Further tighten the adjusting screw one additional turn.

3   Back off the adjusting screw until the spring washer is against the stop of the injector, as shown in Fig 21.21.

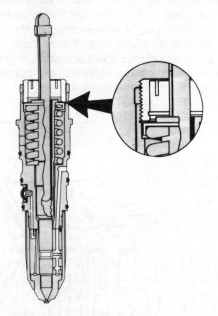

Fig 21.21 The top stop injector with washer and spring against the stop

4   Tighten the adjusting screw to 0.56–0.68 Nm (5–6 lb), using a torque wrench (Fig 21.22). If a torque wrench is not available, tighten the screw until there is light pressure

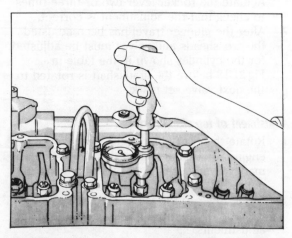

Fig 21.22 Tensioning the adjusting screw—top stop injector

against the injector link. The link must be free enough so that it can be rotated by hand. Finally hold the adjusting screw and tension the locknut.

Note   The figure quoted above is included as a typical figure only. Refer to the workshop manual for the correct setting for any required application.

5   After the injector plunger has been adjusted, the crossheads and valves must be adjusted for the cylinder shown in the table before rotation of the crankshaft to the next valve set mark.

Note   Top stop injectors need two adjustments the first as mentioned earlier in the chapter is the plunger travel distance, which is set at the factory or after reconditioning, and the second is as mentioned above.

The Detroit Diesel Allison fuel injection system

The fuel injection system used on Detroit Diesels features a simple layout, which has some considerable advantages. As can be seen by reference to Fig 21.23, the system consists of the fuel injectors, the fuel manifolds, the fuel pump, the fuel strainer, the fuel filter, and the necessary connecting fuel lines. The fuel

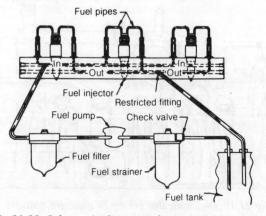

Fig 21.23 Schematic diagram of a typical fuel system

injectors are basically fuel pump–injector combinations, and are operated from the engine camshaft via a pushrod and rocker system. This necessitates the injectors being located under the rocker cover, together with a supply and a return line to each and an injector control tube (or shaft), which links to each injector's rack.

The fuel pump is a constant displacement gear pump—a lift pump—which circulates fuel at (relatively) low pressure through the system. The filter between the fuel tank and the pump is termed a fuel strainer; in other instances it would be known as a primary filter.

The fuel manifolds (or fuel galleries, perhaps) run the length of the cylinder head, and each injector connects to both. A restricting elbow is fitted at the end of the outlet manifold to maintain pressure in the system, particularly in the injectors. In installations in which the engine stands without running for some time, the fuel could drain back into the tank causing starting difficulties; a check valve may be fitted between the fuel strainer and the tank to prevent this.

Fuel passes from the supply tank through the fuel strainer, then enters the gear-type fuel pump, which raises the fuel pressure to an intermediate level, which changes with engine speed, but is somewhat below 500 kPa. From the fuel pump, fuel flows through the fuel filter into the fuel inlet manifold, then through the fuel pipes to the injectors where the fuel required is metered, has its pressure raised to injection pressure, and is passed into each engine cylinder at exactly the required point in the engine operating cycle. Surplus fuel returns from the outlet side of the injectors, through the outlet pipes into the outlet manifold, and then returns to the fuel tank via a return line.

The continuous flow of fuel through the injectors, prevents the buildup of air pockets in the fuel system, and cools those injector parts subject to high combustion temperatures.

The unit fuel injector

While it is simply known as a 'fuel injector', the Detroit Diesel Allison fuel injector does much more than the typical injector described in Chapter 17—it combines a complete pumping element and a closed injector in one compact, pushrod-operated unit. There are two types— the early 'crown valve' type and the later model known as a 'needle valve' injector.

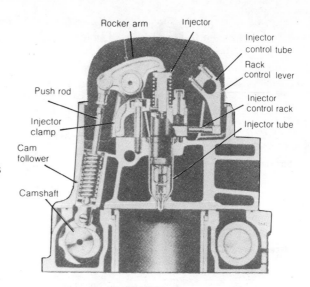

Fig 21.24 Fuel injector mounted in head

The GM unit fuel injector is a lightweight, compact unit, which combines with an open-type combustion chamber to give quick, easy starting directly on diesel fuel without the need for cold starting aids. The design and operation of the unit injector permits the use of simplified

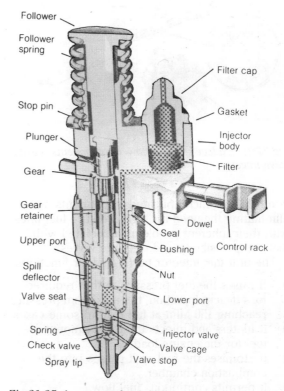

Fig 21.25 A crown valve injector

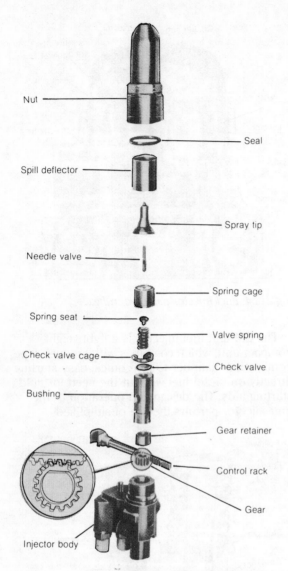

Fig 21.26 *Exploded view of lower components, needle valve injector*

Labels (top to bottom, left then right):
Nut — Seal
Spill deflector — Spray tip
Needle valve — Spring cage
Spring seat — Valve spring
Check valve cage — Check valve
Bushing — Gear retainer
Injector body — Control rack — Gear

controls and uncomplicated adjustments, while eliminating the need for high-pressure fuel lines with their inherent problems associated with fuel compressibility and pressure waves.

The unit fuel injector performs four functions:

- It raises the fuel pressure to that required for efficient injection, the maximum pressure reaching 150 MPa at full load in some cases.
- It meters and injects the exact amount of fuel for the engine load.
- It atomises the fuel as it passes into the combustion chamber.
- It permits continuous fuel flow.

To fulfil the dual role of both fuel pump and injector, this injector assembly has to be a little more complex than a simple injector. Essentially, it consists of the following parts:

- the injector body and (retaining) nut
- the plunger and bushing
- the nozzle assembly
- the check valve, valve spring, cage, etc
- the follower and spring
- the control rack
- numerous lesser components including the filter caps (or inlet and outlet unions), filters and springs, spill deflector, stop pin, etc.

The injector body is, of course, just what its name implies. It carries the necessary drillings to carry fuel both from the fuel inlet—one of the filter caps—and to the fuel outlet—the other filter cap. It houses the two filters, and is threaded on its lower end to accept the nut. Further, the body is drilled to carry the control rack, formed to accept a mounting clamp, and provides the guide in which the follower reciprocates.

The plunger and bushing may well be likened to a jerk-pump element, for there is similarity in both design and function. The bushing is located so that its lower end, carrying two funnel-shaped ports, lies within a fuel supply chamber. Two helices are milled on the plunger, creating an area of considerably reduced diameter a small distance up from its lower end. This area, with a helix at each end, is known as the **metering recess.**

The upper end of the plunger carries the gear (or pinion) which has a 'D'-shaped axial hole to correspond with the 'D'-section upper end of the plunger. This arrangement allows the plunger to reciprocate in the gear (and in the bushing) but causes the plunger to be rotated in the bushing should the gear be turned by the control rack under governor action.

Nozzle assembly design differs between injector types. Crown valve injectors, no longer fitted to current-production engines, utilise a very simple nozzle (or spray tip) without any form of needle valve, while needle valve injectors feature a more conventional needle and spray tip (nozzle body) combination.

Located between the bushing and the spray tip in both injector types are a number of components, including the check valve, valve spring, and (in crown valve types) the crown valve, valve seat, valve stop, and valve cage (or the spring cage in the case of the needle valve design).

In a crown valve injector, a flat, centrally drilled disc lying against the end of the bushing forms the seat for the crown valve, which is held against its lower side by the valve spring inside the valve cage. The valve cage itself is a cylindrical cage/spacer, closed at its lower end but for a smaller axial drilling for the passage of fuel. The disc-type check valve is located below the valve cage in a counterbore in the top of the spray tip, where it can seal against the valve cage.

The position of these components in the current production needle valve system is quite different. The flat check valve, in its own cage in this case, is located immediately below the bushing; next is the spring cage (similar to the crown-valve-type valve cage) containing the valve spring and spring seat, and in contact with the spray tip. The needle valve spigot passes through the small central drilling in the spring cage to contact the spring seat and compress the spring.

The follower, somewhat like a mushroom-type tappet in appearance, reciprocates in the injector housing, and is so designed on its lower end to engage positively with the top of the plunger so that they reciprocate as one.

A round-ended keyway is cut vertically on the body of the follower. It is engaged by the spigot end of a pin (the stop pin) to prevent rotation of the follower and, since the keyway does not extend to the lower end of the follower, to retain the follower in the assembly prior to installation in the engine.

The follower spring—a standard type of compression spring—seats on the injector housing and against the follower flange.

The control rack is basically round in cross-section, with a flat side where the gear teeth are cut, and a square 'U'-end at right angles to its axis. The rack is fitted directly to the injector housing in mesh with the plunger gear. It is moved axially to turn the plunger by the injector rack control lever on the injector control tube, which engages with the 'U'-end.

Of the small parts, most have been mentioned previously—the filters, filter caps, stop pin etc. The stop pin retains the follower in the injector body, and is itself simply retained by the lowest coil of the follower spring engaging with a flat or groove on the body of the pin.

The drilling in which the pin is housed is so positioned that, with the flat of the pin uppermost, it lies flush with the shoulder of the injector against which the spring seats.

The spill deflector is simply a sacrificial tube surrounding the bushing to prevent erosion of the injector body resulting from the high-velocity escape of fuel from the bushing ports.

Operation of the unit fuel injectors

Fuel from the fuel pump enters each injector at the inlet side, passing through the filter cap and filter element. From this filter, the fuel passes through a drilling into the supply chamber, that is, the area between the plunger bushing and the spill deflector. All the fuel entering the injector moves towards the outlet connection until the plunger rises to open the bushing ports to the supply chamber, when fuel can flow into the bushing.

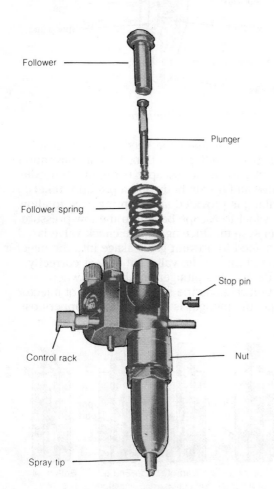

Fig 21.27 Exploded view of upper components, needle valve injector

Follower

Plunger

Follower spring

Stop pin

Control rack

Nut

Spray tip

As the plunger begins its downward stroke, some of the fuel under the plunger is displaced back into the supply chamber through the lower port until that port is closed off by the lower end of the descending plunger. Some of the remaining fuel trapped below the plunger then passes up through the central passage in the end of the plunger and into the metering recess. This fuel continues to escape into the supply chamber through the upper bushing port until this port is closed off by the edge of the upper helix as the plunger descends and the fuel remaining under the plunger is trapped. **Thus it is the upper helix covering the upper port that establishes the beginning of injection**.

Once the fuel is trapped in the bushing, it can only escape by passing (eventually) into the combustion chamber, and it must escape because the plunger continues to move downwards, driving the fuel out.

When the edge of the lower helix uncovers the lower port in the bushing, the metering recess is again in connection with the supply chamber, fuel below the plunger is no longer trapped and injection ends. **Hence it is the uncovering of the lower bushing port as the plunger descends that determines the end of injection.**

As far as the action of the plunger and bushing is concerned, both injectors operate in the same way. However, since the nozzles differ in construction, so too does their operation.

In the case of the crown valve injector, the injector valve is forced downwards from its seat by the sudden pressure rise above, in the bushing, and fuel passes into the spray tip and through the spray holes into the combustion chamber. When the lower port is uncovered allowing the escape of fuel, the fuel pressure below the plunger is relieved and the injector valve is snapped closed by its spring. The check valve, below the injector valve in this type, prevents gas leakage from the combustion chamber into the fuel injector should the injector valve be held open by a dirt particle.

With the needle valve injector, the non-return (or check) valve opens as fuel pressure is built up by the descending plunger, and pressure builds up as far as the needle valve seat. Like a conventional injector, the needle valve lifts from its seat, against spring pressure, when fuel pressure builds up sufficiently, and fuel passes through the spray holes into the combustion chamber. As soon as the lower bushing port is uncovered and fuel pressure is relieved, the

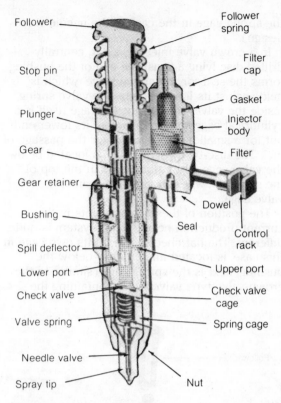

Fig 21.28 Needle valve injector

valve spring snaps the needle onto its seat, positively ending injection. As in a conventional nozzle, some fuel escapes between the needle valve and nozzle body and a pressure relief drilling is provided in the spring cage to allow this fuel to escape back into the low-pressure fuel system. Once again, the check valve is provided to prevent gas leakage into the injector should the needle valve fail to seat correctly.

Back to the plunger and bushing where the action remains the same regardless of injector type, the plunger is returned to its uppermost

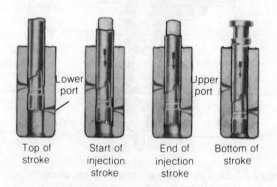

Fig 21.29 Phases of injector operation

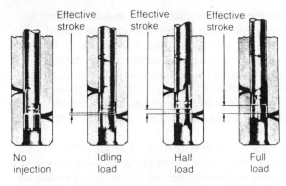

Effective stroke   Effective stroke   Effective stroke

No injection   Idling load   Half load   Full load

Fig 21.30 Fuel metering from no load to full load

position, after completing its downstroke, by the follower spring. Thus the operating cycle begins again as the plunger rises and fuel passes into the bushing from the supply chamber through the bushing ports.

As in most other systems, the quantity of fuel delivered is related directly to the amount of time for delivery. For increased fuel delivery the period of injection is extended, and vice versa. In this instance, injection begins when the upper port is covered, and ends when the lower port is uncovered. And since the upper port is covered by a helical shoulder, as the plunger is rotated to provide more fuel so the upper port is covered sooner and injection is advanced.

In some instances, the lower 'helix' is not a true helix but is a shoulder, square to the plunger. In such a case, the instant at which injection ceases does not vary as the plunger is rotated, and increased fuelling simply means that injection starts earlier but ends at the same time. However, in other instances a true helix is used, and increased fuelling is associated with an earlier start and a later finish to injection.

To stop the engine, the control rack is moved fully to the no-fuel position, rotating the plunger to the point where the upper port is not closed during downward plunger movement until after the lower port is uncovered by the edge of the lower helix. For maximum fuel delivery the plunger is rotated in the opposite direction to a point where the upper port is closed shortly after the lower port, thus producing the longest possible stroke before the lower port is again uncovered, so producing maximum fuel injection. From the no-fuel position to the maximum-fuel position, the contour of the upper helix advances the closing of the upper port, and hence the beginning of injection.

The power output of a Detroit Diesel Allison engine can be varied, within limits, by the use

Metal identification tag pressed into recess in injector body

GM N60

Identification mark on plunger

6N

Identification mark on end of spray tip

Injector	Spray tip *	Plunger
71N5	8.0055−165A	5N
N55	8.0055−165A	$\frac{5}{5}$N
N60	8.0055−165A	6N
N65 (White Tag)	8.0055−165A	$\frac{6}{5}$N
N65 (Brown Tag)	8.006 −165A	N$\frac{6}{5}$
HN65	7.006 −165A	7N
N70	7.006 −165A	7N
N75	7.006 −165A	$\frac{6}{5}$N
N80	7.006 −165A	8N

*First numeral indicates number of spray holes, followed by size of holes and angle formed by spray from holes

Fig 21.31 Needle valve injector identification markings

of injectors having different fuel output capacities. The fuel output of the various injectors is governed by two factors—the helix angle of the plunger, and the type of spray tip used. Refer to Fig 21.31 for the location of identification markings on needle valve injectors; crown valve injectors are similarly marked.

Each fuel injector has a circular disc pressed into a recess on the side of the body for identification purposes—this can be seen in Fig 21.31. The number on the disc indicates the nominal output of the injector in cubic millimetres. On needle valve injectors, the identification disc carries a horizontal bar, or line between the GM and the type code to distinguish them from the earlier crown valve types, which have no such line.

The numbers stamped on the plungers and spray tips aid in assembly of the correct parts when reconditioning the injectors. As is usual with pumping plungers and barrels, Detroit Diesel Allison plungers and bushings are supplied as mated pairs, but additionally are marked with corresponding numbers for ease of identification. Of course these components must never be intermixed with others, or be replaced as anything but an assembly.

Since the plunger helix angle controls the output and operating characteristics of a particular type of injector (while the injectors

Fig 21.32 Injector with offset body

are externally similar), care must be taken to ensure that the correct injectors are used for each engine application. If injectors of different types are used in the same engine, erratic operation will result and serious damage may be caused to the equipment that it powers. Further, it is important not to mix needle valve injectors and crown valve injectors in the same engine.

When space is restricted around the exhaust valve mechanism in engines with a four-valve-per-cylinder cylinder head, an injector with an offset injector body is used (Fig 21.32). A narrower injector clamp is required with the offset injector body but may be used with standard injectors.

On certain offset body injectors, designated **S-type injectors**, the clamp seat is positioned lower on the body and requires a different clamp. S-type injectors also feature a heavier follower spring.

Detroit Diesel Allison electronic unit injectors

With the development of electronic engine control systems, an interface between the electronic control and the mechanical injection is necessary to translate the electronic signals to fuel injection rate, injection timing and duration of injection. To meet this need, Detroit Diesel Allison have developed an electronic unit fuel injector, which utilises an electric solenoid to control the injection functions. The solenoid is itself controlled by the electronic engine control system, as described in Chapter 19.

Construction and operation

The electronic unit injector (Fig 21.33) is basically a two-piece unit consisting of a main body assembly and a nozzle assembly.

As its name suggests, the main body provides the main housing for the assembly and carries a fuel manifold, which connects the fuel supply system to the injector, the control solenoid and poppet control valve as well as the barrel and pumping plunger. The nozzle assembly is a conventional needle valve design.

The poppet control valve, as shown in Fig 21.34, is a simple valve, which operates across port openings to control fuel flow. The mechanical pumping components of the injector

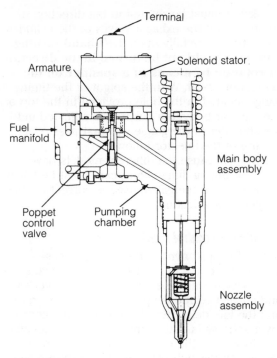

Terminal

Armature

Solenoid stator

Fuel
manifold

Main body
assembly

Poppet
control
valve

Pumping
chamber

Nozzle
assembly

Fig 21.33 Electronic unit injector

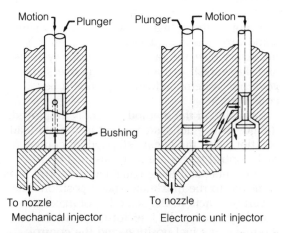

Motion — Plunger

Plunger — Motion

Bushing

To nozzle
Mechanical injector

To nozzle
Electronic unit injector

Fig 21.34 Helix–port injection control—mechanical injector (left); poppet valve injection control—electronic injector (right)

consists of a simple plunger and bore without the normal helices and internal plunger drillings that are synonomous with the mechanical unit injector.

Like the mechanical unit injector, the electronic unit injector is operated by the engine camshaft and a rocker arm. However, the functional difference is the way in which the fuel is metered and injection is timed.

In an electronic unit injector, the poppet control valve, operated by the solenoid, determines the fuel metering and injection timing functions, in a somewhat similar manner to that of the ports and helices in the mechanical unit injector. When the solenoid closes the control valve during the pumping plunger downstroke, fuel is trapped in the pumping chamber and injection begins. Opening of the poppet valve by the solenoid opens a dump circuit allowing the trapped fuel to escape, ending injection. The period of time of poppet valve closure determines the quantity of fuel injected into the engine.

By using an independent valve assembly under the control of the electronic engine control system to control the fuel metering and timing functions in preference to the previous plunger and bushing control, a previously unattainable degree of flexibility in fuel metering, fuel injection duration and injection timing is achieved.

Service requirements of the Detroit Diesel Allison fuel system

Like any other fuel injection system, the Detroit Diesel Allison system components require service periodically. The service procedures are not difficult to follow, and can readily be undertaken if the necessary special equipment is available. However, it is not likely that this equipment will be available, except of course in workshops engaged in Detroit Diesel Allison sales and service. In such workshops the required manufacturer's manuals are sure to be available, and so no advantage can be gained by persuing detailed service procedures here.

Further, at the time of printing, service information on electronic unit injectors was unavailable.

There are, however, two service procedures on the conventional system that can be performed with little special equipment—setting the injector timing and adjusting the control racks—that will allow the general serviceman to fit exchange injectors.

Timing unit injectors

Each injector is timed by setting the injector follower to a specified height in relation to the injector body, a special tool, a **timing gauge**, being used to measure this height. Adjustment is made by rotating the threaded pushrod, which

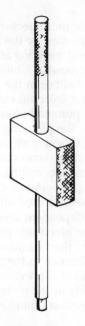

Fig 21.35 Injector timing gauge

is screwed into a clevis (or yoke) attached to a rocker arm. The effect of this adjustment may be likened to adjusting the cam followers in a jerk pump.

The adjustment is made with the governor control lever in the no-fuel position. The

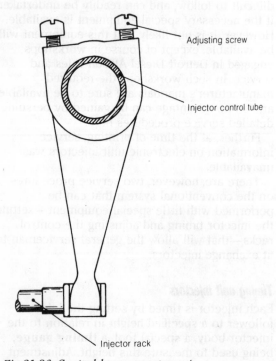

Adjusting screw

Injector control tube

Injector rack

Fig 21.36 Control lever arrangement

crankshaft must be rotated in the direction of rotation until the exhaust valves of the cylinder being timed are fully open. **Left-hand turning engines should not be turned in the direction of rotation by means of a spanner on the crankshaft bolt**. With the spigot of the timing gauge located in the hole provided in the top of the injector body, the pushrod is adjusted until the shoulder of the gauge will **just** pass over the top of the injector follower.

Note   Because the different models have different timing dimensions, care must be taken to ensure that the gauge used is correct for the particular application.

Adjusting the control racks

Accurate adjustment of the control racks— adjustment of the injector rack control levers—is necessary to ensure equality of fuel deliveries and correct relationship between the governor and the fuel deliveries. Alteration of the control lever settings is quite simple, as each is located and locked on the injector control tube by two lock screws, which act against corresponding recesses. If one screw is released and the other tightened, the control level will turn, within limits, in relation to the control tube.

Although the actual adjustment is very simple, some variation in approach and sequence is necessary over the range of governor systems, and the workshop manual, which covers all governor systems individually, should be consulted.

Basically, in the method generally employed, the governor–fuel system is set in the full-load position, with the engine stopped, of course. All rack control levers are first loosened on the control tube, and the engine speed control lever is moved to the maximum speed position. The governor reacts to the selection of maximum speed by moving the injector control tube to the maximum fuel position, and the control levers are adjusted so that the racks reach the full-fuel position.

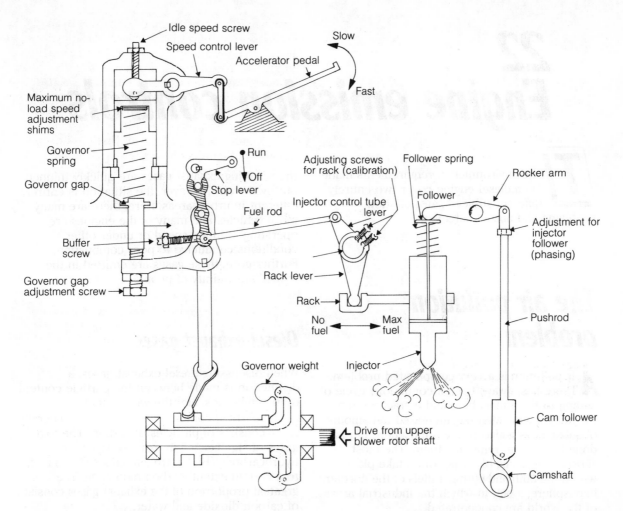

Fig 21.37 Detroit Diesel Allison variable speed governor and injector (schematic)

22
Engine emission controls

The environmental requirements made on a diesel engine cover two entirely different subjects, air pollution problems and noise abatement.

The air pollution problem

Air pollution is a very complicated problem. There has always been a continuous cycle of adding to the atmosphere and extracting from the atmosphere. Man has increased the amount of substances added to the atmosphere but has done little about removing them. The most serious emissions of air pollution take place within the 30th and 60th parallels of the northern hemisphere, a belt in which the industrial areas of the world are concentrated.

Motor vehicles account for only a small proportion of the total amount of emissions—diesel powered trucks account for less than 2 per cent. In principle, diesel exhaust gases are no problem on the highways where the engine is operating under suitable conditions and where

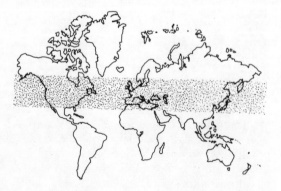

Fig 22.1 Zone of greatest air pollution (shaded)

the resulting exhaust gases are quickly thinned out by the air. However, the situation is entirely different in urban areas. Here, there are many more vehicles and many of the engines are operating at idling speed or under other conditions conducive to poor combustion. Furthermore, these gases are emitted in the immediate vicinity of people.

Diesel exhaust gases

When discussing diesel exhaust gases, a distinction is made between the particle content of the emissions, and the gas content.

The particle content—or soot, as we generally say—consists of particles of carbon. The toxic gases of significance in exhaust gases are hydrocarbons (HC), nitrogen oxides (NO_x) and, to a certain extent, carbon monoxide. The greatest proportion of the exhaust gases consists of carbon dioxide and water.

Soot

Soot consists of particles of carbon that have not been completely burnt. The injection of fuel into the combustion chamber of an engine mixes it with air but the mixture is not complete in all areas, which means there are local surpluses of fuel, and high temperature. Due to the lack of oxygen (rich mixture) dry distillation of the hydrocarbons in the fuel takes place and due to the separation of hydrogen, the result is carbon. As combustion continues, this carbon mixes with oxygen to form carbon monoxide if there is sufficient oxygen and if the temperature is not below 1000°C.

If the supply of oxygen is poor and/or the temperature is too low, the particles of carbon remain in particle form and are seen in the exhaust gases as black smoke.

The particles join together to form small granules. This soot is not injurious to health in itself, but it is thought that it can possibly act as a condensation nucleus for more injurious substances. However, the soot is a source of irritation since it reduces visibility and blackens the surroundings.

Hydrocarbons

The formula HC is used to designate a group of more than 100 different hydrocarbons. The hydrocarbon content in exhaust gases produced by motor vehicles is primarily the result of incomplete fuel combustion as well as of fuel that has evaporated from the fuel system. The greater proportion of HC emissions are not injurious to health, but some of them smell foully and others irritate the eyes and throat.

The amount of HC in exhaust gases is dependent on a number of factors related to the combustion process. High temperature, for example, gives a low HC content. One of the measures taken to combat HC emissions is therefore to design the engine to give a higher temperature during the combustion process under conditions of low load. The HC content of direct-injection diesels is also directly dependent on the so-called sack volume of the nozzle.

HC emissions are being given special study in the United States because they contribute to the formation of photochemical smog. Smog is the result of various reactions in the atmosphere started by the ultraviolet radiation of the sun. These reactions produce substances that, among other things, reduce visibility, irritate the eyes and damage vegetation. Smog can also form where there is a combination of heavy exhaust emissions, poor circulation and strong sunlight. Smog conditions occurred first in Los Angeles but can now be found in most of the larger cities of the USA. In fact, there was smog in Los Angeles long before the motor vehicle, due to air pollution from natural sources.

Nitrogen oxides

NO_x is the collective term used for nitrogen oxides. The oxides present to any extent in diesel exhaust gases are nitric oxide (NO) and nitrogen dioxide (NO_2).

NO is colourless and without smell. In air, NO oxidizes to NO_2, which is brownish red and has a pungent, irritating smell. Excessively high concentrations of NO_2 are injurious to the lungs. Furthermore, NO_2 combines easily with the haemoglobin of the blood, preventing the blood from absorbing and transporting oxygen from the lungs and through our bodies. As is the case with carbon monoxide poisoning, the ultimate consequence of this is suffocation.

The nitrogen oxides in exhaust gases are formed through the reaction between nitrogen and oxygen. This reaction is affected by the conditions prevailing in the combustion chamber, thus high temperature, high pressure and a good supply of oxygen give a high NO_x content.

Carbon monoxide

Carbon monoxide (CO) has no smell and is colourless. Carbon monoxide is poisonous because it can easily combine with the haemoglobin of the blood, preventing the blood from absorbing and transporting oxygen through our bodies. Again, the ultimate consequence of this is suffocation.

CO is formed during an intermediary stage of combustion. It later combines with oxygen to form the non-toxic carbon dioxide, CO_2. CO only occurs in exhaust gases if the supply of oxygen is insufficient. Since diesel engines work with a considerable surplus of air, the content of CO in diesel exhaust gases is therefore low.

Exhaust emission controls

*T*here is no single, simple solution to the complicated problem of air pollution by engine exhaust emissions. However, engine and fuel injection manufacturers have developed some effective methods of reducing the exhaust emissions from diesel engines.

Turbocharging

One of the many benefits of turbocharging is that it provides a greater surplus of air for combustion than is possible in a naturally aspirated engine. This increased air available in the cylinders ensures more complete combustion overall, with resultant reduced concentrations of hydrocarbons, soot and carbon monoxide in the exhaust emissions. However, during acceleration of a turbocharged engine from low

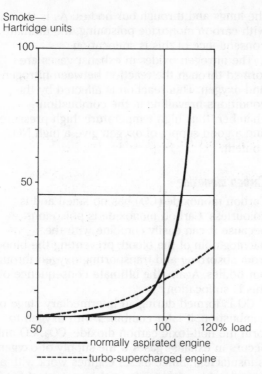

Fig 22.2 Exhaust smoke comparison—turbocharged and naturally aspirated engine

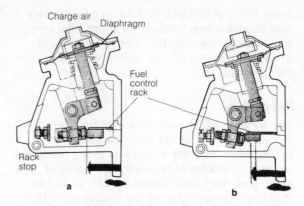

Fig 22.3 Typical air–fuel ratio control unit—rack travel at low charge air pressure (**a**), and rack travel at maximum charge air pressure (**b**)

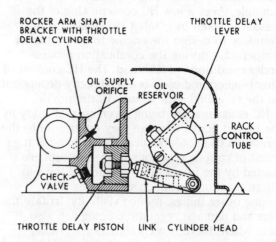

Fig 22.4 Throttle delay cylinder as fitted to a Detroit Diesel Allison engine

speeds, low initial turbocharger boost causes the fuel-to-air ratio to be much higher than under normal operating conditions. As a result, incomplete combustion, caused by more fuel being injected than can be burned with the available air, produces dense black exhaust smoke.

This incomplete combustion is only of short duration for, as the engine gains speed and exhaust gas energy increases, turbocharger output also increases, resulting in a suitable fuel-to-air ratio for efficient combustion.

Manufacturers have overcome this acceleration smoke problem by fitting an air–fuel ratio control, such as the typical example shown in Fig 22.3, to the injection pump. This device limits the quantity of fuel injected during acceleration relative to the turbocharger boost pressure to ensure that there is always enough air available for complete combustion of the fuel charge.

Another alternative that has been used is the throttle delay cylinder, something like a shock absorber, which allows the fuelling control mechanism to move toward maximum at a controlled (slow) rate, irrespective of how fast the throttle is being opened.

Injector nozzle design

The multi-hole injection nozzle is designed so that the spray holes open into a cavity beneath the point of the needle valve called the dome cavity or sack, as shown in Fig 22.5. When the needle valve closes, a small amount of fuel remains in the sack and enters the combustion chamber, too late for efficient combustion. Experiments have shown that the volume of this sack is directly related to the unburned hydrocarbon content of the exhaust gas. To minimise this effect, injection equipment manufacturers have designed the needle and

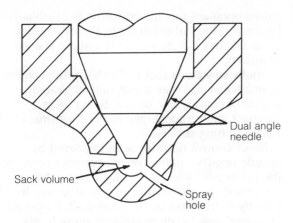

Fig 22.5 Emission control nozzle tip featuring a dual-angle needle and reduced sack volume

seat angles and the positioning so that the needle travels as far down into the sack as possible, thereby reducing the sack volume to a minimum.

Electronic engine control systems

With the advent of electronics to monitor and control engine performance, manufacturers are now able to precisely control fuel metering, the duration of injection, and the timing of injection relative to the changing engine load and speed conditions. By using an electronic engine management system, the quantity of fuel injected is always accurately matched to the available air for efficient combustion throughout the entire engine operating range. Thus, by using

electronics to manage diesel engines' performance, exhaust emissions that have previously been high in the low-to-medium speed range with conventionally governed engines are now as low as when the engine operates under optimal conditions.

Combustion chamber design

In the two-stage combustion system designed by Deutz Engines, fuel is injected into a swirl chamber containing about half the volume of hot compressed air for combustion. This low air-to-fuel ratio that exists at the beginning of the combustion phase limits the formation of oxides of nitrogen, thereby reducing the

Fig 22.6 Deutz dual-stage combustion chamber

Non-toxic content (in concentrations present)

	Direct injection	Indirect injection
Nitrogen gas (N_2)	76%	74%
Carbon dioxide (CO_2)	9%	11%
Water vapour (H_2O)	9%	11%
Oxygen (O_2)	6%	4%
Toxic contents		
Carbon monoxide (CO)	0.10%	0.05%
Nitrogen oxide (NO)	0.12%	0.10%
Nitrogen dioxide (NO_2)	0.02%	0.01%
Hydro carbons (HC)	0.02%	0.01%

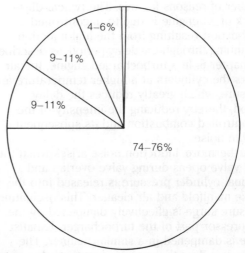

Fig 22.7 Exhaust emissions of a turbocharged engine operating under load (by volume)

hazardous exhaust emissions. As combustion continues (as shown in Fig 22.6), the rising pressure pushes the partly burned combustion products into the double swirl combustion chamber recessed in the piston crown, where the after-burning process takes place in the presence of excess air, under relatively low temperature and high turbulence conditions. These conditions again limit the formation of oxides of nitrogen, and minimise the output of hydrocarbon emissions into the environment.

Noise control

Diesel engine manufacturers in recent years have made determined efforts to control and even lower the noise levels of their engines.

A rigid, strongly built basic engine, with castings made largely from cast iron, appears to be an effective first design step for quiet running. The ribbed convex surfaces of the engine blocks are examples of this design, as are timing covers that have slightly convex surfaces and are stiffened by an irregular network of ribs. As a further noise reduction measure, engine sumps have been strengthened in material and construction to prevent unnecessary vibration.

It would seem reasonable to assume that a turbocharged diesel engine would be noisier than a naturally aspirated model, since it provides greater power output and higher peak operating pressures. However, in practice, the turbocharged engine is much quieter than its naturally aspirated counterpart. There are a number of reasons for this. The typical diesel knock of an engine is due to uncontrolled combustion resulting from the fuel injection continuing through the delay period, as described in Chapter 6. In a turbocharged engine, the air enters the cylinders at a higher temperature and pressure, which greatly reduces the delay period, thereby reducing the intensity of the uncontrolled combustion and its subsequent audible noise.

Furthermore, induction noise arises when the inlet valve opens during valve overlap and residual cylinder pressure is released into the intake manifold and air cleaner. This oscillating pressure surge is effectively dampened by the compressor part of the turbocharger. Exhaust noise is dampened in a similar manner. The pressure pulses that occur when the exhaust

valves open are modulated within the turbine housing of the turbocharger to the extent that some engines can be operated without the aid of a silencer.

Another factor is that the turbocharged engine is smaller in size than a naturally aspirated engine of comparable output. Being more compact, the turbocharged engine has smaller noise-emitting surfaces.

Noise control has also been improved by redesigning the engine cooling system to enclose the cooling fan and its ducting. Further, by increasing the size of the fan and attaching it directly to the front of the crankshaft, it operates at slower speeds and at reduced noise levels.

23
Alternative fuels

Today, the majority of internal combustion (IC) engines operate on either the compression ignition or spark ignition principle. In both cases, suitable fuels have been developed in parallel with the development and improvement of the engines—petrol for use in spark ignition engines, diesel fuels (Chapter 9) for compression ignition engines.

These fuels are subject to exacting standards to ensure constant quality for the consumer. A vast network of distribution points throughout the world ensures the supply of both petrol and diesel fuel. Both situations are important when considering fuel alternatives.

However, with the shortage of fossil fuels predicted during the past decade, together with rising fuel costs, research has been undertaken worldwide into fuel alternatives for existing engines.

Fuels today, both proven and experimental, are either liquid or gaseous.

The liquid fuel options include:

- the diesel fuels—automotive distillate and industrial diesel fuel
- petrol
- alcohols (methanol, ethanol). vegetable oils, and vegetable oil derivatives.

Gaseous fuels that are either in use or are considered to have potential include:

- liquified petroleum gas, LPG (propane butane)
- natural gas (primarily methane)
- biogas.

Liquid fuels

At the present time, the main fuels used for internal combustion engines are the diesel fuels and petrol. As both have a high energy content and are easy to store in non-pressurised tanks, these fuels are most suitable for mobile application. Liquid fuels that are considered as alternatives are the alcohol fuels (methanol ethanol) and vegetable oils.

Diesel fuels

Diesel fuels are mixtures of various hydrocarbons and are produced by distillation of mineral oil and/or by 'cracking' and hydrogenation of the residual products of the distillation process. Future diesel fuels are predicted to have a higher density and viscosity than present-day fuels, as well as a lower cetane number. This will be the result of production by a secondary refining process, which basically converts the heavy oil residuals, left behind after the initial refining, into usable diesel fuel. However, the secondary refining process does result in some degradation of the quality of the diesel fuel produced, although it still can be used in the more efficient diesel engines. Fuel produced by the secondary refining process usually has a lower cetane number and consequently a reduced readiness to burn. Automotive distillate is produced by either distillation or cracking, and its properties have been defined in Chapter 9.

Industrial diesel fuel is to a great extent produced from residues obtained during the oil refining process. It is cheaper than the more refined fuel but requires higher investment and operating costs regarding storage, pre-heating and preparation. The higher percentage of ash, sulphur, sodium, vanadium, asphaltenes and carbon residues in this diesel fuel results in combustion products with corrosive and abrasive action, which increase the rate of engine wear. Despite this, it is used for large engines, particularly for marine propulsion, since the low price makes it economical to use regardless of any initial high expenditure and increased rates of engine wear. Industrial diesel fuel must not be used in high-speed diesel engines .

Petrol

Petrol is also produced from mineral oil and is almost entirely used as fuel for spark ignition engines.

Alcohols

Alcohol fuels, especially methanol and ethanol, have been shown to have some potential as alternative fuels. Methanol (methyl alcohol) can be produced from brown coal, bituminous coal, wood and natural gas. Ethanol (ethyl alcohol) is formed by alcoholic fermentation of plant materials with a high sugar or starch content. As the properties of ethanol are similar to those of methanol, engines suitable for methanol operation can also be operated on ethanol. Due to its higher calorific value (1.36 times that of methanol), the consumption of ethanol is lower than that of methanol.

However, both alcohols have a calorific value roughly half that of the petroleum-based fuels, thus necessitating approximately double the rate of fuel admission for similar engine performance, and increased fuel tank capacity. On the positive side, the exhaust gases emitted have a relatively small percentage of pollutants and a low concentration of nitrogen oxides (NO_x), and are free from black smoke. Both fuels have a high octane rating (which would suit a petrol engine) but a low cetane rating.

Alcohol fuels can be used in a diesel engine by using a dual fuel injection system. In the

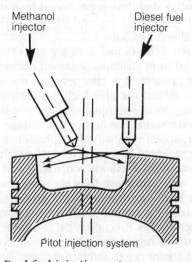

Fig 23.1 Dual-fuel injection system

Methanol injector

Diesel fuel injector

Pitot injection system

system shown in Fig 23.1, immediately prior to the injection of the alcohol fuel, a small amount of diesel fuel is injected, via a second spray nozzle, into the cylinder, where it is ignited normally by the heat of compression. During this combustion, the alcohol is injected into the combustion chamber and is ignited by the burning diesel fuel.

When the alcohol fuels are used in petrol engines, it is desirable to increase the compression ratio to gain comparable power, and to change synthetic fuel system components. However, the major modifications necessary involve increasing the fuelling rate and overcoming the inherent cold starting difficulties.

Vegetable oils

Vegetable oils are not clearly defined chemically. Not only are there differences between oils of the same species, but also of different plants and origins. As the ignition quality and the calorific value of vegetable oils are generally similar to those of commercial diesel fuels, they would seem to provide the alternative. Indeed, some tests have proved similar fuel consumption and engine power.

However, although vegetable oils are all virtually sulphur-free, they create other problems in that they are generally too viscous, often acidic, they choke injector nozzles with carbon and leave a high carbon residue on burning. In addition, the unburned residues form lacquer, leading to ring stick. These problems manifest themselves in long-term use and have yet to be solved. However, a great deal of research has been done into blends with petroleum-derived fuel, and into the use of vegetable oil esters—chemically modified vegetable oils—and there is no doubt that vegetable oils will provide an alternative fuel source for diesel engines in the years to come.

Gaseous fuels

Generally, all combustible gases can be used in internal combustion engines. The most popular gases are:

- **Liquified petroleum gas (LPG)**—This is predominantly propane and is produced from petroleum during the normal distillation process. Depending on the ambient temperature, it is stored in a liquified state at pressures between 511 kPa and 1540 kPa. In this physical state it has an energy content slightly lower than that of liquid fuels. LPG is easier to handle and store as a liquid but for mixture formation and combustion within the engine, it must be transformed into a gaseous state.

- **Natural gas**—This is a petroleum-related gas, which occurs naturally in underground gas fields. Although it is most widely used as a domestic fuel, it is also used as an engine fuel.
- **Biogas**—This gas, which is often referred to as methane, is produced in biogas systems during the decomposition of biomass materials such as plants, liquid manure, and manure in the absence of air and preferably at constant temperature levels. Its composition is principally methane, with hydrogen sulphide and carbon dioxide the major impurities.

Gas ignition systems

Gases readily mix with air to produce a homogeneous mixture, and provided that the gases are relatively free of impurities, burn evenly with almost no black smoke emission. As the energy content of gases, even in compressed state, is lower than that of liquid fuels, the specific fuel consumption of a standard engine is higher. However, with LPG this can be offset to some extent in petrol engines by increased compression ratios.

In the engine (whether it be a former petrol, or a diesel engine designed to run on gas), an ignitable mixture is produced in a gas–air mixing carburettor, and led via the intake manifold into the combustion chamber, where it is compressed and ignited.

Ignition of the gas–air mixture is by an electric spark at a spark plug. This ensures reliable ignition of the mixture with all types of gas. The main disadvantage of gas as an alternative fuel for diesel engines lies in the cost of conversion.

Index